AF598043

METHODS IN MOLECULAR BIOLOGY™

Series Editor
John M. Walker
School of Life Sciences
University of Hertfordshire
Hatfield, Hertfordshire, AL10 9AB, UK

For further volumes:
http://www.springer.com/series/7651

Peptide Modifications to Increase Metabolic Stability and Activity

Edited by

Predrag Cudic

Torrey Pines Institute for Molecular Studies, Port St. Lucie, FL, USA

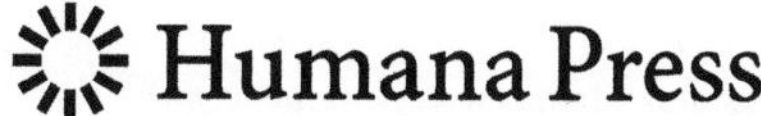

Editor
Predrag Cudic
Torrey Pines Institute for Molecular Studies
Port St. Lucie, FL, USA

ISSN 1064-3745 ISSN 1940-6029 (electronic)
ISBN 978-1-62703-651-1 ISBN 978-1-62703-652-8 (eBook)
DOI 10.1007/978-1-62703-652-8
Springer New York Heidelberg Dordrecht London

Library of Congress Control Number: 2013947403

Printed on acid-free paper

Humana Press is a brand of Springer
Springer is part of Springer Science+Business Media (www.springer.com)

Preface

Historically, natural products have served as important sources of pharmacologically active compounds or lead structures for the development of new drugs. Among natural products, peptides are particularly interesting because of the key roles they play in biological processes. Peptides' potential for high efficacy and their minimal side effects combined with advances in recombinant DNA technology, solid-phase synthetic chemistry, purification technology, and new strategies for peptide drug delivery made them widely considered as lead compounds in drug development. At present around 67 peptides are in the world market for clinical applications, some 270 are in clinical phases, and more than 400 are in advanced preclinical trials worldwide. Peptide-based therapeutics exist for a variety of human diseases, including osteoporosis (calcitonin), diabetes (insulin), infertility (gonadorelin), carcinoid tumors and acromegaly (octreotide), hypothyroidism (thyrotropin-releasing hormone [TRH]), and bacterial infections (vancomycin, daptomycin). However, despite their great potential, there are still limitations for peptides as drugs per se. Major disadvantages are short half-life, rapid metabolism, and poor permeation across biological barriers such as the blood–brain barrier (BBB) and intestinal mucosa. Nevertheless, pharmacokinetic properties of peptides can be improved and optimized through synthetic modifications. Peptidomimetic modifications, cyclization of linear peptides, or incorporation of D- and non-proteinogenic amino acids are traditionally used, both in academia and in industry, as an attractive method to provide more stable and bioactive peptides. In addition, linear peptide sequence modification by cyclization, glycosylation, and incorporation of non-proteinogenic amino acids have been widely used to enhance the potential of peptides as therapeutic agents. *Peptide modifications to increase metabolic stability and activity* is the first volume of a series that summarizes methods for preparation and purification of these peptides, and assessment of their biochemical activity. Readers of this volume will find detailed synthetic protocols that lead to modifications of the peptide backbone, side chains chapter, and terminal residues. Among these are protocols for preparation of conformationally constrained peptides (Chapters 1 and 2), modification of peptide bonds (Chapters 3 and 4), introduction of non-proteinogenic amino acids (Chapters 5–7), and alteration of peptides' physical and biological properties by modification of the amino acid side chains and/or terminal residues (Chapters 8–12). Last chapter (Chapter 13) describes a new experimental approach for the detection of exogenous peptides within living cells using peptides labeled with heavy isotopes and confocal Raman microscopy. This method allows peptide structure–activity relationships and metabolism to be explored directly within the targeted cellular environment. Of course, there are many other ways to improve peptides' metabolic stability and activity (e.g., peptide PEGylation or *N*-methylation of peptide bond and/or incorporation of D-amino acids) and they are well documented in the literature. However, my goal in this volume is to provide alternative approaches to peptide modification that many researchers may find applicable to their specific research requirements.

I believe that the readers will find protocols collected in this volume beneficial and helpful for their own research. At the end, I would like to thank all the authors and coauthors for their generous and enthusiastic contributions to this book. Their efforts and time are much appreciated.

Port St. Lucie, FL, USA *Predrag Cudic*

Contents

Contributors

CHRISTOPHER J. ARMISHAW • *Torrey Pines Institute for Molecular Studies, Port St. Lucie, FL, USA*
JAYATI BANERJEE • *Torrey Pines Institute for Molecular Studies, Port St. Lucie, FL, USA*
MANISHABRATA BHOWMICK • *Torrey Pines Institute for Molecular Studies, Port St. Lucie, FL, USA*
NINA BIONDA • *Torrey Pines Institute for Molecular Studies, Port St. Lucie, FL, USA*
JIANFENG CAI • *Department of Chemistry, University of South Florida, Tampa, FL, USA*
YI-PIN CHANG • *Torrey Pines Institute for Molecular Studies, Port St. Lucie, FL, USA*
MARE CUDIC • *Torrey Pines Institute for Molecular Studies, Port St. Lucie, FL, USA*
PREDRAG CUDIC • *Torrey Pines Institute for Molecular Studies, Port St. Lucie, FL, USA*
DEGUO DU • *Department of Chemistry and Biochemistry, Florida Atlantic University, Boca Raton, FL, USA*
GREGG B. FIELDS • *Torrey Pines Institute for Molecular Studies, Port St. Lucie, FL, USA*
RUŽA FRKANEC • *Institute of Immunology, Inc., Zagreb, Croatia*
REENA GYANDA • *Torrey Pines Institute for Molecular Studies, Port St. Lucie, FL, USA*
YAOGANG HU • *Department of Chemistry, University of South Florida, Tampa, FL, USA*
ANDREJA JAKAS • *Laboratory for Carbohydrate, Peptide and Glycopeptide Research, Division of Organic Chemistry and Biochemistry, Rudjer Boskovic Institute, Zagreb, Croatia*
HAIYANG LIU • *Department of Chemistry and Biochemistry, Florida Atlantic University, Boca Raton, FL, USA*
ADEL NEFZI • *Torrey Pines Institute for Molecular Studies, Port St. Lucie, FL, USA*
YOUHONG NIU • *Department of Chemistry, University of South Florida, Tampa, FL, USA*
BIMLESH OJHA • *Department of Chemistry and Biochemistry, Florida Atlantic University, Boca Raton, FL, USA*
LASZLO OTVOS JR. • *College of Science and Technology, Temple University, Philadelphia, PA, USA*
JEAN-PHILIPPE PITTELOUD • *Department of Chemistry, New York University, New York, NY, USA*
MARIA C. RODRIGUEZ • *Torrey Pines Institute for Molecular Studies, Port St. Lucie, FL, USA*
MACIEJ J. STAWIKOWSKI • *Torrey Pines Institute for Molecular Studies, Port St. Lucie, FL, USA*
ANDREW C. TERENTIS • *Department of Chemistry and Biochemistry, Florida Atlantic University, Boca Raton, FL, USA*
SRDJANKA TOMIĆ • *Faculty of Science, Department of Chemistry, University of Zagreb, Zagreb, Croatia*
BRANKA VRANEŠIĆ • *Institute of Immunology, Inc., Zagreb, Croatia*
HAIFAN WU • *Department of Chemistry, University of South Florida, Tampa, FL, USA*
JING YE • *Department of Chemistry, Salem College, Winston-Salem, NC, USA*

Chapter 1

Hantzsch Based Macrocyclization Approach for the Synthesis of Thiazole Containing Cyclopeptides

Adel Nefzi

Abstract

An innovative macrocyclization approach via high-yielding solid-phase intramolecular thioalkylation reaction is described. The reaction of *S*-nucleophiles with newly generated N-terminal 4-chloromethyl thiazoles leads to the desired cyclic products in high purities and good yields.

Key words Cyclic peptides, Thioalkylation, Solid-phase synthesis, Parallel synthesis, 4-Chloromethyl thiazoles

1 Introduction

Cyclic peptides have been difficult to prepare using traditional synthetic methods. In order for macrocyclization to occur, the activated peptide must adopt an entropically disfavored pre-cyclization state before forming the desired product. Conformational constraint by cyclization is a common approach used to restrict the flexibility of peptides and therefore is a valuable approach to study topographical requirements of receptors [1–6]. Cyclization of peptides can provide potent and selective ligands for receptors when appropriate conformational constraints are incorporated. Furthermore cyclic peptides are often more stable to peptidases, and therefore they can have improved pharmacokinetic profiles and serve as promising lead compounds for further development [7–13]. Macrocycles are known for their broad range of activities including antitumor activities and antibiotic activities such as the structurally complex vancomycin family [14, 15]. Of the various methods of synthesizing cyclic peptides, most often the final ring-closing reaction is a lactamization, a lactonization (depsipeptides), or the formation of a disulfide bridge. Reported approaches on the solid-phase synthesis of macrocyclic compounds include intramolecular nucleophilic substitutions [16, 17], intramolecular amide formations [18–20], disulfide formations [21–23], intramolecular Suzuki

Predrag Cudic (ed.), *Peptide Modifications to Increase Metabolic Stability and Activity*, Methods in Molecular Biology, vol. 1081, DOI 10.1007/978-1-62703-652-8_1, © Springer Science+Business Media New York 2013

reactions [24–26], ring closing metathesis reactions [27–29], and S_NAr displacement reactions [30–33]. Of particular interest, thioalkylation reactions offer a facile and versatile approach to the synthesis of cyclic peptides [16, 17, 34–37]. Examples of described macrocyclizations via thioalkylation include the reaction of the thiol group of a C-terminal cysteine with N-terminal acetyl bromide or N-terminal benzyl bromide [16, 17, 34–37]. A conceptually different approach, wherein thioalkylation proceeds via Michael addition of a thiolate anion to an α,β-unsaturated ester, has been reported for the synthesis of cyclic thioether dipeptides [38].

Many reagents and techniques have been developed to facilitate the synthesis of cyclic peptides, for which the yield-limiting step is generally the cyclization reaction. Particularly, the cyclization of tetra-, penta-, and hexapeptides in the all L-configuration can be problematic, especially in the absence of beta-turn promoting structures such as glycine, proline, or a D-amino acid [6, 39–42]. Our approach outlined in Fig. 1, was tested by performing the parallel synthesis of various thiazole containing cyclic tetrapeptides and pentapeptides from all L-amino acids.

An innovative thioalkylation approach toward the generation of macrocyclic peptides following the intramolecular nucleophilic substitution (S_N2) of N terminus 4-chloro methyl thiazole peptides with the thiol group of cysteine was described. The final products are not entirely peptidic and the described newly generated macrocyclic compounds contain the thiazole ring, a pharmacophore present in many natural and synthetic products with a wide range of pharmacological activities that can be well illustrated by the large numbers of naturally occurring thiazole containing macrocyclic compounds [43–45] and drugs in the market containing this function group [46–48]. We have also performed comparative computational studies of the chemical distribution of different cyclic peptides in the chemical space. This studies show that the prepared thiazole containing cyclic peptides occupy a different region in chemical space as compared to other cyclic forms.

2 Materials

1. All reagents such as 1,3-dichloroacetone, and solvents such as dichloromethane (DCM), Dimethylformamide (DMF) were obtained from Sigma-Aldrich (St. Louis, MO).
2. FmocNCS was obtained from ChemImpex (Wood Dale, IL).
3. Amino acids, Fmoc-isothiocyanate, piperidine, Cs_2CO_3, trifluoroacetic acid (TFA), $(Bu^t)_3SiH$, diisopropylethylamine (DIEA), *p*-methylbenzhydrylamine hydrochloride (MBHA·HCl) resin (100–200 mesh, cross-linked with 1 % divinylbenzene), and peptide coupling reagents such as diisopropylcarbodiimide (DICI), hydroxybenzotriazole (HOBt) were

1) Individual coupling of the first amino acid to resin-bound cysteine. Four different amino acids (Ala, Val, Phe, Tyr) are selected. Each bottle contains 6 bags.

Perform Kaiser test to ensure complete reaction

2) The 24 bags are combined in a single polyethylene bottle to perform the deprotection and washing steps

3) Individual coupling of the second amino acid. Three different amino acids (Phe, Tyr, Leu) are selected. Each bottle contains 8 bags.

Perform Kaiser test to ensure complete reaction

4) The 24 bags are combined in a single polyethylene bottle to perform the deprotection and washing steps

5) Individual coupling of the third amino acid. Two different amino acids (Gly, Ala) are selected. Each bottle contains 12 bags.

Perform Kaiser test to ensure complete reaction

6) The 24 bags are combined in a single polyethylene bottle to perform the deprotection and washing steps.
- All bags are treated with FmocNCS
- All bags are treated with piperidine
- All bags are washed
- All bags are treated with 1,3-dichloroacetone
- All bags are treated with TFA to cleave the Trt group
- All bags are treated with Cs_2CO_3 in DMF

7) The 24 resin packets are separately cleaved at the same time with hydrogen fluoride (HF) using a 24 vessel HF cleavage apparatus.

8) The 24 individual compounds are extracted by sonicating with 50% aqueous acetonitrile (3 x 5 ml). The resulting solutions are lyophilized.

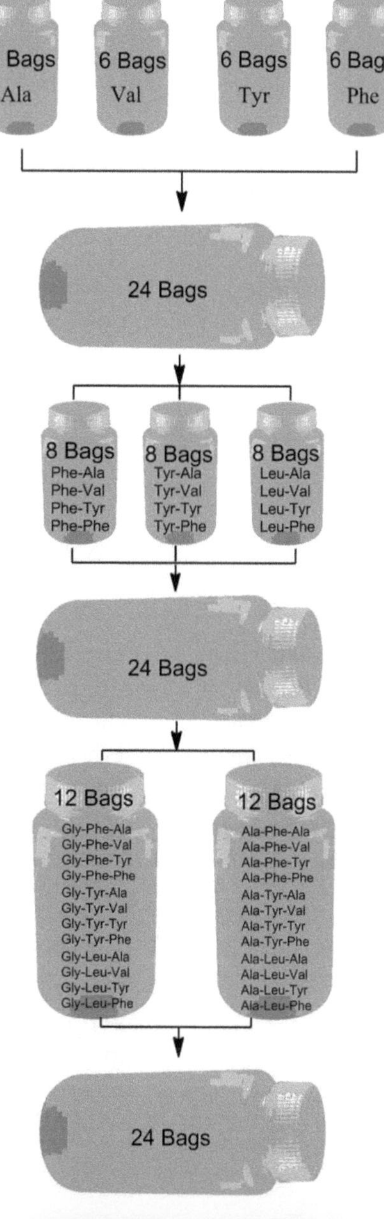

Fig. 1 Hantzsch based macrocyclization strategy for the parallel synthesis of 24 thiazole containing cyclopeptides

obtained from Fisher Scientific (Waltham, MA), ChemImpex and Novabiochem (San Diego, CA).

4. The ninhydrin test kit was obtained from AnaSpec (Fremount, CA).

3 Methods

3.1 The T-Bag Method

T-bags (Fig. 2) are prepared by containing solid phase resins within polypropylene mesh material [49]. Polypropylene is chemically inert and fairly thermally stable (to 150 °C), allowing a wide range of chemical reactions to be used for solid phase synthesis without affecting the bag material. Polystyrene cross-linked with 1 % divinylbenzene, 100–200 mesh, is mainly used as the solid support. It is very important that the size of the resin beads exceeds the size of the pores of the polypropylene mesh material of the T-bags to avoid resin loss during synthesis. Syntheses are carried out manually using polyethylene bottles.

3.2 Simultaneous Multiple Peptide Synthesis: Parallel Solid Phase Peptide Synthesis of the Resin-Bound Linear Peptides

Figure 1 illustrates the applicability of the proposed approach for the parallel synthesis of 24 different thiazole containing macrocyclic peptides. Starting from resin-bound orthogonally protected Fmoc-Cys-(Trt)-OH **1**, the thiomethyl thiazolyl macrocyclic peptidomimetics **6** were synthesized following stepwise Fmoc deprotection [50] and standard repetitive Fmoc-amino-acid couplings yielding the resin-bound linear tetrapeptide **2**. The resulting N-terminal free amine is treated with Fmoc-isothiocyanate. Following Fmoc deprotection, the thioureas are treated with 1,3-dichloroacetone to afford following Hantzsch's cyclocondensation [51–54] the resulting resin-bound chloro methyl thiazolyl peptide **5**. The Trt group is deprotected in the presence of 5 %TFA in DCM and the resin is treated with a solution of Cs_2CO_3 in DMF to undergo an S_N2 intramolecular cyclization. The resin is cleaved with HF/anisole and the desired thiazolyl thioether cyclic peptides

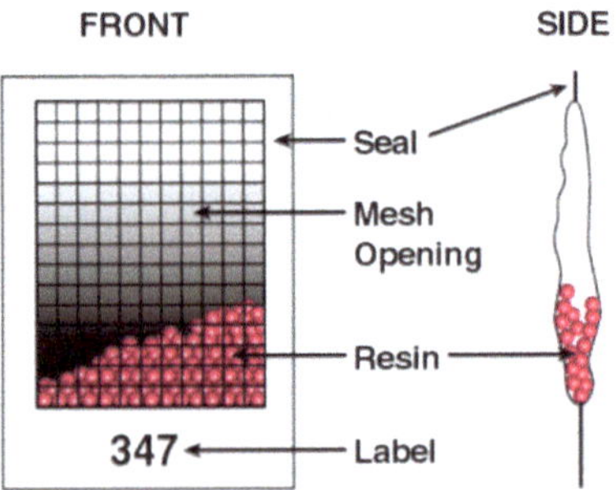

Fig. 2 Simultaneous parallel synthesis "Tea Bag"

6 are obtained in good yield and high purity. The identity of the final products was confirmed by LC-MS and NMR spectroscopy (*see* **Note 1**).

Preparation of T-bags for solid phase synthesis: All syntheses were performed using *p*-methylbenzhydrylamine hydrochloride (MBHA·HCl) resin (1.15 meq/g), and starting with 100 mg resin per bag. Synthesis using the T-bag method can be performed using either Boc [55, 56] or Fmoc [49] synthetic strategies. For all manipulations, enough solvent should be used to cover the T-bags (about 3–4 ml per bag containing 100 mg of resin). To enable efficient washings and reactions, the reaction vessels (polyethylene bottles) should be shaken vigorously, preferably through the use of a reciprocating shaker. Thus, during a T-bag synthesis of various sequences in parallel, the deprotection and washing steps can be performed with all bags combined in a single polyethylene bottle (Fig. 1). For the amino acid couplings, the bags are separated depending on the different corresponding sequences to be prepared. Following the coupling reactions, two washing cycles are done separately before combining all the bags again for subsequent washing and deprotection steps (*see* **Note 2**).

3.3 Synthesis of Resin-Bound Cysteine 1

As outlined in Fig. 3, a 100 mg sample of MBHA·HCl resin (1.15 meq/g) was contained within a sealed polypropylene mesh bag. For the parallel synthesis of 24 different compounds:

1. Prepare 24 separate bags (24×100 mg resin, 2.76 mmol).
2. Put all bags in a polyethylene bottle.
3. Neutralize the resin with 500 ml of 5 % DIEA in DCM.
4. Decant the solution.
5. Couple L-Fmoc-Cys(Trt)-OH (3 eq, 4.85 g, 8.28 mmol) using the conventional reagents HOBt (1.07 g, 8.28 mmol) and DIC (1.16 ml, 8.28 mmol) in 300 ml anhydrous DMF overnight at room temperature.
6. Decant the solution.
7. Wash the resin-bound dipeptide with DMF (3×) and DCM (3×).
8. Monitor the completion of the coupling by the ninhydrin test [57].

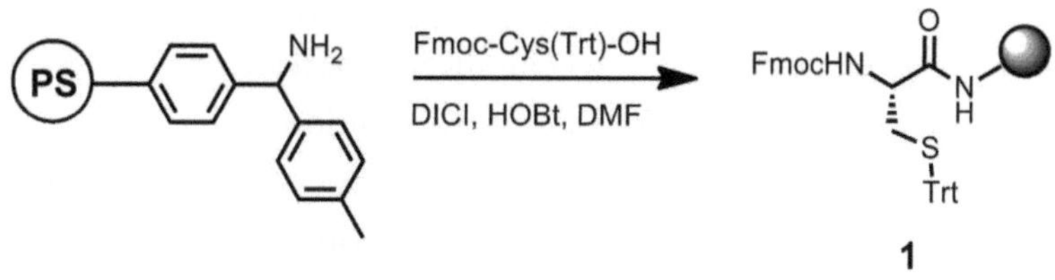

Fig. 3 Synthesis of *p*-methylbenzhydrylamine resin-bound cysteine

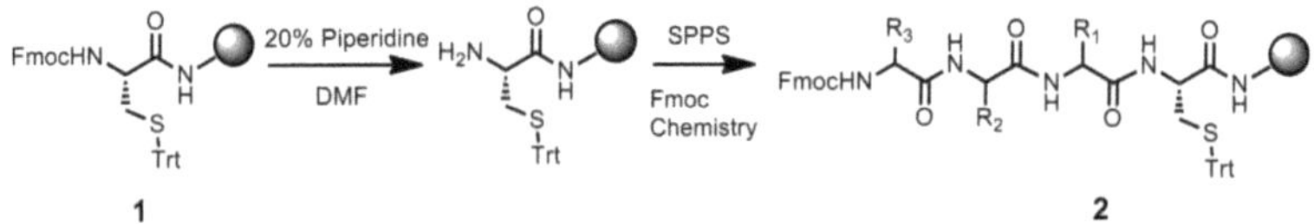

Fig. 4 Solid phase peptide synthesis of the resin-bound linear peptides using Fmoc-chemistry

3.4 General Procedure for the Solid-Phase Synthesis of Resin-Bound Linear Peptide 2

Starting from resin-bound orthogonally protected Fmoc-Cys-(Trt)-OH **1**, the linear peptides **2** are synthesized following step-wise Fmoc deprotection [14] and standard repetitive Fmoc-amino-acid couplings (Fig. 4).

Following is the procedure for the preparation of one resin-bound linear peptide:

1. Prepare one bag of resin **1** (100 mg, 0.115 mmol) in a small polyethylene bottle.
2. Deprotect the Fmoc group with 10 ml 20 % piperidine in DMF (2 × 10 min).
3. Wash the resin with 10 ml DMF (3×) and 10 ml DCM (3×).
4. Couple the first amino acid L-Fmoc-Xaa1-OH (6 eq, 0.69 mmol) in the presence of HOBt (6 eq, 0.094 g, 0.69 mmol) and DIC (6 eq, 0.101 ml, 0.69 mmol) in 10 ml of anhydrous DMF for 2 h at room temperature.
5. Decant the solution.
6. Wash the resin-bound dipeptide with DMF (3×) and DCM (3×).
7. Monitor the completion of the coupling by the ninhydrin test.
8. Deprotect the Fmoc group with 10 ml 20 % piperidine in DMF (2 × 10 min).
9. Wash the resin-bound dipeptide with DMF (3×) and DCM (3×).
10. Proceed with the coupling of the second amino acid L-Fmoc-Xaa2-OH (6 eq, 0.69 mmol) using the same reaction conditions.
11. Monitor the completion of the coupling by the ninhydrin test.
12. Deprotect the Fmoc group with 10 ml 20 % piperidine in DMF (2 × 10 min).
13. Wash the resin-bound dipeptide with DMF (3×) and DCM (3×).
14. Couple L-Fmoc-Xaa3-OH to the resin-bound tripeptide in the same conditions to yield the corresponding resin-bound protected linear peptide **2**.
15. Decant the solution.
16. Wash the resin-bound dipeptide with DMF (3×) and DCM (3×).
17. Monitor the completion of the coupling by the ninhydrin test [57].

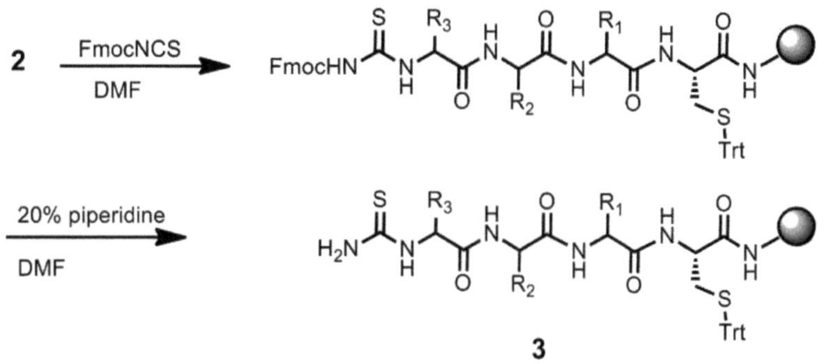

Fig. 5 Synthesis of resin-bound N-terminal thiourea linear peptide **3**

3.5 General Procedure for the Synthesis of Resin-Bound N-Terminal Thiourea Linear Peptide 3

The generation of the N-terminal thiourea (Fig. 5) is performed according to the following steps:

1. Deprotect the Fmoc group from the N-terminal amino acid with 10 ml 20 % piperidine in DMF (2 × 10 min).
2. Decant the solution.
3. Wash the resin-bound tetrapeptide with DMF (3×) and DCM (3×).
4. Treat the resulting N-terminal free amine of resin-bound linear peptide **2** with Fmoc-isothiocyanate (6 eq, 0.193 g, 0.69 mmol) in 10 ml of anhydrous DMF overnight at room temperature.
5. Decant the solution.
6. Wash the resin-bound tetrapeptide with DMF (3×) and DCM (3×).
7. Deprotect the Fmoc group with 10 ml 20 % piperidine in DMF (2 × 10 min).
8. Decant the solution.
9. Wash the resin-bound N-thiourea tetrapeptide with DMF (3×) and DCM (3×).

3.6 Generation of N-Terminal 4-Chloromethyl Thiazole 4

Following Fmoc deprotection, the synthesis of resin-bound N-terminal 4-chloromethyl thiazole peptides **4** (Fig. 6) is performed as follow:

1. Treat the resin-bound N-terminal thiourea with 1,3-dichloroacetone (10 eq, 0.145 g, 1.15 mmol) in DMF anhydrous overnight at 70 °C to afford following Hantzsch's cyclocondensation the resulting resin-bound chloro methyl thiazolyl peptide **4**.
2. Decant the solution.
3. Wash the resin-bound N-thiourea tetrapeptide with DMF (3×) and DCM (3×).

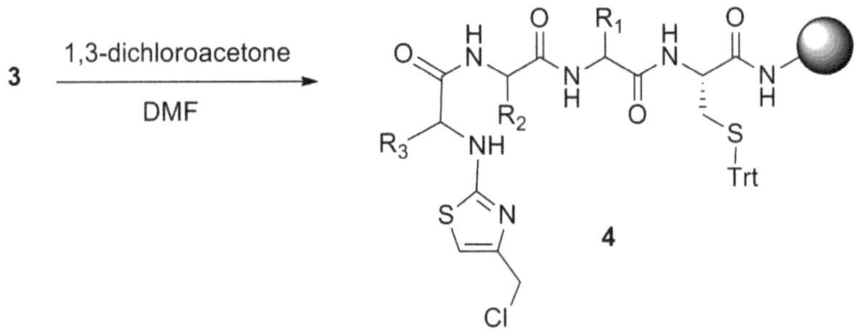

Fig. 6 Synthesis of N-terminal 4-chloromethyl thiazole **4**

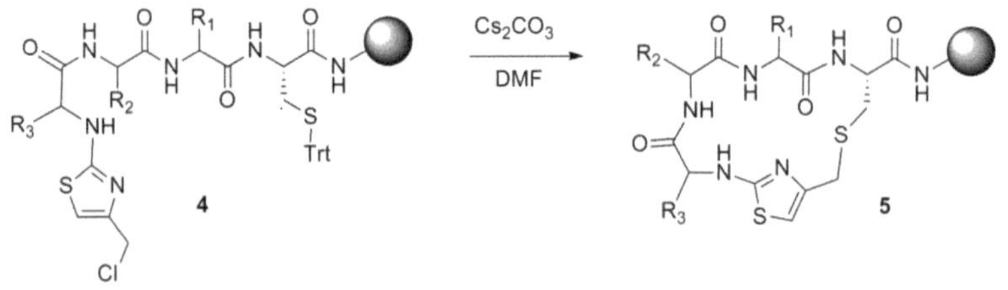

Fig. 7 Intramolecular thioalkylation cyclization

3.7 Intramolecular Thioalkylation Cyclization

The reaction of *S*-nucleophiles with newly generated N-terminal 4-chloromethyl thiazoles (Fig. 7) leads to the resin-bound cyclic products as follow:

1. The Trt group is deprotected in the presence of 10 ml TFA/$(Bu^t)_3$SiH/DCM (5:5:90) for 30 min.
2. The resin is washed with DCM (5×) and DIEA/DCM (5:95).
3. The resin is treated overnight with a solution of Cs_2CO_3 (10 eq) in 10 ml DMF to undergo an S_N2 intramolecular cyclization (*see* **Note 3**).

3.8 Cleavage of the Cyclic Compounds from the Resin

The resin is cleaved with HF/anisole and the desired thiazolyl thioether cyclic peptides **6** is obtained. Following the parallel synthesis of all individual cyclic peptides, the cleavage of the compounds from the resin packets is performed 24 at a time with hydrogen fluoride (HF; approximately 5 ml of HF per resin packet containing up to 0.225 mmol of resin-bound compound with 0.35 ml anisole added as a scavenger; 90 min, 0 °C) by using a 24 vessel HF cleavage apparatus (Fig. 8). The hydrogen fluoride is removed from the apparatus with nitrogen (*see* **Note 4**). The resulting individual compounds are extracted by sonicating with 50 % aqueous acetonitrile (3 × 5 ml). The resulting solutions are lyophilized twice from 50 % aqueous acetonitrile.

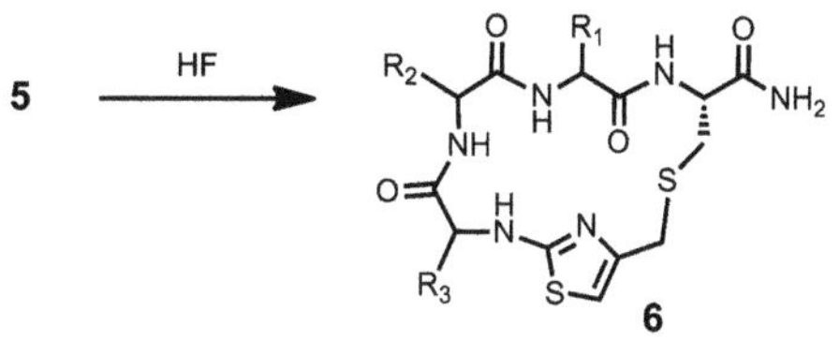

Fig. 8 Cleavage of the solid support

4 Notes

1. We selected different amino acids for each of the positions of diversity R_1, R_2, and R_3 for the synthesis of various tetrapeptides with side chains having different physicochemical properties including hydrophobic, hydrophilic, polar, apolar, basic, acidic, aliphatic, and aromatic properties. We also tested the effect of the stereochemistry by the incorporation of D-amino acids at each position. High purities were obtained for all compounds. In all cases, the intramolecular thioalkylation reaction led to the desired cyclic monomers with negligible traces of dimerization. The NMR data show a clear singlet at 6.4 ppm which is specific to the proton on C-5 of the aminothiazole ring.
2. The presented approach can be extended toward the synthesis of macrocyclic libraries where the cysteine residue can be placed anywhere in the peptide sequence, allowing for extension of the peptide beyond the cyclic link.
3. The intramolecular macrocyclization reaction is performed under anhydrous conditions in a nitrogen atmosphere. A small portion of the resin (5–10 mg) is cleaved by HF/anisole to ensure completion of the cyclization by LC-MS.
4. The HF is trapped by in-line traps containing solid CaO.

References

1. Lambert JN, Mitchell JP, Roberts KD (2001) The synthesis of cyclic peptides. J Chem Soc Perkin Trans 1:471–484
2. Hruby VJ, Balse PM (2000) Conformational and topographical considerations in designing agonist peptidomimetics from peptides leads. Curr Med Chem 7:945–970
3. Hruby VJ, Agnes RS (1999) Conformation activity relationships of opioid peptides with selective activities at opioid receptors. Biopolymers 51:391–410
4. Jones RM, Bulaj G (2000) Combinatorial chemistry at a cone snail's pace. Curr Opin Drug Discov Dev 3:141–154
5. Vagner J, Qu H, Hruby VJ (2008) Peptidomimetics, a synthetic tool of drug discovery. Curr Opin Chem Biol 12:292–296
6. Hruby VJ (2002) Designing peptide receptor agonists and antagonists. Nat Rev Drug Discov 1:847–858
7. Schiller PW (1993) Development of opioid peptides analogs as pharmacological tools and as potential drugs. Handb Exp Pharmacol 104(1) (Opioids I):681–710
8. Fang W-J, Cui Y, Murray TF, Aldrich JV (2009) Design, synthesis, and pharmalogical activities of dynorphin A analogues cyclized by ring-closing metathesis. J Med Chem 52:5619–5625

9. Berezowska I, Lemieux C, Chung NN, Wilkes BC, Schiller PW (2009) Dicarba analogues of the cyclic enkephalin peptides H-Tyr-c-[D-Cys-Gly-Phe-D9or L)-Cys]NH_2 retain high opioid activity. Chem Biol Drug Des 74: 329–334
10. Purington LC, Pogozheva ID, Traynor JR, Mosberg HI (2009) Pentapeptides displaying mu opioid receptor agonist and sigma opioid receptor partial agonist/antagonist properties. J Med Chem 52:7724–7731
11. Mollica A, Guardiani G, Davis P, Ma S, Porreca F, Lai J, Manina L, Sobolev AP, Hruby VJ (2007) Synthesis of stable and potent sigma/mu opioid peptides: analogues of H-Tyr-c[D-Cys-Gly-Phe-D-Cys]-OH by ring closing metathesis. J Med Chem 50:3138–3142
12. Weltrowska G, Lu Y, Lemieux C, Chung NN, Schiller PW (2004) A novel cyclic enkephalin analogue with potent opioid antagonist activity. Bioorg Med Chem Lett 14:4731–4733
13. Mollica A, Davis P, Ma S, Porreca F, Lai J, Hruby VJ (2006) Synthesis and biological activity of the first cyclic biphalin analogues. Bioorg Med Chem Lett 16:367–372
14. Driggers EM, Hale SP, Lee J, Terrett NK (2008) The exploration of macrocycles for drug discovery—an underexploited structural class. Nat Rev Drug Discov 7:608–624
15. Blout ER (1981) Cyclic peptides: Past, present, and future. Biopolymers 20:1901–1912
16. Feng Y, Pattarawarapan M, Wang Z, Burgess K (1999) Solid-phase S_N2 macrocyclization reactions to form β-turn mimics. Org Lett 1: 121–124
17. Roberts KD, Lambert JN, Ede NJ, Bray AM (2006) Efficient methodology for the cyclization of linear peptide libraries via intramolecular S-alkylation using multipin solid phase peptide synthesis. J Pept Sci 12:525–532
18. Dixon MJ, Nathubhai A, Andersen OA, van Aalten DMF, Eggleston IM (2009) An efficient synthesis of argifin: a natural product chitinase inhibitor with chemotherapeutic potential. Org Biomol Chem 7:259–268
19. Romanovskis P, Spatola AF (1988) Preparation of head-to-tail cyclic peptides via side-chain attachment: implications for library synthesis. J Pept Res 52:356
20. Alsina J, Jensen KJ, Albericio F, Barany G (1999) Solid-phase synthesis with tris(alkoxy) benzyl backbone amide linkage (BAL). Chem Eur J 5:2787–2795
21. Craik DJ, Cemazar M, Daly NL (2007) The chemistry and biology of cyclotides. Curr Opin Drug Discov Dev 10:176–184
22. Pons M, Albericio F, Royo M, Giralt E (2000) Disulfide bonded cyclic peptide dimers and trimers: an easy entry to high symmetry peptide frameworks. Synlett 2:172–181
23. Annis I, Chen L, Barany G (1998) Novel solid-phase reagents for facile formation of intramolecular disulfide bridges in peptides under mild conditions. J Am Chem Soc 120:7226–7238
24. Feliu L, Planas M (2005) Cyclic peptides containing biaryl and biaryl ether linkages. Int J Pept Res Ther 11:53–97
25. Li P, Roller PP, Xu J (2002) Current synthetic approaches to peptide and peptidomimetic cyclization. Curr Org Chem 6:411–440
26. Kaiser M, Siciliano C, Assfalg-Machleidt I, Groll M, Milbradt AG, Moroder L (2003) Synthesis of a TMC-95A ketomethylene analogue by cyclization via intramolecular Suzuki coupling. Org Lett 5:3435–3437
27. Blackwell HE, Grubbs RH (1988) Highly efficient synthesis of covalently cross-linked peptide helices by ring-closing metathesis. Angew Chem Int Ed 37:3281–3284
28. Reichwein JF, Versluis C, Liskamp RMJ (2000) Synthesis of cyclic peptides by ring-closing metathesis. J Org Chem 65:6187–6195
29. Boyle TP, Bremner JB, Coates J, Deadman J, Keller PA, Pyne SG, Rhodes DI (2008) New cyclic peptides via ring-closing metathesis reactions and their anti-bacterial activities. Tetrahedron 64:11270–11290
30. Feng Y, Burgess K (1999) Solid phase SNAr macrocyclizations to give turn-extended-turn peptidomimetics. Chem Eur J 5:3261–3272
31. Grieco P, Cai M, Liu L, Mayorov A, Chandler K, Trivedi D, Lin G, Campiglia P, Novellino E, Hruby VJ (2008) Design and microwave-assisted synthesis of novel macrocyclic peptides active at melanocortin receptors: discovery of potent and selective hMC5R receptor antagonists. J Med Chem 51:2701–2707
32. Derbal S, Ghedira K, Nefzi A (2010) Parallel synthesis of 19-membered ring macro-heterocycles via intramolecular thioether formation. Tetrahedron Lett 51:3607–3609
33. Giulianotti M, Nefzi A (2003) Efficient approach for the diversity-oriented synthesis of macro-heterocycles on solid-support. Tetrahedron Lett 44:5307–5309
34. Jung G (1991) Lantibiotics—ribosomally synthesized biologically active polypeptides containingsulfidebridgesanda,b,-didehydroamino acids. Angew Chem Int Ed Engl 30: 1051–1068
35. Campiglia P, Gomez-Monterrey I, Longobardo L, Lama T, Novellino E, Grieco P (2004) A novel route to synthesize Freidinger lactams by microwave irradiation. Tetrahedron Lett 45:1453–1456
36. Jack RW, Jung G (2000) Lantibiotics and microcins: polypeptides with unusual chemical diversity. Curr Opin Chem Biol 4:310–317

37. Kaiser D, Jack RW, Jung G (1998) Lantibiotics and microcins: novel posttranslational modifications of polypeptides. Pure Appl Chem 70:97–104
38. Crescenza A, Botta M, Corelli F, Santini A, Tafi A (1999) Cyclic dipeptides. Synthesis of methyl (R)-6-[(tert-butoxycarbonyl)amino]-4,5,6,7- tetrahydro-2-methyl-5-oxo-1,4-thiazepine-3-carboxylate and its hexahydro analogues: elaboration of a novel dual ACE/NEP inhibitor. J Org Chem 64:3019–3025
39. Olson GL, Bolin DR, Bonner MP, Bos M, Cook CM, Fry DC, Graves BJ, Hatada M, Hill DE, Kahn M, Madison VS, Rusiecki VK, Sarabu R, Sepinwall J, Vincent GP, Voss ME (1993) Concepts and progress in the development of peptide mimetics. J Med Chem 36:3039–3046
40. MacDonald M, Aube J (2001) Approaches to cyclic peptide beta-turn mimics. Curr Org Chem 5:417–421
41. Suat Kee K, Jois SDS (2003) Design of β-turn based therapeutic agents. Curr Pharm Des 9:1209–1212
42. Zhang J, Xiong C, Ying J, Wang W, Hruby V (2003) Stereoselective synthesis of novel dipeptide β-turn mimetics targeting melanocortin peptide receptors. J Org Lett 5: 3115–3118
43. Jin Z (2003) Muscarine, imidazole, oxazole, and thiazole alkaloids. Nat Prod Rep 20: 584–605
44. Bertram A, Blake AJ, de Turiso F, Hannam JS, Jolliffe KA, Pattenden G, Skae M (2003) Concise synthesis of stereodefined, thiazole-containing cyclic hexa- and octapeptide relatives of the Lissoclinums, via cyclooligomerisation reactions. Tetrahedron 59:6979–6990
45. Jin Z (2006) Imidazole, oxazole and thiazole alkaloids. Nat Prod Rep 23:464–496
46. Aulakh VS, Ciufolini MA (2011) Total synthesis and complete structural assignment of thiocillin I. J Am Chem Soc 133:5900–5904
47. Sanfilippo PJ, Jetter MC, Cordova R, Noe RA, Chourmousis E, Lau CY, Wang E (1995) Novel thiazole based heterocycles as inhibitors of LFA-1/ICAM-1 mediated cell adhesion. J Med Chem 38:1057–1059
48. Suzuki S, Yonezawa Y, Shin C (2004) Useful synthesis of fragment A–C–D of a thiostrepton-type macrocylic antibiotic, thiocilline I. Chem Lett 33:814–815
49. Houghten RA (1985) General method for the rapid solid-phase synthesis of larger numbers of peptides: specificity of antigen–antibody interaction at the level of individual amino acids. Proc Natl Acad Sci USA 82:5131–5135
50. Fields GB, Noble RL (1999) Solid phase peptide synthesis utilizing 9-fluorenylmethoxycarbonyl amino acids. Int J Pept Protein Res 35:161–214
51. Hantzsch AR, Weber JH (1987) Ueber verbindungen des thiazols pyridins der thiophenreihe. Ber Dtsch Chem Gen 20: 3118–3132
52. Garcia-Egido E, Wong SYF, Warrington BH (2002) A Hantzsch synthesis of 2-aminothiazoles performed in a heated microreactor system. Lab Chip 2:31–33
53. Lin PY, Hou RS, Wang HM, Kang IJ, Chen LC (2009) Efficient synthesis of 2-aminothiazoles and fanetizole in liquid PEG-400 at ambient conditions. J Chin Chem Soc 56:455–458
54. Arutyunyan S, Nefzi A (2010) Synthesis of chiral polyaminothiazoles. J Comb Chem 12:315–317
55. Gunnarsson K, Grehn L, Ragnarsson U (1988) Synthesis and properties of N'-di-Ter-butoxycarbonyl and N-benzyloxycarbonyl tertbutoxycarbonyl amino acids. Angew Chem Int Ed Engl 27:400–401
56. Gunnarsson K, Ragnarsson U (1990) Preparation and properties of N'-di-tertbutoxycarbonyl amino acids. Applicability in the synthesis of Leu-enkephalin. Acta Chem Scand 44:944–951
57. Kaiser E, Colescott RL, Bossinger CD, Cook PI (1970) Color test for detection of free terminal amino groups in the solid-phase synthesis of peptides. Anal Biochem 34:595–598

Chapter 2

The Chemical Synthesis of α-Conotoxins and Structurally Modified Analogs with Enhanced Biological Stability

Jayati Banerjee, Reena Gyanda, Yi-Pin Chang, and Christopher J. Armishaw

Abstract

α-Conotoxins are peptide neurotoxins isolated from the venom ducts of carnivorous marine cone snails that exhibit exquisite pharmacological potency and selectivity for various nicotinic acetylcholine receptor subtypes. As such, they are important research tools and drug leads for treating various diseases of the central nervous system, including pain and tobacco addiction. Despite their therapeutic potential, the chemical synthesis of α-conotoxins for use in structure–activity relationship studies is complicated by the possibility of three disulfide bond isomers, where inefficient folding methods can lead to a poor recovery of the pharmacologically active isomer. In order to achieve higher yields of the native isomer, especially in high-throughput syntheses it is necessary to select appropriate oxidative folding conditions. Moreover, the poor biochemical stability exhibited by α-conotoxins limits their general therapeutic applicability in vivo. Numerous strategies to enhance their stability including the substitution of disulfide bond with diselenide bond and N-to-C cyclization via an oligopeptide spacer have successfully overcome these limitations. This chapter describes methods for performing both selective and nonselective disulfide bond oxidation strategies for controlling the yields and formation of α-conotoxin disulfide bond isomers, as well as methods for the production of highly stable diselenide-containing and N-to-C cyclized conotoxin analogs.

Key words Conotoxins, Disulfide bonds, Diselenide bonds, Cyclized conotoxins

1 Introduction

Disulfide rich polypeptides isolated from venom sources have provided researchers with a vast array of research probes for studying a variety of neuropathological conditions such as pain, depression, schizophrenia, and drug addiction, with enormous potential as drug leads [1]. Of increasing interest are the conotoxins, which are isolated from marine gastropods that inhabit tropical reef ecosystems [2]. Conotoxins exhibit a small number of conserved disulfide bond frameworks, which give rise to very rigid and well defined three dimensional scaffolds that project hypervariable amino acid

Predrag Cudic (ed.), *Peptide Modifications to Increase Metabolic Stability and Activity*, Methods in Molecular Biology, vol. 1081, DOI 10.1007/978-1-62703-652-8_2, © Springer Science+Business Media New York 2013

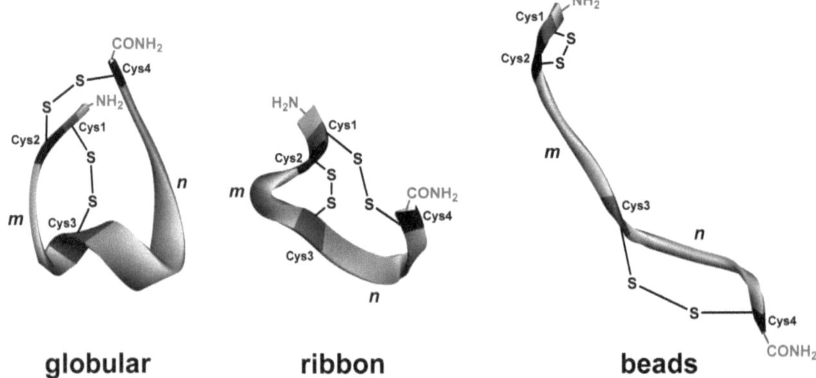

Fig. 1 Three possible disulfide bond isomers of α-conotoxins

residues thus accounting for their exquisite selectivity for different ion-channels and receptor classes [3]. Among these, the α-conotoxins can target different subtypes of nicotinic acetylcholine receptors (nAChRs) with a high degree of specificity [4]. As such, they can be used as novel subtype specific probes to map the role that nAChRs play in the central nervous system and in the development of drug leads for treating pain and tobacco addiction. The χ-conotoxins are a structurally related class, which are selective norepinephrine transporter inhibitors [5]. Importantly, a synthetic χ-conotoxin derivative, Xen2174, is currently undergoing phase II clinical trials as an analgesic for treating chronic neuropathic pain [6].

α-Conotoxins consist of 12–20 amino acids that exhibit a highly conserved cysteine framework consisting of two disulfide bonds (Fig. 1). Residues Cys1 and Cys2 are always adjacent to each other, with Cys4 occurring at or near the C-terminal. The two intervening loops of amino acids are projected from the scaffold between Cys2-Cys3 and Cys3-Cys4, denoted by letters "*m*" and "*n*" respectively. The predominant disulfide bond connectivity of naturally occurring α-conotoxins occurs between Cys[1–3, 2–4] and is commonly referred to as the "globular" isomer. However, two additional disulfide bond isomers are also possible, namely, the "ribbon" (Cys[1–4, 2–3]) and "beads" (Cys[1–2, 3–4]) isomers. While in most cases the globular isomer is the predominant bioactive isomer in α-conotoxins, χ-conotoxins exhibit the ribbon isomer in their natural bioactive form [5]. Nonetheless, nonnative isomers of α-conotoxins have been reported to exhibit novel pharmacological profiles. For example, the ribbon isomer of α-AuIB has been shown to be several times more potent at rat parasympathetic nAChR than the globular isomer [7]. As such, access to all three synthetic conotoxin isomers is a valuable tool for performing structure–activity relationship studies. With increasing use of combinatorial chemistry for performing accelerated structure–activity relationship studies of α-conotoxins, high-throughput methods for the production of synthetic analogs and their respective disulfide bond isomers are essential [8].

Like most other classes of peptides, α-conotoxins exhibit poor biochemical stability and resistance to proteolytic degradation, resulting in a short biological half-life that limits their general applicability as therapeutics. Furthermore, the disulfide bonds in α-conotoxins are inherently unstable and can undergo reduction or scrambling to other isomers under biological reducing conditions encountered in vivo [9]. Engineering conotoxin analogs with higher biochemical stability has proven to be very effective in slowing down the process of degradation in human serum and extending their biological half-life in vivo.

This chapter describes common methods for accessing synthetic isomers of α-conotoxins, as well as engineering highly stable analogs for use in structure–activity relationship studies. Such strategies that will be addressed include substitution of disulfide bonds with non-reducible diselenide bonds, and N-to-C backbone cyclization. Although the methods here describe the synthesis of α-conotoxins, they can be readily applied to any class of disulfide rich peptides.

2 Materials

2.1 General Requirements for Peptide Synthesis

1. All syntheses are performed manually as previously described using either a glass peptide synthesis vessel with a fritted filter, screw cap, and PTFE stop-cock (VWR, Radnor PA) [10], or tea bags prepared from 74 μM polypropylene mesh (Spectrum, Houston TX) using an impulse sealer as previously described (*see* **Note 1**) [11].
2. Unless otherwise indicated, 4-methylbenzyl-1-yl (MBHA) polystyrene resin (Chem-Impex, Wood Dale IL) is used for all syntheses.
3. 2-(1H-benzotriazole-1-yl)-1,1,3,3-tetramethyluroniumhexafl uorophosphate (HBTU) (ChemPep, Miami FL).
4. Dichloromethane (DCM), dimethylformamide (DMF), isopropanol (IPA), methanol, acetonitrile and trifluoroacetic acid (TFA) (Sigma-Aldrich, St. Louis MO). All solvents are reagent grade and are used without further purification.
5. N^{α}-*tert*-butyloxycarbonyl (Boc) amino acids with the following side chain protecting groups: Asn and Gln, xanthanyl (Xan); Asp and Glu, *O*-cyclohexyl (OcHxl); Arg and His, *p-toluenesulfonyl* (Tos); Cys, 4-methylbenzyl (MeBzl) or acetomidomethyl (Acm); Lys, 2-chlorobenzyloxycarbonyl (ClZ); Ser, Hyp and Thr, benzyl (Bzl); Tyr, 2-bromobenzyloxycarbonyl (BrZ); Trp, *N*-formyl (For) (Chem-Impex).

6. *N,N*-Diisopropylethylamine (DIEA) (Chem-Impex).
7. Anhydrous hydrogen fluoride (HF) (Airgas, La Porte TX) and a specialized HF cleavage apparatus constructed of corrosion resistant material (Peptides International, Louisville KY) [12, 13].
8. Scavengers for HF cleavage as follows: *p*-cresol, *p-thiocresol*, dimethylsulfide (DMS) and 1,2-ethanedithiol (EDT) (Sigma-Aldrich).

2.2 Analysis and Purification

1. A liquid chromatography mass spectrometer (LC-MS) (Shimadzu, Kyoto, Japan) is used to assess the molecular weight and purity of crude and purified peptide products, and to assess the completion of oxidation reactions. Analytical LC-MS is performed using a Jupiter, 50 mm×4.6 mm ID reversed phase C_{18} HPLC column (Phenomenex, Torrance CA). Buffer "A" 0.05 % aqueous formic acid; Buffer "B" 95 % acetonitrile, 5 % water, 0.05 % formic acid; Linear gradient, 0–60 % over 12 min; Flow rate 0.5 mL/min.
2. Peptides are purified using a preparative HPLC system (Waters, Milford MA) using a Luna, 150 mm×21.2 mm ID reversed phase C_{18} HPLC column (Phenomenex); Buffer "A" 0.1 % aqueous TFA; Buffer "B" 95 % acetonitrile, 5 % water, 0.1 % TFA; Linear gradient; 0–40 % or 0–60 % "B" over 40 min. Flow rate 20 mL/min. Detection wavelength 214 nm. The product peak is fractionated and the purity analyzed by LC-MS.
3. CD spectra are recorded using a J-720 spectropolarimeter (Jasco, Easton MD) using a 400 μL photometer cell with a 1 mm path length. Spectra are recorded between 190 and 260 nm, with an average of 4 scans.

2.3 Oxidation of Conotoxins

2.3.1 Preparation of Oxidation Buffers

1. 0.1 M Ammonium bicarbonate. Prepared by dissolving ammonium bicarbonate (7.91 g) in 1 L deionized water and then adjusting the pH with 1 M HCl or NH_4OH as required.
2. 0.1 M Ammonium acetate. Prepared by dissolving ammonium acetate (7.71 g) in 1 L deionized water and then adjusting the pH with glacial acetic acid or NH_4OH as required.

2.3.2 Iodine Mediated Oxidation of Conotoxins

1. 80 % Methanol in deionized water (v/v).
2. 0.1 M HCl solution.
3. 0.1 M $Na_2S_2O_3$ solution. Prepared by dissolving $Na_2S_2O_3$ (1.58 g) in 100 mL deionized water.
4. 0.1 M Iodine in methanol. Prepared by dissolving elemental iodine (0.253 g) in 10 mL of methanol.

2.4 Selenocysteine Directed Folding

2.4.1 Synthesis of Boc–Sec[MeBzl]OH

1. Metallic selenium powder, $NaBH_4$, $NH_4NH_2{\cdot}HCl$, celite, α-bromo-*p*-xylene, K_2CO_3, $MgSO_4$ and *tert*-butyloxycarbonyl dicarbonate (Sigma-Aldrich).
2. L-β-chloroalanine (Bachem, Bubendorf, Switzerland).

2.4.2 Synthesis of Diselenide Containing Conotoxins

1. Fmoc-SCAL Linker (Chem Impex) and aminomethyl ChemMatrix resin (BioMatrix, Saint-Jean Sur-Richelieu, Quebec).
2. NH_4I (Sigma-Aldrich).
3. 0.1 M Ammonium formate, pH 4.2. Prepared by dissolving ammonium formate (6.31 g) in 1 L deionized water, then adjusting the pH with formic acid or NH_4OH as required.

2.5 Cyclized Conotoxin Analogs

1. *S*-trityl-β-mercaptopropionyl MBHA resin (Peptides International).
2. 0.1 M Phosphate buffer, pH 8.2. Prepared by slowly adding 0.1 M sodium phosphate (dibasic) to 0.1 M sodium phosphate (monobasic) while monitoring with a pH meter as required. 0.1 M Phosphate buffer (dibasic) is prepared by dissolving Na_2HPO_4 (14.2 g) in 1 L deionized water. 0.1 M Phosphate buffer (monobasic) is prepared by dissolving NaH_2PO_4 (11.9 g) in 1 L deionized water.

3 Methods

3.1 General Boc Solid Phase Peptide Synthesis Procedure

All of the procedures described in this chapter use the Boc-chemistry approach. However, the Fmoc-chemistry approach may be used with equal effectiveness where indicated. For detailed procedures on solid phase peptide synthesis using both Boc or Fmoc chemistry, readers are referred to refs. [14, 15].

1. 55 % TFA in DCM (v/v) is used for the stepwise removal of the *N*α-Boc protecting group.
2. For syntheses using tea bags, batch washes with DCM (2×), IPA (2×) and DCM (2×) are used following Boc deprotection. Prior to coupling, the tea bags are neutralized using 5 % DIEA/DCM (v/v) (3×), followed by additional washes with DCM (3×) and DMF (3×).
3. For syntheses using a glass peptide synthesis vessel, in situ neutralization procedures and flow washes with DMF are used as previously described [15, 16].
4. For all syntheses, HBTU and DIEA are used to activate the amino acid prior to coupling. DMF is used as the coupling solvent.
5. Coupling reactions are monitored using the quantitative ninhydrin assay [17].
6. HF cleavage reactions are performed using either a two-step "low-high," or a one step "high" HF cleavage procedure as indicated for each method. For the low HF cleavage, tea bags containing peptide-resin are treated with "low" HF cleavage cocktail (25 % HF, 60 % DMS, 10 % *p*-cresol, and 5 % EDT (v/v/v/v)) for 2 h at 0 °C. The HF cleavage cocktail is

discarded and the tea bags are washed alternately with DCM and IPA (6×), then alternately with DMF and DCM (6×), and finally with methanol (6×) before drying under high vacuum.

7. For the “high” HF cleavage, peptide-resin is cleaved (95 % HF, 5 % *p*-cresol (v/v)) for 2 h at 0 °C. Following cleavage, the HF is evaporated and the peptide is precipitated with cold diethyl ether, centrifuged for 1 min (or filtered) and then washed again with additional diethyl ether. The peptide is then extracted with 95 % acetic acid/5 % H_2O (v/v) and lyophilized (*see* **Note 2**).

3.2 Directed Formation of Disulfide Bonds

The use of orthogonal protecting groups on each pair of cysteine residues can direct the formation of the desired target disulfide bond isomer (Fig. 2). For this approach, the *S*-acetomidomethyl (Acm) in combination with *S*-methylbenzyl (MeBzl) is widely employed for Boc chemistry. Similarly, the *S*-triphenylmethyl (Trityl, Trt) protecting group can be used with equal effectiveness for Fmoc-chemistry. The Acm protecting group is particularly versatile since deprotection and oxidation occurs simultaneously using iodine as the oxidation reagent [18, 19]. However, sensitive side chain residues, in particular Met and Trp are vulnerable to side reactions, which include oxidation of methionine or formation of tryptophan 2-thioether [20]. Depending on the extent of tryptophan 2-thioether formation, the use of Trp[For] with Boc chemistry can prevent this side reaction; however, this requires an additional deprotection and purification step (*see* **Note 3**) and may decrease the final yield [21].

1. Assemble the linear precursor conotoxin using orthogonal Boc-Cys[Acm]-OH and Boc-Cys[MeBzl]-OH on each pair of cysteine residues. Cleave from the resin using the “low-high” HF procedure and lyophilize (for peptides containing Trp[For], *see* **Note 3**).
2. Check the quality of the crude sample using LC-MS. If the sample purity is sufficiently high (>80 %), proceed to the next step. Otherwise, prior purification by preparative RP-HPLC may be required.
3. Dissolve the reduced precursor conotoxin in 100 mL of 0.1 M ammonium bicarbonate, pH 8.2. Up to 50 % organic co-solvent (e.g., isopropanol or methanol) may be added to aid dissolution. For particularly hydrophobic sequences, or those that are sensitive to basic conditions (e.g., sequences containing Asn, Gly, or Trp[For]), dissolve the peptide in 0.1 M ammonium acetate, pH 5.8 containing up to 30 % dimethylsulfoxide (DMSO). Agitate for 24 h using either a magnetic stirrer plate, or an orbital shaker platform (for parallel oxidation of multiple samples) in an open vessel (*see* **Note 4**). Monitor the progress of the reaction by analytical LC-MS.

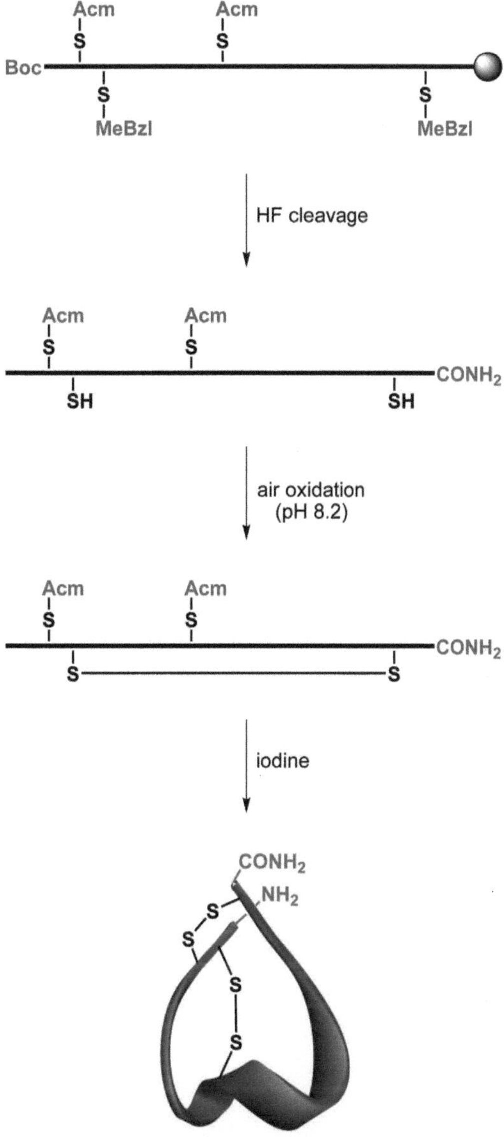

Fig. 2 Directed formation of disulfide bond isomers of α-conotoxins using orthogonal cysteine protecting groups

4. When the oxidation is judged to be complete, acidify the sample to pH 2.0 using TFA. Organic co-solvents should be first evaporated in vacuuo prior to desalting and purification (*see* **Note 5**). For samples containing DMSO, dilute with aqueous 0.1 % TFA such that the final concentration of DMSO is <5 % to ensure that the peptide is retained on the column during the desalting step.
5. Desalt by passing the entire sample through a C_{18} reversed phase HPLC column via direct infusion. Equilibrate the column with buffer A and isolate the partially oxidized/partially

protected peptide using a HPLC gradient. Alternatively, samples can be evaporated to dryness and redissolved in a smaller volume (<5 mL) for injection using an autosampler. Following purification, lyophilize the purified partially oxidized peptide.

6. Dissolve the partially oxidized purified conotoxin in 80 % methanol to a concentration of approximately 2.5 mg/mL. Add 1 M HCl to a final concentration of 10 mM HCl. Add 10 equivalents of I_2 solution per Cys[Acm] group and stir for 5 min using a magnetic stirrer. To quench the reaction, add 0.1 M aqueous sodium thiosulfate solution until the reaction mixture becomes colorless. Dilute the sample to <5 % methanol with 0.1 % aqueous TFA prior to desalting.
7. Desalt by passing the entire sample through a C_{18} reversed phase HPLC column via direct infusion and isolate the fully oxidized peptide using a HPLC gradient.

3.3 Nondirected Formation of Disulfide Bonds

The nondirected formation of disulfide bond isomers is a simplified one step oxidation procedure, which is more generally applicable to the synthesis of native α-conotoxins and their analogs (Fig. 3).

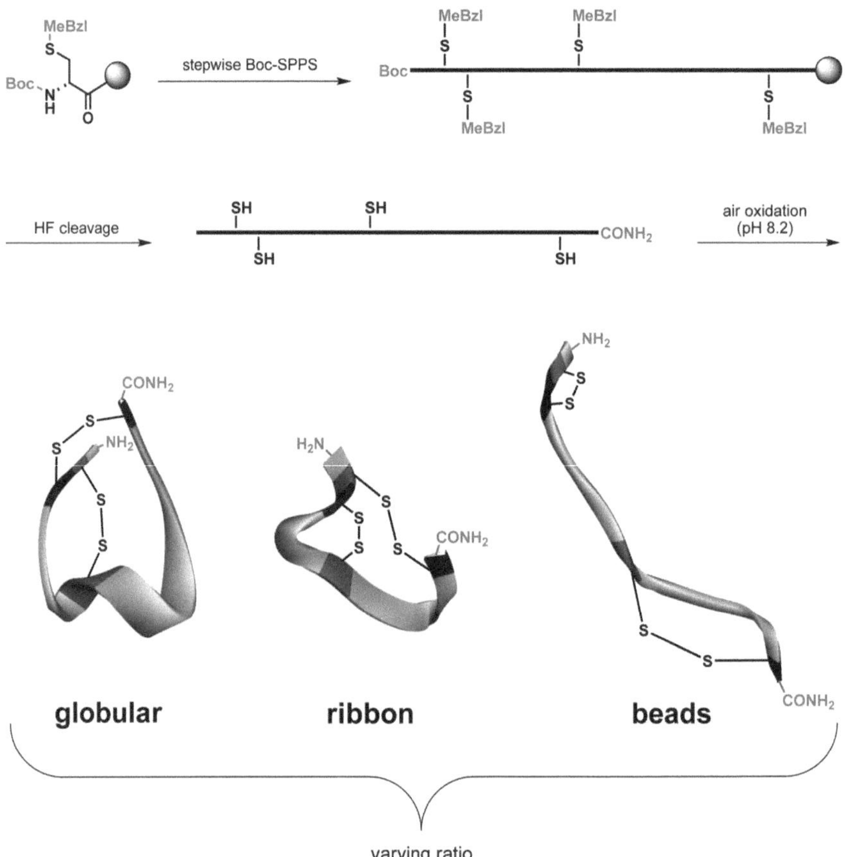

Fig. 3 Nondirected formation of α-conotoxin disulfide bonds using random oxidation of unprotected cysteine residues

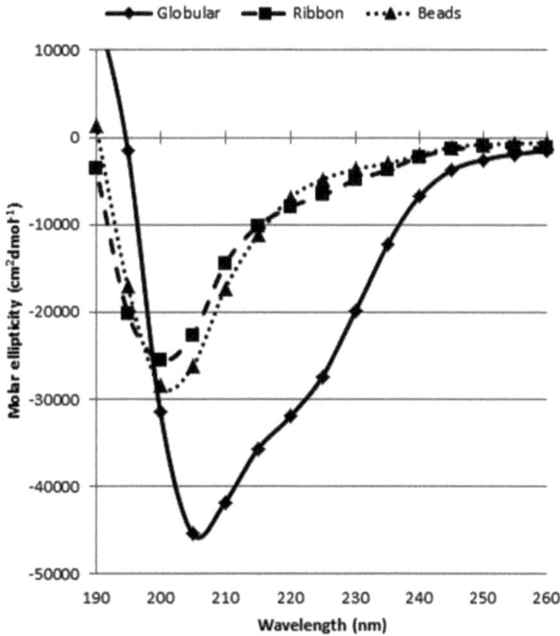

Fig. 4 Circular dichroism spectra of the three disulfide bond isomers of α-conotoxin MII

This strategy may be performed in one step, thus it is less labor intensive and is very useful for accelerating the high-throughput production of conotoxin libraries. Furthermore, it leads to increased product yields due to fewer purification steps and has fewer propensities for side reactions.

However, a mixture of disulfide bond isomers may be obtained in varying ratios, thus it is important to select conditions to maximize the accumulation of the native globular isomer. As such, optimization of oxidation conditions through a series of small scale trial oxidations may be required to obtain the desired isomer in high yield and purity. To this end, directed formation of each disulfide bond isomer can be carried out prior to commencing oxidation trials, which can be used as retention time markers by analytical RP-HPLC. Alternatively, native α-conotoxin isomers exhibit distinct circular dichroism (CD) spectra, with a minima occurring at 222 nm that is characteristic of a helical structure, whereas ribbon and bead isomers exhibit a more random conformation (Fig. 4).

When designing a series of random oxidation trials, one should consider a variety of factors, including the choice of oxidation buffer, organic co-solvents, redox reagents, pH, and temperature. Each of these factors can be investigated to obtain the best optimized conditions for maximization of yield and purity of the final isomer [22]. Generally, 0.1 M ammonium bicarbonate or 0.1 M ammonium acetate is used as oxidation buffers due to their compatibility with performing LC-MS analysis. In many cases, yields of native disulfide bond isomer have been found to increase by adding up to

50 % organic co-solvents such as isopropanol to the reaction mixture [23]. Additionally, inclusion of redox reagents such as a mixture of reduced and oxidized glutathione have been used to mimic physiological conditions to allow intermolecular disulfide interchange to occur more rapidly [22].

3.3.1 Oxidation Trials for Optimizing α-Conotoxin Folding

1. Synthesize each of the globular, ribbon and beads isomers using directed disulfide bond formation as described in Subheading 3.1. Analyze by analytical RP-HPLC and record the retention time of each isomer.
2. Assemble the linear precursor conotoxin using Boc-Cys [MeBzl]-OH on all four cysteine residues (*see* **Note 6**). Cleave from the resin, lyophilize and purify the reduced linear precursor by preparative RP-HPLC.
3. Prepare a series of oxidation buffers to examine the effect of buffer salt (e.g., ammonium bicarbonate or ammonium acetate), organic co-solvents (e.g., isopropanol, methanol, ethanol, or acetonitrile), redox reagents (e.g., reduced and oxidized glutathione), pH (e.g., 6.0–9.0) and temperature (e.g., 4 °C and ambient temperature) and time (24–72 h). Place 990 μL aliquots of each oxidation buffer into individual 3 mL glass vials containing a magnetic stirrer flea.
4. Prepare a 10 mg/mL aqueous stock solution of purified reduced conotoxin in deionized water. Aliquot 10 μL of conotoxin stock solution into each vial containing oxidation buffer (the final concentration of conotoxin in each trial should be 0.1 mg/mL). Agitate the vials for 24–72 h using a magnetic stirrer plate (*see* **Note 3**).
5. At various time points, remove an aliquot and quench the oxidation by acidifying to pH 2.0 with a solution of 10 % aqueous TFA. Analyze the reaction mixture by analytical HPLC and compare the retention times of the products with each isomer obtained using directed disulfide bond formation to identify the proportion of each isomer for each condition.

3.3.2 Large Scale Random Oxidation of α-Conotoxins

1. Assemble the linear precursor conotoxin using Boc-Cys [MeBzl]-OH on all four cysteine residues (*see* **Note 6**). Cleave from the resin using the "low-high" HF procedure and lyophilize.
2. Check the quality of the crude sample using LC-MS. If the sample purity is sufficiently high (>80 %), then proceed to the next step. Otherwise, prior purification by preparative RP-HPLC may be required.
3. Weigh 20–50 mg of the reduced precursor conotoxin into a 125 mL Erlenmeyer or round bottom flask and dissolve in 100 mL of optimized oxidation buffer as determined from small scale trial oxidations (*see* Subheading 3.2, **step 1**).

Agitate for 24–72 h using either a magnetic stirrer plate, or an orbital shaker platform (for parallel oxidation of multiple samples) in an open vessel (*see* **Note 4**). Monitor the progress of the reaction by analytical LC-MS or analytical HPLC.

4. When judged to be complete, acidify to pH 2.0 using a solution of 50 % aqueous TFA (2 mL). Organic co-solvents should be first evaporated in vacuuo prior to desalting and purification (*see* **Note 5**).
5. Desalt by passing the entire sample through a C_{18} reversed phase HPLC column via direct infusion and isolate the fully oxidized α-conotoxin using a HPLC gradient.

3.4 Selenocysteine Directed Folding

Although the presence of multiple disulfide bonds in α-conotoxins is crucial for stabilizing their three dimensional conformations, they are very prone to reduction or scrambling to other isomers by thiol containing molecules usually found in blood plasma, which can decrease their efficacy in vivo. Several strategies have been explored to overcome this problem, including substitution of disulfides with non-reducible moieties such as diselenide, lactam, thioether, or dicarba-linkages [9, 24–26]. Among these approaches, systematic replacement of disulfide bonds with diselenide bonds has been shown to be the most promising for increasing stability in vivo, while retaining pharmacological activity at the target receptors [9, 27].

Selenocysteine (Sec) is a naturally occurring amino acid, which exhibits the propensity to oxidatively form a diselenide bond in analogy to the disulfide bond. Diselenide bonds exhibit very similar bond geometry to disulfide bond and can be viewed as one of the most conservative amino acid substitutions available [28]. Importantly, oxidation of selenocysteine to the corresponding diselenide bond occurs much faster than cysteine at pH 5.0, allowing selenocysteine to be selectively oxidized over cysteine at lower pH. Furthermore, the redox potential for a mixed sulfide/selenide bond is higher than that of a diselenide bond, suggesting that its formation is unfavorable. As such, selenocysteine can be used to selectively control the formation of α-conotoxin disulfide bond isomers in a one-pot reaction without the requirement of multiple isolation steps when appropriately incorporated into the precursor peptide sequence (Fig. 5) [9]. Moreover, diselenide containing α-conotoxin analogs exhibit increased resistance to reduction or scrambling under several biological reducing conditions, including blood plasma [9, 27].

3.4.1 Synthesis of Boc-Sec[MeBzl]-OH

L-Selenocysteine ($[Sec]_2$)

1. Suspend metallic selenium powder (4.5 g, 57 mmol) in H_2O (25 mL) in a two-neck 250 mL round bottom flask with a magnetic stirrer bar. Cool the flask to 0 °C in an ice/salt bath.
2. Dissolve $NaBH_4$ (4.5 g, 119 mmol) in H_2O (25 mL) and transfer to a dropping funnel. Slowly add the $NaBH_4$ solution drop wise to the reaction mixture. After the vigorous exothermic

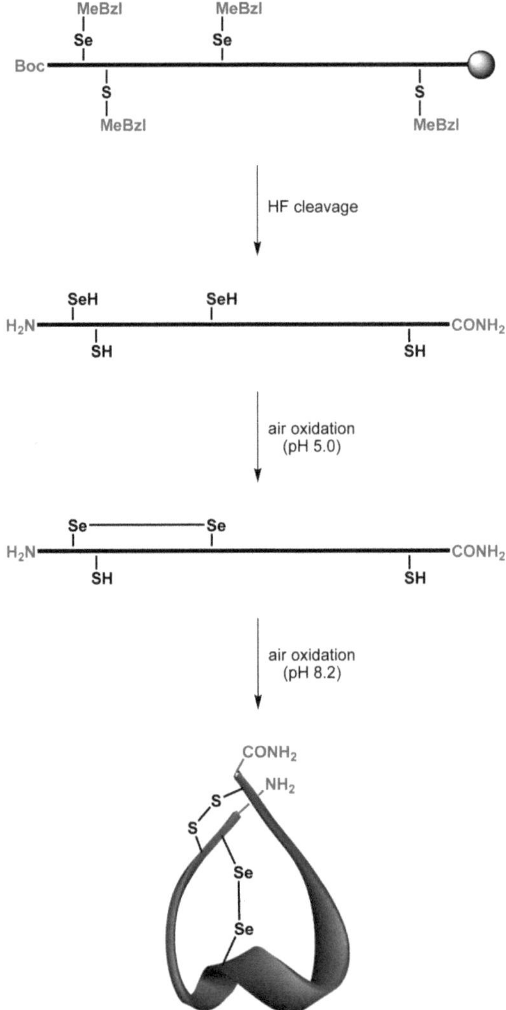

Fig. 5 General synthesis of an α-conotoxin using selenocysteine directed folding

reaction has subsided and the suspension becomes colorless, add additional metallic selenium powder and stir for further 15 min at room temperature until the remaining selenium is dissolved, forming a red/brown colored solution of sodium diselenide.

3. Dissolve L-β-chloroalanine (5.0 g, 31 mmol) in H_2O (40 mL). Adjust the solution to pH 9.0 by adding 0.1 M NaOH and transfer to a dropping funnel and add drop wise over a period of 2 h to the sodium diselenide solution (*see* **Note 7**). Seal the reaction flask and stir the reaction overnight at 40 °C.

4. Acidify the reaction mixture to pH 2.0 using 6 M HCl and add hydroxylamine hydrochloride (0.330 g, 9.7 mmol). Flush the reaction vessel with nitrogen gas for 2 h while passing the

exhaust gas through two successive NaOCl traps. Filter the reaction mixture though a celite plug to remove any excess selenium powder. Flush the yellow filtrate with nitrogen for an additional 1 h using a fritted Drescher bottle.

5. Carefully adjust the pH of the yellow filtrate to 6.5 using 10 M NaOH, allowing the product to precipitate as a yellow solid. Filter using a Buchner funnel and redissolve in minimum volume of 2 M HCl. Filter any residual elemental selenium and again precipitate the product by adjusting the solution to pH 6.5 using 10 M NaOH. Filter the final amorphous yellow product using vacuum filtration (yield 4.62 g, 70 %).
6. Characterize the product using NMR spectroscopy. ^{1}H NMR (300 MHz, D_2O + DCl + DSS) δ 8.2 (d, 2H), 7.9 (d, 2H), 5.1 (m, 1H), 4.8 (s, 2H), 3.9 (m, 1H), 3.1 (s, 3H).

Se-(4-methylbenzyl)-L-selenocysteine

1. Suspend L-selenocysteine (3.80 g, 9.3 mmol) in 0.5 M NaOH (15 mL) with magnetic stirring. Cool the suspension to 0 °C with an ice/salt bath.
2. Dissolve $NaBH_4$ (3.60 g, 95 mmol) in H_2O (15 mL) and transfer to a dropping funnel. Add the $NaBH_4$ solution drop wise to the reaction vessel. After the vigorous exothermic reaction has subsided and the suspension becomes colorless (approximately 30 min), adjust the solution to pH 7.0 under a blanket of argon using glacial acetic acid.
3. Dissolve α-bromo-*p*-xylene (1.11 g, 60 mmol) in ethanol (15 mL) and add dropwise to the reaction mixture over 30 min. Stir the reaction mixture for a further 2 h at 0 °C under argon.
4. Acidify the reaction mixture to pH 2 using 6 M HCl, which will produce the product as a white precipitate. Filter the product under vacuum, wash with water and then diethyl ether. The final product is recrystallized from hot water (yield 3.50 g, 61.1 %).
5. Characterize the product using NMR spectroscopy. ^{1}H NMR (300 MHz, CD_3OD + D_2O + DCl + DSS) δ 8.2 (d, 2H), 7.9 (d, 2H), 5.1 (m, 1H) 4.8, (s, 2H), 3.9 (m, 1H), 3.1 (s, 3H); 13C NMR (75.4 MHz, CD_3OD + D_2O + DCl) δ 170.7, 138.0, 136.8, 130.3, 130.1, 53.9, 28.7, 23.2, 21.4.

Boc-Sec[MeBzl]-OH

1. Dissolve *Se*-(4-methoxybenzyl)-L-selenocysteine (3.24 g, 10 mmol) together with K_2CO_3 (3.4 g, 0.25 mmol) in water (25 mL) with magnetic stirring. Gently heat to aid dissolution.
2. Dissolve *tert*-Butyloxycarbonyl dicarbonate (2.30 g, 11 mmol) in THF (25 mL) and add to the reaction mixture. Stir the reaction for 1 h at room temperature.

3. Add 100 mL of H_2O to the reaction mixture and transfer to a separating funnel. Wash the mixture with diethyl ether (2 × 100 mL) and then separate the aqueous layer.
4. Acidify the aqueous layer to pH 4.0 with solid citric acid and extract with ethyl acetate (3 × 100 mL). Combine the ethyl acetate extracts and wash with 10 % citric acid (3 × 100 mL) and then brine (100 mL). Dry over solid $MgSO_4$, filter and evaporate the solvent using a rotary evaporator. Recrystallize the final product using petroleum spirits/diethyl ether (yield 2.8 g, 75.3 %).
5. Characterize the product using NMR spectroscopy. ^{1}H NMR (300 MHz, $CDCl_3$ + TMS) δ 7.14 (d, 2H), 7.02 (d, 2H), 5.29 (d, 2H), 3.79 (m, 2H), 2.87 (s, 2H), 2.32 (s, 3H), 1.45 (s, 9H); ^{13}C NMR (75.4 MHz, $CDCl_3$) δ 175.6, 155.4, 136.6, 135.4, 129.3, 128.8, 80.5, 53.3, 28.3, 27.8, 25.3, 21.1.

3.4.2 General Synthesis of Diselenide Containing Conotoxins

1. Assemble the linear precursor conotoxin using Boc-solid phase peptide synthesis with combinations of Boc-Sec[MeBzl]-OH and Boc-Cys[MeBzl]-OH to achieve the desired disulfide/diselenide bond connectivity.
2. Cleave the peptide from the resin using the high HF procedure for 2 h at 0 °C.
3. Following evaporation of HF, precipitate the conotoxin with cold ethyl acetate degassed with nitrogen. Filter and wash the precipitated peptide with additional cold ethyl acetate.
4. The crude reduced conotoxin can be immediately dissolved in 0.1 M ammonium formate buffer (pH 4.2) containing 50 % isopropanol to a concentration $<$10 mmol and stirred for 2 h at room temperature in an open vessel.
5. Carefully adjust the pH of the solution to 8.2 using ammonium hydroxide solution and continue stirring overnight at room temperature in an open vessel. Monitor the oxidation progress using LC-MS.
6. When the oxidation is judged to be complete, acidify the sample to pH 2.0 using TFA. Remove the isopropanol in vacuo (*see* **Note 4**). Desalt by passing the entire sample through a C_{18} reversed phase HPLC column via direct infusion and isolate fully oxidized selenoconotoxin using a HPLC gradient.

3.4.3 On-Resin Oxidation of Diselenide Containing Conotoxins

On resin supported oxidation of selenocysteine containing conotoxins can potentially be used in the parallel production of synthetic combinatorial libraries of α-conotoxins [27]. In this synthesis, the peptide is assembled on an amphiphilic resin containing a HF stable safety catch amide linker (SCAL), thus allowing deprotection of all side chain protecting groups using HF. This facilitates on-resin oxidation of the disulfide/diselenide bond framework prior to cleavage of the fully oxidized conotoxin from the resin (Fig. 6).

1. Couple the Fmoc-SCAL linker to aminomethyl ChemMatrix® resin containing three glycine residues as a spacer. Deprotect with 50 % piperidine/DMF (v/v).
2. Transfer 100–500 mg of resin to 74 μm mesh polypropylene tea bags. Assemble the linear precursor conotoxin using Boc-solid phase peptide synthesis with combinations of Boc-Sec[MeBzl]-OH and Boc-Cys[MeBzl]-OH to achieve the desired disulfide/diselenide bond connectivity.
3. Deprotect the side-chain protecting groups using the "high" HF procedure (90 % HF, 5 % *p*-cresol, 5 % *p*-thiocresol) for 2 h at 0 °C. After evaporating the HF, wash the tea bags TFA (2×), DCM (6×), DMF (6×) and H_2O (6×).

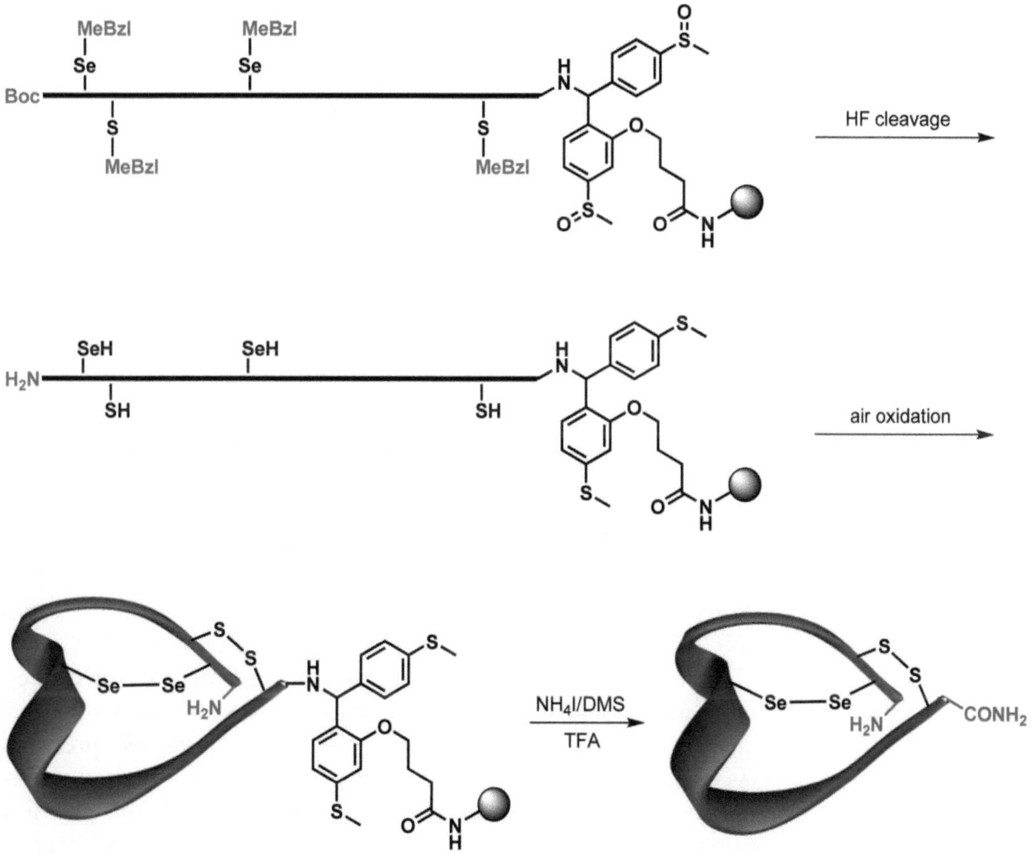

Fig. 6 On-resin selenocysteine directed folding of a α-conotoxin using a safety catch amide linker

4. Place the tea bags containing resin into a solution of 0.1 M NH_4HCO_3, pH 8.4 (10 mL/100 mg of resin). Shake overnight in a Nalgene container at room temperature and then wash with methanol (6×) and dry under vacuum.
5. Remove the resin from the tea bags and weigh approximately 300 mg into a 25 mL round bottom flask. Add a solution of NH_4I (100 mg), dimethyl sulfide (200 μL), and TFA (5 mL) to the resin and stir for 4 h at room temperature.
6. Evaporate the TFA under a stream of nitrogen. Precipitate the peptide product with cold ethyl acetate, filter and wash with additional ethyl acetate. Redissolve the peptide in 50 % aqueous acetonitrile containing 0.1 % TFA and lyophilize. Purify the final product by preparative C_{18} RP-HPLC.

3.5 Cyclized Conotoxin Analogs

Cyclization is an effective modification strategy to increase overall biochemical stability of therapeutically relevant peptides in vivo. Numerous classes of disulfide rich cyclic peptides are found in nature which exhibit compact three-dimensional structures and remarkable stability. These include the plant cyclotides, sunflower trypsin inhibitor, and mammalian theta-defensins [29–31]. In view of the superior stability offered by cyclic disulfide rich peptides, reengineering of stable conotoxins through backbone cyclization serves as a useful synthetic modification strategy to enhance their in vivo stability.

N-to-C cyclization of α-conotoxins via an inert oligopeptide spacer unit has yielded analogs with vastly improved stability than the native toxin under biological conditions, while retaining the pharmacological activity of the native conotoxin (Fig. 7) [32]. This has been attributed to the fact that cyclization leads to an overall tightening of the peptide structure resulting in a loss of flexibility, while preserving key structural characteristics that are crucial for maintaining pharmacological activity [32]. As such, cyclized α-conotoxins exhibit increased resistance towards proteolytic degradation and improved stability in human serum. Significantly, a cyclized analog of α-conotoxin Vc1.1 has recently shown promise as an orally available analgesic in rodent neuropathic pain models [33].

Cyclization of conotoxins is usually achieved in mildly basic aqueous buffer through intramolecular native chemical ligation (NCL) reaction between an N-terminal cysteine residue and a C-terminal thioester and subsequent rearrangement to form a peptide amide bond with the regeneration of N-terminal cysteine side chain (Fig. 8) [34]. As such, the linear thioester precursor α-conotoxin is assembled as a cyclic permutant such that one of the four cysteine residues occurs at the N-terminal. While there are four potential ligation sites to consider, a convenient Gly-Cys occurs in many α-conotoxins, which offers minimal steric hindrance and increased reaction rates. The presence of internal disulfide bonds has also been proposed to accelerate the cyclization process via a

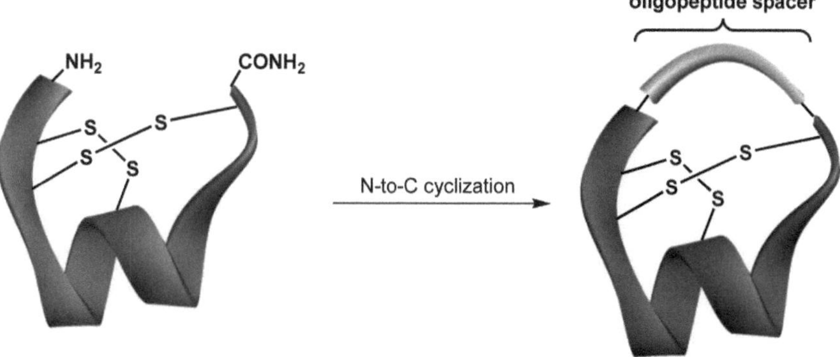

Fig. 7 N-to-C cyclization of an α-conotoxin through an inert spacer

thiazip mechanism, with complete conversion to the cyclized product occurring within minutes [35]. As such, cyclization and oxidation can be performed using a one-pot reaction, since oxidative formation of the disulfide bonds occurs at a much slower rate than cyclization. Although numerous methods have been reported for the production of C-terminal thioesters using Fmoc SPPS [36–38], Boc chemistry represents the most widely applicable and robust approach for synthesizing cyclized conotoxin analogs. Disulfide bond formation can be achieved using either the two-step directed strategy, or a one-pot nondirected strategy. However, varying disulfide bond isomers may be obtained by using a nondirected approach, particularly if the length of the spacer is not optimized.

When designing cyclic α-conotoxin analogs, the length and nature of the oligopeptide spacer spanning the N- and C-termini must be considered. Spacers consisting of consecutive arrangements of functionally inert Gly-Ala residues have proven to be effective for enhancing the biological stability of numerous α-conotoxins. Moreover, the use of an inert Gly-Ala linker allows for the possibility of introducing additional functional groups to further enhance the physical properties of cyclized conotoxins without interfering with the pharmacophore [39]. The spacer length must be optimized in order to achieve the correct orientation of key binding residues that form the α-conotoxin pharmacophore. Additionally, the spacer length also plays an important role in dictating the formation of disulfide bond isomers in one pot nonselective oxidation and cyclization reactions. For example, cyclization of α-conotoxins ImI and AuIB with three or fewer amino acid spacers showed greater preferences towards formation of the ribbon isomer [21, 40]. In such cases directed oxidation strategy implementing orthogonal cysteine protection was used for increasing the yield of the globular isomer. Therefore an important criteria for choosing linker size will depend on the three dimensional conformation of the native conotoxins and the directed

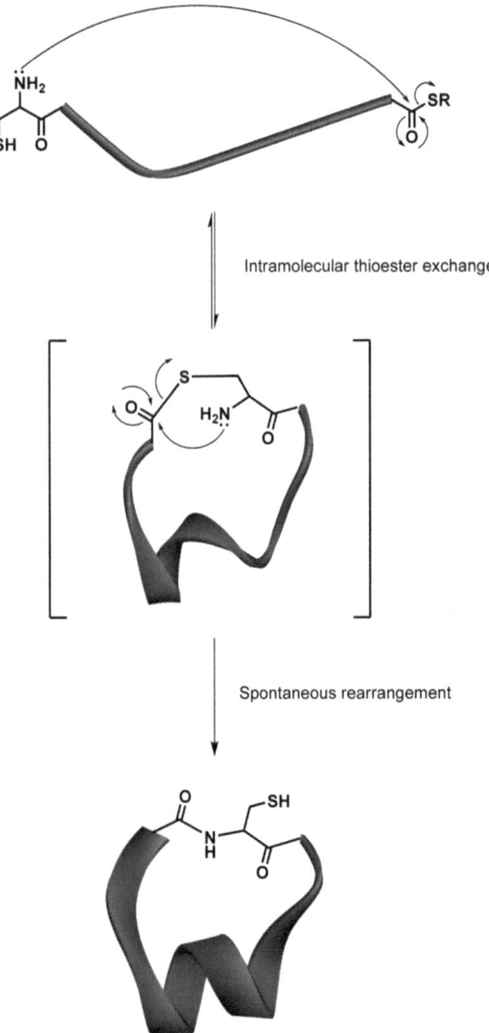

Fig. 8 Intramolecular native chemical ligation reaction in the synthesis of cyclized α-conotoxin analogs

oxidation approach employing orthogonal protection of two cysteine residues can be used to obtain desired disulfide bond isomers in major proportion.

3.5.1 Preparation of Cyclized Conotoxin Analogs (Directed Folding Approach)

1. Deprotect *S*-trityl-β-mercaptopropionyl MBHA the resin by treating with a cocktail of 90 % TFA, 5 % H_2O and 5 % triisopropylsilane (2 × 10 min treatments).
2. Assemble the conotoxin sequence as a cyclic permutant using pairs of Boc-Cys[MeBzl]-OH and Boc-Cys[Acm]-OH to achieve the desired disulfide bond connectivity (*see* **Note 8**).
3. Cleave the linear thioester peptide precursor from the resin using the "high" HF procedure using 90 % HF, 10 % *p*-cresol

(*see* **Note 9**). Check the quality of the crude sample using LC-MS. If the sample quality is sufficiently good, proceed to the cyclization step. Otherwise, prior purification by preparative RP-HPLC may be required.

4. Dissolve 30–50 mg of crude sample in 180 mL of 0.1 M phosphate buffer, pH 8.2 and shake or stir vigorously for 24–48 h in an open vessel (*see* **Note 4**). An organic co-solvent may be used to aid dissolution of hydrophobic peptides. Monitor the progress of the cyclization using LC-MS.
5. When the oxidation is judged to be complete, acidify the sample to pH 2.0 using TFA. For samples containing organic co-solvent, remove the solvent in vacuo (*see* **Note 5**). Desalt by passing the entire sample through a C_{18} reversed phase HPLC column via direct infusion. Equilibrate the column with buffer A and isolate the partially oxidized/partially protected peptide using a HPLC gradient.
6. Dissolve the partially oxidized peptide in 80 % methanol containing 10 mM HCl to a concentration of 2.5 mg/mL. Prepare a 0.1 M solution of I_2 in methanol. Add 10 equivalents of the I_2 solution per Cys[Acm] group and stir for 5 min using a magnetic stirrer. Add 0.1 M aqueous sodium thiosulfate solution until the reaction becomes colorless. Dilute the sample to <5 % methanol prior to desalting.
7. Desalt by passing the entire sample through a C_{18} reversed phase HPLC column via direct infusion and isolate the fully oxidized cyclized conotoxin using a HPLC gradient.

4 Notes

1. Tea bags may be used to efficiently prepare large synthetic combinatorial libraries of α-conotoxins and cyclic analogs as previously described [8, 40]. For detailed procedures regarding synthetic combinatorial libraries, readers are referred to [41].
2. Peptides can appear oily after lyophilization with 95 % acetic acid. Further lyophilization with 50 % aqueous acetonitrile containing 0.1 % TFA may be required to obtain crude solid peptide material.
3. For sequences containing a protected Trp[For], HF cleavage procedures using thiol scavengers should be avoided to prevent premature removal of the formyl protecting group. As such, the "low" HF cleavage is not compatible. To remove the formyl protecting group following oxidation, redissolve the oxidized conotoxin (approximately 10 mg) in 6 M guanidine hydrochloride (16 mL) and cool to 0 °C. Add 1 mL of ethanolamine and stir for 5 min at 0 °C. Acidify to pH 2 using 1 M

HCl, dilute to >100 mL and purify by C_{18} preparative RP-HPLC.

4. The mouth of each vessel can be sealed with plastic laboratory film and pierced with a pipette tip to allow air to permeate the oxidation mixture while preventing foreign objects from entering.
5. A rotary solvent evaporator can be used to evaporate co-solvents for single samples, or an automated system can be used for multiple samples (e.g., Genevac Rocket or Biotage V10 evaporators).
6. When using tea bags, assembly and cleavage can be performed in parallel with the synthesis of individual isomer via the directed synthesis strategy to increase efficiency.
7. Dropwise addition over 2 h is required to prevent selenomethionine formation.
8. It is crucial that a free cysteine side chain is present as the N-terminal residue in the linear thioester precursor to allow the internal NCL reaction to occur. As such, the location of Cys[Acm] residues should be considered carefully to achieve the desired disulfide bond connectivity.
9. HF cleavage procedures using thiol scavengers should be avoided to prevent premature cleavage of the peptide via thioester exchange. As such, the "low" HF cleavage is not compatible with thioester peptides.

Acknowledgment

The authors acknowledge financial support by the James and Esther King Biomedical Research Program (New Investigator Grant, 1KN02-33990), the Arthritis and Chronic Pain Research Institute, and the State of Florida.

References

1. Lewis RJ, Garcia ML (2003) Therapeutic potential of venom peptides. Nat Rev Drug Discov 2:790–802
2. Han TS, Teichert RW, Olivera BM, Bulaj G (2008) Conus venoms—a rich source of peptide-based therapeutics. Curr Pharm Des 14:2462–2479
3. Woodward SR, Cruz LJ, Olivera BM, Hillyard DR (1990) Constant and hypervariable regions in conotoxin peptides. EMBO J 9:1015–1020
4. Armishaw CJ (2010) Synthetic α-conotoxin mutants as probes for studying nicotinic acetylcholine receptors and in the development of novel drug leads. Toxins (Basel) 2: 1470–1498
5. Sharpe IA, Gehrmann J, Loughnan ML, Thomas L, Adams DA, Atkins A, Palant E, Craik DJ, Alewood PF, Lewis RJ (2001) Two new classes of conopeptides inhibit the α1-adrenoreceptor and noradrenaline transporter. Nat Neurosci 4:902–907
6. Brust A, Palant E, Croker DE, Colless B, Drinkwater R, Patterson B, Schroeder C, Wilson D, Nielsen CK, Smith MT, Alewood D, Alewood PF, Lewis RJ (2009) χ-Conopeptide pharmacophore development: Toward a novel

class of norepinephrine transporter inhibitor (Xen2174) for pain. J Med Chem 52: 6991–7002

7. Dutton JL, Bansal PS, Hogg RC, Adams DJ, Alewood PF, Craik DJ (2002) A new level of conotoxin diversity, a non-native disulfide bond connectivity in α-conotoxin AuIB reduces structural definition but increases biological activity. J Biol Chem 277: 48849–48857
8. Armishaw CJ, Singh N, Medina-Franco J, Clark RJ, Scott KCM, Houghten RA, Jensen AA (2010) A synthetic combinatorial strategy for developing α-conotoxin analogs as potent α7 nicotinic acetylcholine receptor antagonists. J Biol Chem 285:1809–1821
9. Armishaw CJ, Daly NL, Nevin ST, Adams DJ, Craik DJ, Alewood PF (2006) α-Selenoconotoxins: a new class of potent α7 neuronal nicotinic receptor antagonists. J Biol Chem 281:14136–14143
10. Schnölzer M, Alewood P, Jones A, Alewood D, Kent SBH (1992) In situ neutralization in Boc-chemistry solid phase peptide synthesis. Int J Protein Pept Res 40:180–193
11. Houghten RA (1985) General-method for the rapid solid-phase synthesis of large numbers of peptides—specificity of antigen-antibody interaction at the level of individual amino-acids. Proc Natl Acad Sci USA 82:5131–5135
12. Houghten RA, Bray MK, Degraw ST, Kirby CJ (1986) Simplified procedure for carrying out simultaneous multiple hydrogen fluoride cleavages of protected peptide resins. Int J Pept Protein Res 27:673–678
13. Stewart JM (1997) Cleavage methods following boc-based solid phase peptide synthesis. In: Fields GB (ed) Methods in enzymology. Academic, New York, pp 29–44
14. Pennington MW, Dunn BM (eds) (1994) Peptide synthesis protocols. Methods in molecular biology. Humana, Totowa, NJ
15. Fields GB (ed) (1997) Solid-phase peptide synthesis. Methods in enzymology, vol 289. Academic, New York, NY
16. Schnölzer M, Alewood P, Jones A, Alewood D, Kent SBH (2007) In situ neutralization in Boc-chemistry solid phase peptide synthesis. Int J Pept Res Ther 13:31–44
17. Sarin VK, Kent SBH, Tam JP, Merrifield RB (1981) Quantitative monitoring of solid phase peptide synthesis by the ninhydrin reaction. Anal Biochem 117:147–157
18. Kamber B, Hartmann A, Eisler K, Riniker B, Rink H, Sieber P, Rittel W (1980) The synthesis of cystine peptides by iodine oxidation of S-trityl-cysteine and S-acetamidomethyl-cysteine peptides. Helv Chim Acta 63:899–915
19. Ponsati B, Giralt E, Andreu D (1990) Solid-phase approaches to regiospecific double disulfide bond formation. Application to a fragment of bovine pituitary peptide. Tetrahedron 46:8255–8266
20. Sieber P, Kamber B, Riniker B, Rittel W (1980) Iodine oxidation of S-trityl- and S-acetomidomethyl-cysteine-peptides containing tryptophan: conditions leading to the formation of tryptophan-2-thioethers. Helv Chim Acta 63:2358–2362
21. Armishaw CJ, Dutton JL, Craik DJ, Alewood PF (2010) Establishing regiocontrol of disulfide bond isomers of α-conotoxin ImI via the synthesis of N-to-C cyclic analogs. Biopolymers 94:307–313
22. Bulaj G (2005) Formation of disulfide bonds in proteins and peptides. Biotechnol Adv 23: 87–92
23. Neilsen JS, Buczek P, Bulaj G (2004) Cosolvent-assisted oxidative folding of a bicyclic α-conotoxin ImI. J Pept Sci 10:249–256
24. Hargittai B, Solé NA, Groebe DR, Abramson SN, Barany G (2000) Chemical synthesis and biological activities of lactam analogues of α-conotoxin SI. J Med Chem 43:4787–4792
25. Dekan Z, Vetter I, Daly NL, Craik DJ, Lewis RJ, Alewood PF (2011) α-Conotoxin ImI incorporating stable cystathione bridges maintains full potency and identical three-dimensional structure. J Am Chem Soc 133: 15866–15869
26. MacRaild CA, Illesinghe J, van Lierop BJ, Townsend AL, Chebib M, Livett BG, Robinson AJ, Norton RS (2009) Structure and activity of (2,8)-Dicarba-(3,12)-cystino α-ImI, an α-conotoxin containing a nonreducible cystine analogue. J Med Chem 52:755–762
27. Muttenthaler M, Nevin ST, Grishin AA, Ngo ST, Choy PT, Daly NL, Hu S-H, Armishaw CJ, Wang C-I, Lewis RJ, Martin JL, Noakes PG, Craik DJ, Adams DJ, Alewood PF (2010) Solving the α-conotoxin folding problem: efficient selenium-directed on-resin generation of more potent and stable nicotinic acetylcholine receptor antagonists. J Am Chem Soc 132: 3514–3522
28. Muttenthaler M, Alewood PF (2008) Selenopeptide chemistry. J Pept Sci 14: 1223–1239
29. Craik DJ (2006) Seamless proteins tie up their loose ends. Science 311:1563–1564
30. Luckett S, Santiago-Garcia R, Barker JJ, Konarev AV, Shewry PR, Clarke AR, Brady RL (1999) High-resolution structure of a potent, cyclic proteinase inhibitor from sunflower seeds. J Mol Biol 290:525–533

31. Tang Y-Q, Yuan J, Ösapay G, Ösapay K, Tran D, Miller CJ, Ouellette AJ, Selsted ME (1999) A cyclic antimicrobial peptide produced in primary leukocytes by the ligation of two truncated α-defensins. Science 286:498–502
32. Clark RJ, Fischer H, Dempster L, Daly NL, Rosengren KJ, Nevin ST, Meunier FA, Adams DJ, Craik DJ (2005) Engineering stable peptide toxins by means of backbone cyclization: stabilization of the α-conotoxin MII. Proc Natl Acad Sci USA 102:13767–13772
33. Clark RJ, Jensen J, Nevin ST, Callaghan BP, Adams DJ, Craik DJ (2010) The engineering of an orally active conotoxin for the treatment of neuropathic pain. Angew Chem Int Ed 49:6545–6548
34. Camarero JA, Muir TW (1997) Chemoselective backbone cyclisation of unprotected peptides. Chem Commun 1997:1369–1370
35. Tam JP, Lu Y, Yu Q (1999) Thia zip reaction for synthesis of large cyclic peptides: mechanisms and applications. J Am Chem Soc 121: 4316–4324
36. Clippingdale AB, Barrow CJ, Wade JD (2000) Peptide thioester preparation by Fmoc solid phase peptide synthesis for use in native chemical ligation. J Pept Sci 6:225–234
37. Brask J, Albericio F, Jensen KJ (2003) Fmoc solid-phase synthesis of peptide thioesters by masking as trithioortho esters. Org Lett 5: 2951–2953
38. Tofteng AP, Jensen KJ, Hoeg-Jensen T (2007) Peptide dithioethanol esters for in situ generation of thioesters for use in native chemical ligation. Tetrahedron Lett 48: 2105–2107
39. Dekan Z, Paczkowski FA, Lewis RJ, Alewood PF (2007) Synthesis and in vitro biological activity of cyclic lipophilic χ-conotoxin MrIA analogues. Int J Pept Res Ther 13: 307–312
40. Armishaw CJ, Jensen AA, Balle LD, Scott KCM, Sørensen L, Strømgaard K (2011) Improving the stability of α-conotoxin AuIB through N-to-C cyclization: the effect of spacer length on stability and activity at nicotinic acetylcholine receptors. Antioxid Redox Signal 14:65–76
41. Cabilly S (ed) (1998) Combinatorial peptide library protocols. Methods in molecular biology, vol 87. Humana, Totowa, NJ

Chapter 3

Synthesis of AApeptides

Youhong Niu, Yaogang Hu, Haifan Wu, and Jianfeng Cai

Abstract

The creation and development of nonnatural peptidomimetics has become an area of increasing significance in bioorganic and chemical biology. A wide range of new peptide mimics with novel structures and functions are urgently needed to be explored in order to identify potential drug candidates and targeted probes, and to study protein functions. AApeptides are a new class of peptide mimics based on chiral PNA backbone. They are resistant to proteolytic degradation and have limitless potential for diversification. They have been found to have a wide variety of biological applications including cellular translocation, disruption of protein–protein interactions, formation of nanostructures, antimicrobial activity, etc. The synthesis of AApeptides is modular and straightforward. In this chapter, methods for the synthesis of AApeptides (including different subclasses) are described.

Key words α-AApeptides, γ-AApeptides, Solid-phase synthesis, Cyclization, Lipidation

1 Introduction

The creation of novel peptidomimetics with discrete structures and functions has become an area of high significance in chemical biology and biomedical sciences [1]. There has been extensive effort in the past two decades to develop sequence-specific oligomers [2, 3], including α/β-peptides [4, 5], polyamides [6], peptoids [7], β-peptides [8–10], γ- and δ-peptides [11–13], oligoureas [14, 15], azapeptides [16, 17], α-aminoxy-peptides [18], sugar-based peptides, γ- and δ-peptides [11–13], and phenylene ethynylenes [19]. These different classes of peptidomimetics are designed through the modification of α-peptide backbone, or introduction of peptide isosteres. As a result, they can at least mimic peptide primary structure, as well as secondary folding structure. It is noticeable that because of their nature of unnatural backbone, they are often resistant to proteolytic degradation, and are believed to have reduced immunogenicity and improved bioavailability compared to peptides [20]. Indeed, they have begun to

Predrag Cudic (ed.), *Peptide Modifications to Increase Metabolic Stability and Activity*, Methods in Molecular Biology, vol. 1081, DOI 10.1007/978-1-62703-652-8_3, © Springer Science+Business Media New York 2013

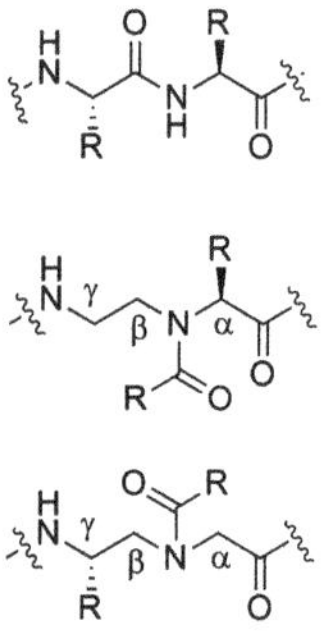

α-peptide

β-AApeptide

γ-AApeptide

Fig. 1 General structures of conventional α-peptides, α-AApeptides, and γ-AApeptides

find some important biological and biomedical applications [21, 22]. Despite great potential, the applications of peptidomimetics are still under developed, partially due to the limited frameworks [22]. New peptide mimics with novel structures and functions are urgently needed to be designed and investigated [5, 22]. Such new classes of peptidomimetics are increasingly important for the generation of chemically diverse library for drug discovery, design of protein/peptide mimics to study their biological functions, and design of novel biological probes, etc. To facilitate the application of peptidomimetics and to advance the field, we recently have developed a new class of peptide mimics termed "AApeptides" (Fig. 1), called so because the residues of this class of peptidomimetics are *N*-acylated-*N-aminoethyl* amino acids [23–30]. The scaffold of AApeptides is organically derived from chiral PNAs; however, unlike PNAs which are used for the mimicry of nucleic acids, AApeptides are developed to mimic the structure and function of peptides. Depending on the position of the side chains, two subclasses of AApeptides, α-AApeptides and γ-AApeptides were designed and synthesized.

As shown in Fig. 1, in a unit of AApeptides, one side chain is connected to either α- or γ-C in relation to the carbonyl group, and the other side chain is linked to the central N through acylation. Compared with natural α-peptides, the repeating unit (building block) of AApeptides is comparable to a dipeptide residue. As a result, AApeptides project an identical number of functional side groups as conventional peptides of the same length [26, 30]. As such, AApeptides can presumably mimic the primary structure and function of peptides. As half of side chains are introduced through acylation by any carboxylic acids, the potential of developing AApeptides with chemically diverse functional groups is limitless. In addition, half of side chains are still chiral, which may pose conformational bias onto AApeptides, and lead to the formation of certain folding structure. Similar to other classes of peptidomimetics, AApeptides are highly resistant to protease degradation [26, 30]. They are not digested by a mixture of enzymes in 24 h [26, 30]. They also have found to have some important biological applications. For instance, they can mimic the structure of p53 and disrupt cancer-related p53/MDM2 protein–protein interactions [26, 30]. One AApeptide is able to bind to RNA with high affinity comparable to the Tat peptide [24], demonstrating the potential of AApeptides to mimic bioactive peptides. Two AApeptides have also shown the capability to translocate the cellular membrane with high efficiency, indicating their potential use as drug carriers for cellular delivery of cargos [28, 29]. In addition, AApeptides have also found applications in nanotechnology. Two AApeptides are able to form novel nanorods and nanovesicles, with the morphology different from those formed from peptides, which shows the potential use of AApeptides in the development of novel nanomaterials [31]. Very recently, a few AApeptides were designed and have shown potent and broad-spectrum activity against a range of gram-positive and gram-negative bacteria by mimicking the structure, function and mechanisms of antimicrobial peptides [25, 27, 32, 33]. There is no doubt that AApeptides will have enormous biological and biomedical applications in the future.

The synthesis of AApeptides is also very straightforward, which is expected to further facilitate their potential applications. So far the synthetic protocols for both α-AApeptides and γ-AApeptides have been well established, including their subclasses, such as lipidated and cyclic forms. A general synthetic strategy can be outlined below:

1. Synthesis of AApeptide building blocks.
2. Synthesis of AApeptide sequences through the incorporation of AApeptide building blocks.
3. Lipidation (for lipidated version) or cyclization (for cyclized form) on solid phase.
4. Cleavage from solid support, purification by HPLC and lyophilization.

In this chapter we will briefly describe the synthetic protocol of α-AApeptides, γ-AApeptides, including their lipidated and cyclized forms. As chemistry is still being developed in our lab, we envision the synthetic procedure of AApeptides will be further evolved and simplified in the near future.

2 Materials

1. All materials and reagents are commercially available and used as received.
2. Organic solvents, such as ethyl acetate (EtOAc), dichloromethane (DCM), *tert*-butanol (*t*-BuOH), tetrahydrofurane (THF), methanol (MeOH), and hexanes were ACS grade and can be obtained from Sigma-Aldrich and Fisher. Acetonitrile was HPLC grade and used for HPLC purification of AApeptides.

3. Peptide coupling reagents such as diisopropylcarbodiimide (DIC), and benzotriazole-1-yl-oxy-tris-pyrrolidino-phosphonium hexafluorophosphate (PyBOP), were obtained from ChemImpex, and Aksci.
4. Amino acids (Fmoc-protected) were obtained from ChemImpex.
5. Resins for solid-support synthesis can be obtained from ChemImpex and Aapptec.

3 Methods

Solid phase synthesis has been the general approach for the preparation of peptides and other natural biomacromolecules including oligonucleotides and polysaccharides. Meanwhile, it is also the routine strategy to generate oligomeric peptidomimetic sequences. Since AApeptides are a new class of peptidomimetics, no synthetic protocols have been summarized systematically. In this chapter, we are trying to highlight some most important tips in the preparation of AApeptides. We will first discuss the synthesis of different subclasses of α-AApeptides (linear and lipo-linear forms), and then we will elaborate the synthetic procedure of γ-AApeptides (linear, cyclic, and lipo-cyclic forms).

3.1 Synthesis of α-AApeptides

Both canonical α-AApeptides and lipidated α-AApeptides are synthesized on a Rink-amide resin using α-AApeptide building blocks. Lipidation was achieved by capping the amino terminus of α-AApeptides with desired fatty acids.

3.1.1 Synthesis of α-AApeptide Building Blocks

The general synthetic scheme of α-AApeptide building blocks is shown in Fig. 2 [30].

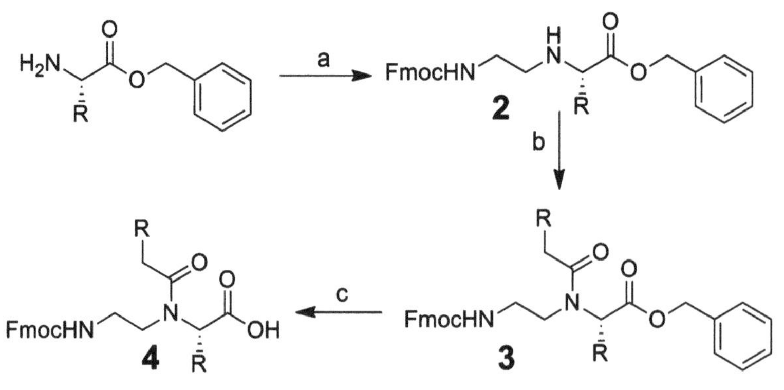

Fig. 2 Synthesis of α-AApeptide building blocks. (*a*) Fmoc-amino ethyl aldehyde, $NaBH_3CN$, CH_3OH; (*b*) RCH_2COOH, DhBtOH/DIC. (*c*) Pd/C, H_2, EtOAc. DhBtOH = Ox ohydroxybenzotriazole, DIC = Diisopropylcarbodiimide, EtOAc = Ethyl acetate

The typical synthetic protocol is shown as follows:

Synthesis of 2:

1. Dissolve the amino acid ester hydrochloride in methanol in a round bottom flask.
2. Add 1.1 eq of triethylamine and stir the solution at 0 °C for 15 min.
3. Add 1 eq Fmoc-glycinaldehyde and stir the solution for another 30 min.
4. Add 2–5 drops of acetic acid, then add 2 eq of $NaBH_3CN$.
5. Stir the solution at 0 °C for 1 h and then at room temperature overnight.
6. Evaporate the solvent, and wash the residue with ethyl acetate and saturated sodium bicarbonate solution.
7. After washing the organic layer with brine for three times, dry it over Na_2SO_4, and concentrate it in vacuo.
8. Purify the residue by flash chromatography.

Synthesis of 3:

1. Add compound **2**, 1.2 eq of DIC, DhBtOH, and RCH_2COOH in DMF in a round bottom flask, and stir overnight.
2. Add ethyl acetate and water to the flask.
3. Separate organic layer and wash with water (3×) and Brine (2×).
4. Dry organic layer anhydrous sodium sulfate, concentrate in vacuo, and purify by Flash chromatography.

Typical synthesis of 4:

1. Dissolve **3** in 20 ml ethyl acetate in a round bottom flask.
2. Add 10 % Pd/C and hydrogenate the solution at atmospheric pressure overnight.
3. Evaporate the solution and purify the residue by flash chromatography (*see* **Note 1**).

3.1.2 Synthesis of Linear α-AApeptides

The general synthetic scheme of α-AApeptides is shown in Fig. 3 [30]. The typical synthetic protocol is shown below.

1. Place Rink-amide resin in a peptide synthesis vessel.
2. Wash the resin three times with DMF and DCM (Dichloromethane) and then swell in DCM.

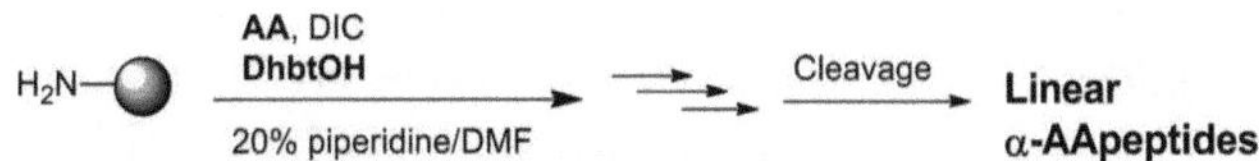

Fig. 3 Synthesis of linear α-AApeptides. DMF = dimethylformamide

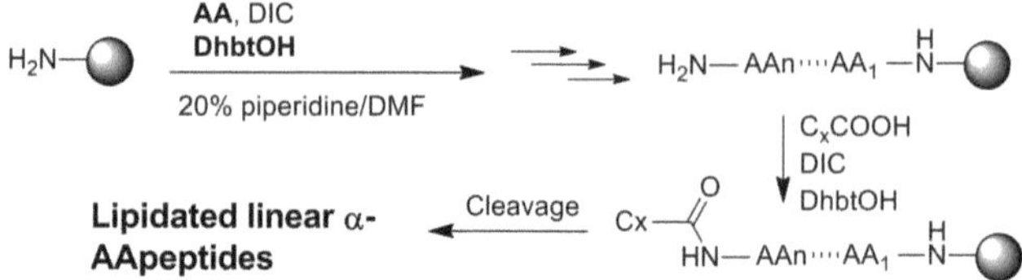

Fig. 4 Synthesis of lipidated linear α-AApeptides. AA = α-AApeptide building blocks. CxCOOH = fatty acids

3. Dissolve 1.5 eq (based on the loading of the resin) α-AApeptide building block in DMF in a vial, and add DhBtOH and DIC.
4. Add the solution to the resin in the peptide synthesis vessel, and stir for 4 h.
5. Wash the resin three times with DMF and DCM (Dichloromethane) and then swell in DCM (*see* **Note 2**).
6. Add 20 % Piperidine in DMF to the vessel and stir for 30 min (*see* **Note 3**).
7. Repeat **steps 2–6** until the desired sequence is synthesized on the resin (*see* **Note 4**).
8. Transfer the resin to a vial, add TFA/TIS/H_2O (95:2.5:2.5), and incubate for 2 h (*see* **Note 5**).
9. Evaporate the solvent and purify the residue by HPLC.

3.1.3 Synthesis of Lipidated Linear α-AApeptides

The general synthetic scheme of lipidated linear α-AApeptides is shown in Fig. 4 [32]. Lipidated linear α-AApeptides have been found to be antimicrobial agents [32].

The typical synthetic protocol is shown below.

1. Follow the **steps 1–7** of Subheading 3.1.2, until the desired residues are assembled on the solid phase.
2. Dissolve the fatty acid in DMF in a vial, and add DIC and DhbtOH.
3. Add the solution to the resin in the peptide synthesis vessel, and stir for 4 h.
4. Wash the resin three times with DMF and DCM (Dichloromethane) and then swell in DCM.
5. Transfer the resin to a vial, add TFA/TIS/H_2O (95:2.5:2.5), and incubate for 2 h (*see* **Note 5**).
6. Evaporate the solvent and purify the residue by HPLC.

3.2 Synthesis of γ-AApeptides

Both canonical γ-AApeptides, lipidated linear γ-AApeptides and cyclic γ-AApeptides, are synthesized on a Rink-amide resin using γ-AApeptide building blocks. Lipidation was achieved by capping the amino terminus of γ-AApeptides with desired fatty acids.

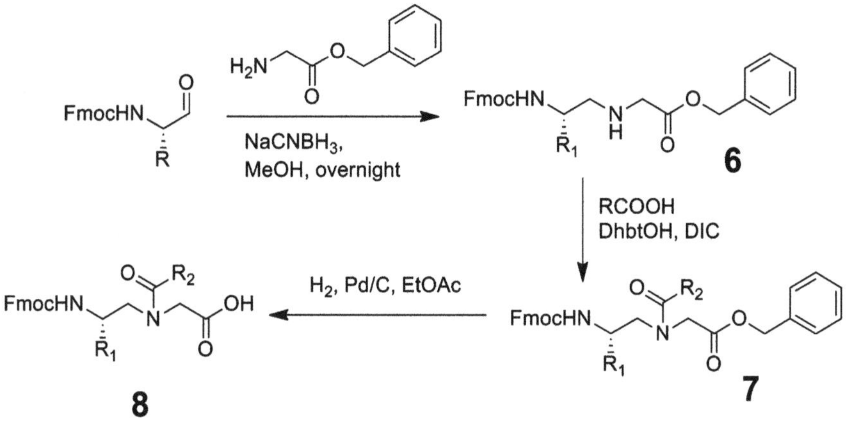

Fig. 5 Synthesis of γ-AApeptide building blocks

Cyclization was achieved on the solid support by utilizing an allyl carboxylate ester-containing γ-AApeptide building block.

3.2.1 Synthesis of γ-AApeptide Building Blocks

The general synthetic scheme of γ-AApeptide building blocks is shown in Fig. 5 [26].

The typical synthetic protocol is shown as follows:

Synthesis of 6:

1. Dissolve glycine benzyl ester hydrochloride in methanol in a round bottom flask.
2. Add 1.1 eq of triethylamine and stir the solution at 0 °C for 15 min.
3. Add 1 eq Fmoc protected amino acid aldehyde [34, 35] and stir the solution for another 30 min.
4. Add 2–5 drops of acetic acid, then add 2 eq of $NaBH_3CN$.
5. Stir the solution at 0 °C for 1 h and then at room temperature overnight.
6. Evaporate the solvent, and wash the residue with ethyl acetate and saturated sodium bicarbonate solution.
7. After washing the organic layer with brine for three times, dry it over Na_2SO_4, and concentrate it in vacuo.
8. Purify the residue by flash chromatography.

Synthesis of 7:

1. Add compound **6**, 1.2 eq of DIC, DhBtOH, and RCH_2COOH in DMF in a round bottom flask, and stir overnight.
2. Add ethyl acetate and water to the flask.
3. Separate organic layer and wash with water (3×) and brine (2×).
4. Dry organic layer anhydrous sodium sulfate, concentrate under the reduced pressure, and purify by flash chromatography.

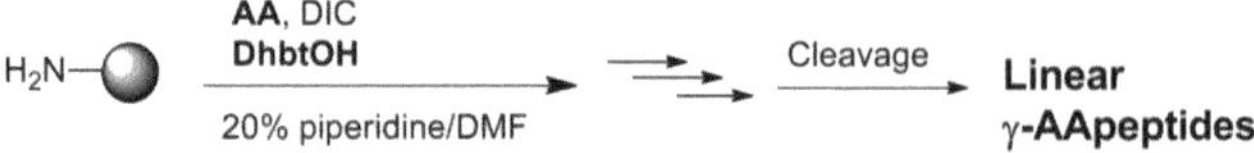

Fig. 6 Synthesis of linear γ-AApeptides. AA = γ-AApeptide building blocks

Typical synthesis of 8:

1. Dissolve 7 in 20 ml ethyl acetate in a round bottom flask.
2. Add 10 % Pd/C and hydrogenate the solution at atmospheric pressure overnight.
3. Evaporate the solution and purify the residue by flash chromatography.

3.2.2 Synthesis of Linear γ-AApeptides

The general synthetic scheme of γ-AApeptides is shown in Fig. 6 [26].
The typical synthetic protocol is shown below.

1. Place Rink-amide resin in a peptide synthesis vessel.
2. Wash the resin three times with DMF and DCM (dichloromethane) and then swell in DCM.
3. Dissolve 1.5 eq (based on the loading of the resin) γ-AApeptide building block in DMF in a vial, and add DhBtOH and DIC.
4. Add the solution to the resin in the peptide synthesis vessel, and stir for 4 h.
5. Wash the resin three times with DMF and DCM, and then swell in DCM.
6. Add 20 % Piperidine in DMF to the vessel and stir for 30 min.
7. Repeat **steps 2–6** until the desired sequence is synthesized on the resin.
8. Transfer the resin to a vial, add TFA/TIS/H_2O (95:2.5:2.5), and incubate for 2 h (*see* **Note 5**).
9. Evaporate the solvent and purify the residue by HPLC.

3.2.3 Synthesis of Lipidated Linear γ-AApeptides

The general synthetic scheme of lipidated linear γ-AApeptides is shown in Fig. 7 [26]. Lipidated linear γ-AApeptides have been found to be antimicrobial agents [27] and novel nanomaterials [31].
The typical synthetic protocol is shown below.

1. Follow the **steps 1–7** of Subheading 3.2.2, until the desired γ-AApeptide building blocks are assembled on the solid phase.
2. Dissolve the fatty acid in DMF in a vial, and add DIC and DhbtOH.
3. Add the solution to the resin in the peptide synthesis vessel, and stir for 4 h.
4. Wash the resin three times with DMF and DCM (Dichloromethane) and then swell in DCM.

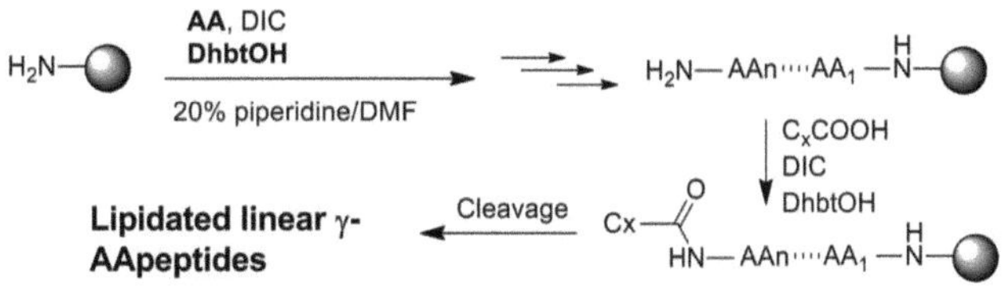

Fig. 7 Synthesis of lipidated linear γ-AApeptides. AA = γ-AApeptide building blocks. CxCOOH = fatty acids

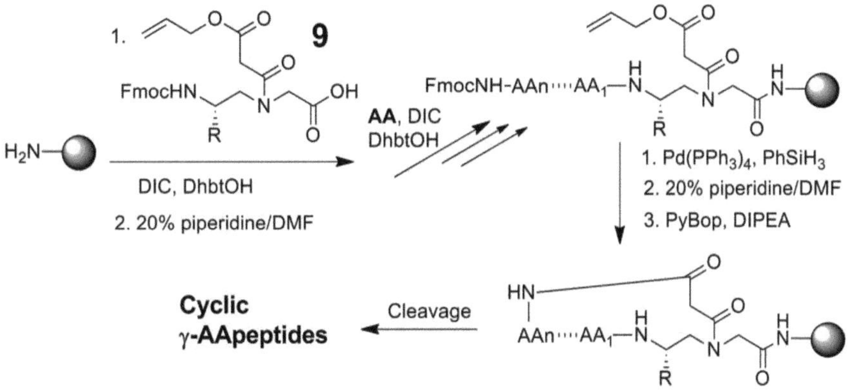

Fig. 8 Synthesis of cyclic γ-AApeptides. AA = γ-AApeptide building blocks. $Pd(PPh_3)_4$ = Tetrakis(triphenylphosphine)palladium(0), $PhSiH_3$ = Phenylsilane, PyBop = benzotriazol-1-yl-oxytripyrrolidinophosphonium hexafluorophosphate

5. Transfer the resin to a vial, add TFA/TIS/H_2O (95:2.5:2.5), and incubate for 2 h (*see* **Note 5**).
6. Evaporate the solvent and purify the residue by HPLC.

3.2.4 Synthesis of Cyclic γ-AApeptides

The general synthetic scheme of cyclic γ-AApeptides is shown in Fig. 8 [33]. Cyclic γ-AApeptides have shown broad-spectrum activity against both gram-positive and gram-negative bacteria [33]. Cyclization on the solid phase was achieved by utilizing an allyl carboxylate ester-containing γ-AApeptide building block **9**.

The typical synthetic protocol is shown below.

1. Prepare γ-AApeptide building block **9** following the protocol of synthesizing general γ-AApeptide building blocks.
2. Place Rink-amide resin in a peptide synthesis vessel.
3. Wash the resin three times with DMF and DCM (Dichloromethane) and then swell in DCM.
4. Dissolve 1.5 eq (based on the loading of the resin) γ-AApeptide building block **9** in DMF in a vial, and add DhBtOH and DIC.

5. Add the solution to the resin in the peptide synthesis vessel, and stir for 4 h.
6. Wash the resin three times with DMF and DCM and then swell in DCM.
7. Add 20 % Piperidine in DMF to the vessel and stir for 30 min.
8. Wash the resin three times with DMF and DCM and then swell in DCM.
9. Dissolve 1.5 eq (based on the loading of the resin) γ-AApeptide building block in DMF in a vial, and add DhBtOH and DIC.
10. Add the solution to the resin in the peptide synthesis vessel, and stir for 4 h.
11. Wash the resin three times with DMF and DCM (Dichloromethane) and then swell in DCM.
12. Repeat **steps 8–12** until the desired building blocks are assembled on the resin.
13. Add $Pd(PPh_3)_4$ (0.2 eq) and $PhSiH_3$ (10 eq) in CH_2Cl_2 to the resin and stir for 2 h.
14. Add 20 % Piperidine in DMF to the vessel and stir for 30 min.
15. Add 1.5 eq PyBop, 1.5 eq HOBt, 5 eq DIEA in DMF to the synthesis vessel and stir for 4 h.
16. Transfer the resin to a vial, add TFA/TIS/H_2O (95:2.5:2.5), and incubate for 2 h (*see* **Note 5**).
17. Evaporate the solvent and purify the residue by HPLC.

4 Notes

1. The purity of AApeptide building blocks should be at least >98 %. Otherwise the yield on the solid phase will be significantly dropped down.
2. Coupling of building blocks onto solid phase can be repeated once. In other words, double coupling can be used to increase the yield.
3. Deprotection of Fmoc group by 20 % piperidine in DMF can also be repeated one more time to ensure the complete removal of Fmoc group.
4. Acetic anhydride can be used to cap the failure sequences in each step to minimize the side products.
5. TFA = Trifluoroacetic acid, TIS = Triisopropylsilane.

References

1. Wu Y-D, Gellman S (2008) Peptidomimetics. Acc Chem Res 41:1231–1232
2. Tuwalska D, Sienkiewicz J, Liberek B (2008) Synthesis and conformational analysis of methyl 3-amino-2,3-dideoxyhexopyranosiduronic acids, new sugar amino acids, and their diglycotides. Carbohydr Res 343:1142–1152
3. Risseeuw MD, Mazurek J, van Langenvelde A, van der Marel GA, Overkleeft HS, Overhand M (2007) Synthesis of alkylated sugar amino acids: conformationally restricted L-Xaa-L-Ser/Thr mimics. Org Biomol Chem 5:2311–2314
4. Horne WS, Johnson LM, Ketas TJ, Klasse PJ, Lu M, Moore JP, Gellman SH (2009) Structural and biological mimicry of protein surface recognition by alpha/beta-peptide foldamers. Proc Natl Acad Sci USA 106: 14751–14756
5. Horne WS, Gellman SH (2008) Foldamers with heterogeneous backbones. Acc Chem Res 41:1399–1408
6. Dervan PB (1986) Design of sequence-specific DNA-binding molecules. Science 232: 464–471
7. Simon RJ, Kania RS, Zuckermann RN, Huebner VD, Jewell DA, Banville S, Ng S, Wang L, Rosenberg S, Marlowe CK et al (1992) Peptoids: a modular approach to drug discovery. Proc Natl Acad Sci USA 89: 9367–9371
8. Cheng RP, Gellman SH, DeGrado WF (2001) beta-Peptides: from structure to function. Chem Rev 101:3219–3232
9. Seebach D, Ciceri PE, Overhand M, Jaun B, Rigo D, Oberer L, Hommel U, Amstutz R, Widmer H (1996) Probing the helical secondary structure of short-chain beta-peptides. Helv Chim Acta 79:2043–2066
10. Kritzer JA, Stephens OM, Guarracino DA, Reznik SK, Schepartz A (2005) beta-Peptides as inhibitors of protein–protein interactions. Bioorg Med Chem 13:11–16
11. Kumbhani DJ, Sharma GV, Khuri SF, Kirdar JA (2006) Fascicular conduction disturbances after coronary artery bypass surgery: a review with a meta-analysis of their long-term significance. J Card Surg 21:428–434
12. Arndt HD, Ziemer B, Koert U (2004) Folding propensity of cyclohexylether-delta-peptides. Org Lett 6:3269–3272
13. Trabocchi A, Guarna F, Guarna A (2005) gamma- and delta-Amino acids: synthetic strategies and relevant applications. Curr Org Chem 9:1127–1153
14. Violette A, Petit MC, Rognan D, Monteil H, Guichard G (2005) Oligourea foldamers as antimicrobial peptidomimetics. Biopolymers 80:516
15. Boeijen A, van Ameijde J, Liskamp RMJ (2001) Solid-phase synthesis of oligourea peptidomimetics employing the Fmoc protection strategy. J Org Chem 66:8454–8462
16. Lee HJ, Song JW, Choi YS, Park HM, Lee KB (2002) A theoretical study of conformational properties of N-methyl azapeptide derivatives. J Am Chem Soc 124:11881–11893
17. Graybill TL, Ross MJ, Gauvin BR, Gregory JS, Harris AL, Ator MA, Rinker JM, Dolle RE (1992) Synthesis and evaluation of azapeptide-derived inhibitors of serine and cysteine proteases. Bioorg Med Chem Lett 2: 1375–1380
18. Li X, Wu YD, Yang D (2008) Alpha-aminoxy acids: new possibilities from foldamers to anion receptors and channels. Acc Chem Res 41:1428–1438
19. Nelson JC, Saven JG, Moore JS, Wolynes PG (1997) Solvophobically driven folding of non-biological oligomers. Science 277:1793–1796
20. Patch JA, Barron AE (2002) Mimicry of bioactive peptides via non-natural, sequence-specific peptidomimetic oligomers. Curr Opin Chem Biol 6:872–877
21. Gellman S (2009) Structure and function in peptidic foldamers. Biopolymers 92:293
22. Goodman CM, Choi S, Shandler S, DeGrado WF (2007) Foldamers as versatile frameworks for the design and evolution of function. Nat Chem Biol 3:252–262
23. Niu Y, Padhee S, Wu H, Bai G, Harrington L, Burda WN, Shaw LN, Cao C, Cai J (2011) Identification of gamma-AApeptides with potent and broad-spectrum antimicrobial activity. Chem Commun (Camb) 47: 12197–12199
24. Niu Y, Jones AJ, Wu H, Varani G, Cai J (2011) gamma-AApeptides bind to RNA by mimicking RNA-binding proteins. Org Biomol Chem 9:6604–6609
25. Padhee S, Hu Y, Niu Y, Bai G, Wu H, Costanza F, West L, Harrington L, Shaw LN, Cao C, Cai J (2011) Non-hemolytic alpha-AApeptides as antimicrobial peptidomimetics. Chem Commun (Camb) 47:9729–9731
26. Niu Y, Hu Y, Li X, Chen J, Cai J (2011) [gamma]-AApeptides: design, synthesis and evaluation. New J Chem 35:542–545
27. Niu Y, Padhee S, Wu H, Bai G, Qiao Q, Hu Y, Harrington L, Burda WN, Shaw LN, Cao C, Cai J (2012) Lipo-gamma-AApeptides as a new class of potent and broad-spectrum antimicrobial agents. J Med Chem 55: 4003–4009
28. Bai G, Padhee S, Niu Y, Wang RE, Qiao Q, Buzzeo R, Cao C, Cai J (2012) Cellular uptake

of an alpha-AApeptide. Org Biomol Chem 10:1149–1153
29. Niu Y, Bai G, Wu H, Wang RE, Qiao Q, Padhee S, Buzzeo R, Cao C, Cai J (2012) Cellular translocation of a gamma-AApeptide mimetic of Tat peptide. Mol Pharm 9:1529–1534
30. Hu Y, Li X, Sebti SM, Chen J, Cai J (2011) Design and synthesis of AApeptides: a new class of peptide mimics. Bioorg Med Chem Lett 21:1469–1471
31. Niu Y, Wu H, Huang R, Qiao Q, Constanza F, Wang X, Hu Y, Amin MN, Naguyen A, Zhang J, Haller E, Ma S, Li X, Cai J (2012) Nanorods formed from a new class of peptidomimetics. Macromolecules. doi:10.1021/ma3015992
32. Hu Y, Amin MN, Padhee S, Wang R, Qiao Q, Ge B, Li Y, Mathew A, Cao C, Cai J (2012) Lipidated peptidomimetics with improved antimicrobial activity. ACS Med Chem Lett 3:683–686
33. Wu H, Niu Y, Padhee S, Wang RE, Li Y, Qiao Q, Ge B, Cao C, Cai J (2012) Design and synthesis of unprecedented cyclic gamma-AApeptides for antimicrobial development. Chem Sci 3:2570–2575
34. Debaene F, Da Silva JA, Pianowski Z, Duran FJ, Winssinger N (2007) Expanding the scope of PNA-encoded libraries: divergent synthesis of libraries targeting cysteine, serine and metallo-proteases as well as tyrosine phosphatases. Tetrahedron 63: 6577–6586
35. Debaene F, Mejias L, Harris JL, Winssinger N (2004) Synthesis of a PNA-encoded cysteine protease inhibitor library. Tetrahedron 60:8677–8690

Chapter 4

Peptoids and Peptide–Peptoid Hybrid Biopolymers as Peptidomimetics

Maciej J. Stawikowski

Abstract

Peptoids (oligomers of *N*-substituted glycine residues) and peptide–peptoid hybrid polymers (peptomers) are interesting classes of compounds mimicking structure and function of biologically active peptides. The oligomeric peptidomimetics such as peptoids are particularly important compounds since they provide access to an enormous molecular diversity, by variation of the building blocks. The modular structure of peptoids, ease of synthesis, and high compatibility with existing peptide chemistry synthetic protocols, make peptoids and peptoid-containing peptidomimetics ideal tools for structure–activity and drug discovery related studies.

Key words Peptoids, Solid-phase peptide synthesis, Peptide–peptoid hybrid polymers, Foldamers, Peptomers

1 Introduction

Peptide research on drug design and discovery is a very exciting field in the development of the new therapeutics. Peptides are an important class of biopolymers responsible for molecular recognition and biological activities. Their effectiveness and minimal side effects make them ideal candidates for drug discovery. However, peptides are limited in that they are metabolically unstable due to proteolytic susceptibility and may have poor bioavailability, partly because of low bio-membrane transport characteristics. The engineering and design of peptide based drug candidates are related to advances in understanding of biological activity of given parent peptides. Thus structure–activity studies are indispensable in the search for better peptidomimetics. At present there is a wide range of synthetic methods allowing for modification of native peptides. In particular of interest are those which would be compatible with classic methods for peptide synthesis. The improvement of a peptide's pharmacokinetic properties could involve the modification of the peptide backbone, particular amino acid residues or cyclization.

Predrag Cudic (ed.), *Peptide Modifications to Increase Metabolic Stability and Activity*, Methods in Molecular Biology, vol. 1081,
DOI 10.1007/978-1-62703-652-8_4, © Springer Science+Business Media New York 2013

The very potent and promising classes of peptidomimetics at the peptide backbone level are peptoids and the peptide–peptoid hybrids. The oligomeric peptidomimetics such as peptoids are particularly interesting compounds since they provide access to an enormous molecular diversity by variation of the building blocks. Peptoids represent a class of polymers that are not found in nature. They differ from peptides in the manner of side chain attachment, and can be considered as peptide mimics in which the side chains of particular monomeric building blocks has been shifted from the chiral α-carbon atom to the achiral nitrogen atom. In polypeptoids *N*-substituted glycine residues are connected through polyimide bonds. Peptoids lack secondary amide hydrogen atoms, and thus are incapable of forming hydrogen bonds characteristic to the secondary structure found in peptides and proteins. Since peptoid monomers are derived from *N*-substituted glycine residues the peptoid backbone is achiral. However, chirality can be incorporated into side-chains to obtain fairly stable secondary structures [1].

Peptoid monomer nomenclature abbreviation is somehow tentative. When a *N*-substituted glycine residue monomer mimics a proteinogenic amino acid, the four letter code *N*Xaa is used; for example *N*Leu, *N*Lys. In the case of proteinogenic amino acid homologs, four or five letter abbreviation is used: e.g., homoserine—*N*hse or *N*hser (a four-letter format is preferred). Very often abbreviations are formed based on the name of the amine building block used to create the peptoid monomer (e.g., *N*spe = (*S*)-*N*-(1-phenylethyl)-glycine; *N*aeg = *N*-aminoethylglycine). Due to such ambiguity in the nomenclature of abbreviations of peptoid monomers clear definitions are required for clarity in each peptoid study.

Polypeptoids lack conformational rigidity in comparison with peptides. Despite their inability to form hydrogen bond networks, peptoids are known to adopt stable three-dimensional structures, which often localize in backbone conformational spaces not readily accessible to peptides [2]. For example, the *cis* and *trans* amide bond geometries of peptoids are generally iso-energetic, allowing peptoids to form polyproline type I helices featuring repeating *cis* amide bonds that are rarely observed in peptides. The backbone structural preferences in peptoids are predominantly driven by local interactions, including steric, stereoelectronic, and bond-resonance effect [3, 4]. Additionally, unlike the Ramachandran plots for the canonical chiral nonglycine amino acids, the energy surface for the achiral peptoid is center symmetric, i.e., the energy of a structure with a given (φ, ψ) will be identical to its mirror image with ($-\varphi$, $-\psi$) [2, 5].

The secondary structure of peptoids is typically experimentally evaluated by circular dichroism spectroscopy (CD) as this tool allows rapid analysis relative to characterization by NMR or X-ray crystallography. Although CD spectroscopy analysis gives the relative results about the secondary structure it is found to be in good agreement with NMR and X-ray crystallography results. One of the interesting features of peptoids is their ability to adopt stable conformations in a variety of solvent media [1]. Having this feature, peptoids belong to the class of sequence-specific peptidomimetics referred to as foldamers which can be designed to adopt specific backbone conformations [6]. There have been several studies that investigated, by a variety of techniques, the folding behavior of peptoids with diverse chiral, α-branched side chains that had formed stable polyproline type I-like helix [1, 7]. Barron and coworkers developed a set of predictive rules for helix formation in peptoids [8]. First, helical structure is stabilized by the incorporation of at least 50 % α-chiral (in the side chain) monomer units into an oligomer, or if the helix contains one or more aromatic faces running parallel to the helix axis (i.e., aromatic side chains at the i, $i+3$ positions). Second, peptoid helices are stabilized when the C-terminal residue is α-chiral. Third, helical structure is further stabilized at longer peptoid oligomer lengths (e.g., decamers and beyond). Stable β-turn structures were reported by Krishenbaum and coworkers in cyclic hexamer and octamer peptoids [9]. These peptoids possessed an alternating pattern of benzyl and methoxyethyl side chains, and their crystal structures revealed that the aromatic and alkyl side chains were segregated to opposite faces of the macrocycle. The reported peptoids were found to closely resemble both type I and type III β-turns. Ongoing studies seek new side chain types that will generate novel, stable folded structures. Research directed by Zuckermann has pursued the self-assembly of peptoid secondary structures into complex folds called peptoid nanosheets [10]. Peptoid nanosheets sandwich hydrophobic side-chains between surfaces that display alternating strips of positive and negative charge carried by parallel all-*trans* peptoid oligomers bearing 2-aminoethyl and 2-carboxyethyl side-chains. These structures resemble β-sheet secondary structures found in proteins and could potentially play a similar role in peptoid engineering. The potential diversity of peptoid nanosheet variants offers unique opportunities for functional engineering and suggests prospects for extensive applications as robust lipid-membrane analogs.

Due to specific physicochemical properties, peptoids are an attractive type of oligomeric peptidomimetics addressing many objectives in the field of biology and medicine. The spectrum of biological applications of peptoids ranges from biomolecular sensing, drug delivery, antimicrobial, antiviral, antifungal, and cancer therapeutics to self-assembling nanosystems [11]. In the past several years there has been a growing interest in peptoid research area and researchers interested in this field can attend biannual conferences called "Peptoid Summits" [12].

The engineering of transformation from peptide to peptoid often possess challenging task as the simple monomer-to-monomer (amino acid to *N*-substituted glycine) translation will drastically change the folded conformation of the oligomer, and in turn will result in unpredictably different biological activity. The straightforward approach to the design of biologically active peptoid sequences is the systematic modification of an active peptide target sequence. One has to keep in mind that such a simple modification results also in a side chain shift in a sequence ($C\alpha \rightarrow N$) relative to the original position. An alternative approach utilizes combinatorial chemistry by generation of library of compounds in which site specific substitutions are made using a diverse set of peptoid monomers. This approach is called "peptoid scan" which in principle is similar to the alanine scan commonly employed. Characterization of peptoid-containing analogs of peptide ligands can provide valuable information. For example, this process can help elucidate the position of critical residues in the protein target that provide important binding determinants. It can identify molecules with enhanced activity relative to the starting structure and/or with enhanced specificity to a target protein. Finally, it can identify active species with peptoid monomer substitution levels sufficient to grant significantly improved protease stability relative to the original peptide of interest.

Peptide–peptoid hybrids (peptomers) are an interesting class of compounds, and such hybrid polymers can beneficiate from both classes of compounds. Many interesting applications of that approach have been reported in the literature [13]. Goodman and coworkers were among first to report examples of peptide–peptoid hybrids studying collagen mimetics. In this study hydroxyproline residue in a peptide repeat [Gly–Pro–Hyp]$_n$ was replaced with *N*-isobutylglycine (*N*Leu) [14]. An interesting example of a peptide–peptoid hybrid capable of inhibiting a protein–ligand interaction was reported by Hamy and coworkers who discovered an oligomer of nine residues, that specifically inhibits the Tat/TAR RNA interaction, both in vitro (at nanomolar concentrations) and in vivo (at micromolar concentrations) [15]. Moreover, that compound blocked HIV-1 replication in primary human lymphocytes. The successful oligomer was selected from a library of 3.2 million compounds by means of split and mix method.

There are several successful examples of receptor targeting based on peptide–peptoids described in the literature. One of the first examples comes from the Goodman laboratory, which was working on somatostatin analogs [16]. Somatostatin, a tetradecapeptide hormone released by the hypothalamus, plays important physiological roles: it is a potent inhibitor of the release of several hormones (i.e., glucagon, growth hormone, insulin, gastrin) and regulates many other biological activities. Goodman and coworkers investigated analogs of cyclic hexapeptide containing chiral (*S*)- or (*R*)-phenylethylamine [(*R*)- and (*S*)-βMe *N*phe] and *N*-benzylglycine

(*N*Phe) peptoid monomers, which led to a ligand with increased selectivity for human somatostatin 2 receptor, over somatostatin 1 and somatostatin 3–5 receptors. In vivo, these peptoid analogs selectively inhibited the release of a growth hormone but have no effect on the inhibition of insulin [17]. The melanocortin (MC) receptor system belongs to the G-protein coupled receptor (GPCR) superfamily and is involved in many physiological processes. Haskel-Luevano and coworkers reported synthesis of a series of peptoids and peptide–peptoid [18]. These peptidomimetics were synthesized based on the Ac–His–Phe–Arg–Trp–NH_2 tetrapeptide template. The melanocortin-4 receptor (MC4-R) is an excellent drug target for obesity. Mutations of MC4-R in humans and mice have been associated with the development of obesity. Liskamp described the synthesis and pharmacological activity of a library of 32 peptoid–peptide ligands (seven residues long oligomers) of an MC4 receptor agonist that were synthesized and investigated on cells expressing different MC receptor subtypes and for rat grooming behavior [19]. In general, receptor selectivity remained, while affinity and potency were decreased. It was reported that introduction of peptoid monomers resulted in peptoid - peptide hybrid ligands with an increase in EC_{50} values of 1–2 orders of magnitude for every additional peptoid monomer that was introduced into the core sequence. However, selectivity for the MC4 receptor usually remained. Additionally, replacement of the d-Phe residue by the corresponding peptoid monomer (*N*Phe) strongly increased the EC_{50} values of the resulting ligands.

Peptoids and peptide–peptoid hybrids are also promising candidates as antimicrobial agents [20]. The potency of such compounds originates from cationic antimicrobial peptides. Their antimicrobial activity is related to their ability to adopt amphiphilic secondary structures, including α-helices or β-hairpins, in a membrane like environment. Their principal mode of antibiotic action appears to involve depolarization or permeabilization of the bacterial cell membranes, although the detailed mechanisms are still not fully understood. Additionally, peptoids' proteolytic resistance is crucial for development of effective peptidomimetic antimicrobials [21]. Magainins, helical antimicrobial peptides, served as a template for the development of new antibiotics with improved therapeutic properties [22, 23]. Other examples of peptoid containing antimicrobials described in the literature include protegrin I—cyclic β-hairpin [24], tritrpticin and indolicidin [25], melittin [26], apidaecin Ib [27], and analogs of autoinducing peptide I (AIP-I) [28].

The modular structure of peptoids, ease of synthesis and high compatibility with existing peptide chemistry synthetic protocols make peptoids and peptoid-containing peptidomimetics, an ideal tool for structure–activity and drug discovery related studies.

2 Materials

1. All materials and reagents are commercially available and used as received.
2. For manual synthesis of peptoids a standard reaction vessels for manual peptide synthesis are used. They can be purchased from Chemglass Life Sciences (USA), Peptides International (USA) or other commercial sources.
3. Resins for solid-support synthesis can be obtained from Rapp-Polymere (Germany), Novabiochem (USA), ChemImpex International (USA), Aapptec (USA, and other suppliers.
4. Ninhydrin test kit can be purchased from Anaspec (USA) or prepared in-house.
5. Coupling reagents such as *N*,*N'*-diisopropylcarbodiimide (DIC), benzotriazole-1-yl-oxy-tris-pyrrolidino-phosphonium hexafluorophosphate (PyBOP), Bromo-tris-pyrrolidino-phosphonium hexafluorophosphate (PyBrop), bromoacetic acid may be obtained from Sigma-Aldrich, ChemImpex International (USA), EMD Millipore (Novabiochem), or Anaspec (USA).
6. Trifluoroacetic acid (TFA), triisopropylsilane (TIS), and methyl *tert*-butyl ether (MTBE) can be purchased from Sigma-Aldrich, Acros Organics, or other commercial sources.
7. Amines of choice are usually commercially available. They can be purchased from Sigma-Aldrich, TCI America (or TCI Europe), Acros Organics, or other commercial sources. The sources can be found through an online chemical directory search, such as ChemExper Chemical Directory (www.chemexper.com).

3 Methods

The facile synthesis of peptoids, along with its designable structure and advantageous pharmacological properties, make peptoids excellent candidates for peptidomimetics. Moreover, there are many ways of modifications in particular side chains of *N*-substituted glycines which extend the scope of molecular diversification [11].

There are presently several methods for peptoid synthesis. The most common route involves solid-phase synthesis, which is compatible with standard solid-phase peptide synthesis and involves the Fmoc/*t*Bu chemistry. In general, synthetic methods can be divided into two approaches: monomeric and submonomeric (Fig. 1). The monomeric approach requires that the completely protected *N*-substituted glycine unit is coupled to the resin (in the case of first residue) or the amino group of a former peptoid residue.

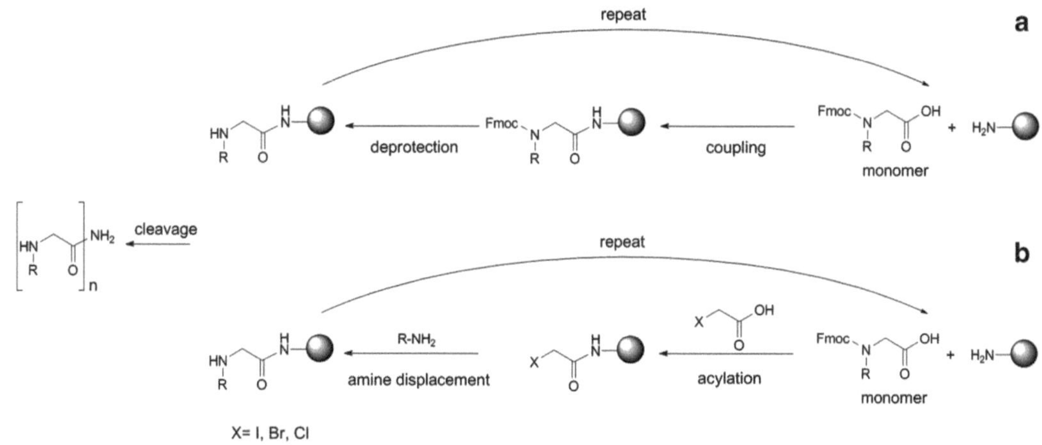

Fig. 1 General methods for solid-phase peptoid synthesis. Monomer approach (**a**) and submonomeric approach (**b**)

After coupling, temporary *N*-protecting (Fmoc) group is removed and the peptoid chain can be elongated (Fig. 1a). A disadvantage of the monomeric approach, however, is the necessity of preparing suitable quantities of a diverse set of protected *N*-substituted glycine monomers. In the submonomeric method the incorporation of each peptoid unit requires two steps: acylation using chloro- or bromoacetic acid followed by substitution with the appropriate amine (Fig. 1b). The first facile solid phase synthesis of peptoids was reported by Simon and coworkers in 1992 [5]. It was based on the monomeric approach, but later on Zuckermann and coworkers had published the submonomeric approach which simplified the solid-phase peptoid and peptide–peptoid hybrid synthesis [29]. Since the classical solid-phase peptide synthesis occurs in the C to the N terminus direction, it is possible that solid-phase peptoid and peptide synthesis could be combined, giving peptide–peptoid hybrid polymers. That approach may be used in the conversion of biologically active peptide ligands to more potent peptide–peptoid peptidomimetics with improved properties. Ostergaard and Holm named such hybrids "peptomers" [30]. The term "peptomer" had also been used earlier by Robey and coworkers to describe peptides that are specifically cross-linked in head-to-tail fashion [31]. The polymerization reaction is performed between bromoacetylated N-termini and thiol group from C-terminal Cys residue [32].

3.1 Solid Support

As a solid support for peptoid/peptomer synthesis, the polyethylene grafted polystyrene (PS-PEG) is widely used (in form of Rink amide MBHA, TentaGel S RAM, or Rink amide AM resins), although there are reports in the literature for use of polystyrene resins (with 2-chlorotrityl chloride linker) [33], cellulose membranes (SPOT synthesis) [34], or modified glass surfaces particularly used for photolithographic synthesis of arrays [35, 36]. The use of Rink amide linker based resins will result in formation of C-terminal amides whereas application of 2-chlorotrityl chloride linker will result in C-terminal acids upon cleavage from the resin.

3.2 Submonomeric Approach: The Acylation Step

The first step in peptoid synthesis by the submonomeric method is the acylation step (Fig. 1b). The typical reaction involves the use of bromoacetic acid together with coupling agent such as *N*, *N-diisopropylcarbodiimide* (DIC) in DMF as a solvent (*see* **Note 1**). Typical reaction times may vary from 30 s (microwave assisted synthesis) to 30 min. during manual or automated synthesis.

3.3 Submonomeric Approach: The Amine Substitution Step

Amine displacement is the second step during synthesis of the peptoid residue, which allows for incorporation of amines that will create the side chain of the peptoid residue and can be further used to extend molecular diversity (e.g., essential for combinatorial chemistry). Standard reaction conditions involve 20 M equivalent excess of amine in DMSO and reaction times vary from 30 s to 2 h, depending mainly on type of the amine used. Alternative solvents for this reaction are DCM or NMP with typical amine concentrations of 1–2 M. Regardless of the synthetic method used, functional groups present in the side chain must be appropriately protected, due to possible side reactions. The successful syntheses of peptoids with unprotected side chain residues bearing different heterocyclic groups such as imidazoles, pyridines, pyrazines, indoles, and quinolones was also developed [37]. Application of chloroacetic acid instead of bromoacetic acid minimized unwanted alkylation side reactions due to reduced reactivity of the chloride during amine displacement step. On the other hand in situ protection (or further modification) of side chain functionalities using submonomeric approach is also possible. For example *N*Lys can be easily incorporated into peptoid/peptomer sequence by using 1,4-diaminobutane and trityl as side chain protecting group conveniently applied after amination step [38]. Similarly *N*-substituted glycines bearing guanidine group (*N*Arg mimetics) can be synthesized on solid support by means of guanylation reactions [18] (for methods of guanidinylation *see* Chapter 10).

1. Prepare all solvents, reagents and prepare the workspace under fume hood.
2. Place resin in dry reaction vessel.
3. Remove the Fmoc protecting group (if necessary) by agitating the resin with 20 % piperidine in DMF (3 × 6 min) using 4–5 times resin volume (*see* **Note 2**).
4. Wash the resin (3 min. each) with 3 × 10 mL of DMF and 3 × 10 mL of DCM.
5. Dissolve bromoacetic acid (20 eq, relative to resin loading), and DIC (20 eq) in amount of DMF which equals 2–3 volumes of resin and add this solution to the resin.

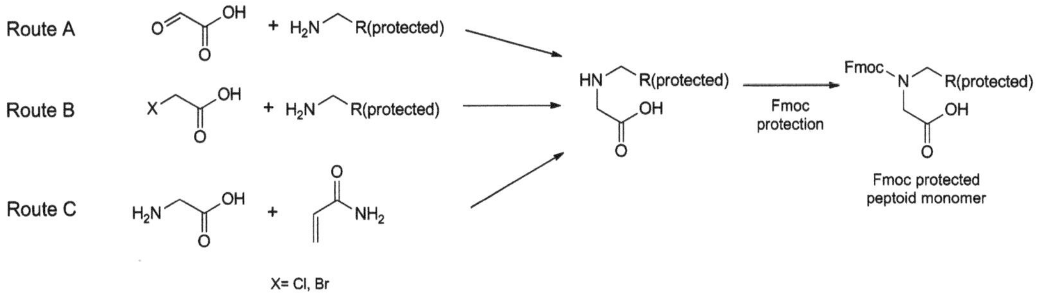

Fig. 2 General synthetic routes for synthesis of peptoid monomers proposed by Simon et al. [5]

6. Allow the resin to agitate at room temperature for 30 min.
7. Wash the resin with 3 × 10 mL of DMF for 3 min. each.
8. Dissolve amine (25 eq) in DMSO, NMP or DMF (*see* **Note 3** for preparation of amines in hydrochloride salt form).
9. Upon addition of amine solution to the resin, allow the resin to agitate at room temperature for 2 h.
10. Wash the resin (3 min each) with 3 × 10 mL of DMF; and 3 × 10 mL of DCM.
11. Perform ninhydrin test when necessary.
12. Keep repeating **steps 3–11** until throughout desired sequence is obtained.
13. Prepare the resin bound peptoid/peptomer for cleavage (*see* Subheading 3.5 below).

    Remarks: In some cases the displacement step (**step 9**) has to be prolonged up to 12 h (overnight).

3.4 Monomeric Approach

Although submonomeric approach is very facile method for peptoid synthesis, the monomeric approach possesses some advantages. It allows for easy implementation of the synthetic protocols for peptides on commercial automated peptide synthesizers, including monitoring during Fmoc removal and permits the use of more powerful coupling reagents during difficult couplings. The peptoid monomer synthesis approach is required when the side chain consists of functional group(s) that are labile and require additional protection during synthesis.

The general synthetic routes for the preparation of peptoid monomers in solution were described by Simon and coworkers in 1992 (Fig. 2) [5]. The synthesis of peptoid residues mimicking natural amino acids, namely, Ala, Arg, Ala, Asp, Glu, Gln, and Leu were described. They have found that the most generally useful synthetic method was reductive amination of a side-chain amine with glyoxylic acid (route A). Alternative routes are alkylation of a side-chain amine with α-haloacetic acid (route B), or Michael addition of glycine to acrylamide (route C). After formation of the *N*-substituted glycines, the α-amino group is protected as the Fmoc derivative. *Tert*-butyl based groups are used to protect side chains containing amino, carboxyl, hydroxyl, and indole moieties; guanidine and imidazole moieties are blocked with 2,2,5,7,8-pentamethylchroman-6-sulfonyl (Pmc) and trityl groups, respectively. This scheme was chosen for its orthogonality to the Fmoc protection of the α-amino group and its compatibility with the global deprotection chemistry implemented on automatic peptide synthesizers.

Liskamp and coworkers have reported slightly modified synthesis of peptoid monomers based on alkylation of amine submonomers using ethyl bromoacetate instead of the free acid (Fig. 2b) which simplified the purification by column chromatography [39]; multi-gram scale synthesis of Fmoc protected *N*Phe, *N*leu, *N*Met, *N*Lys, *N*Gln, and *N*Tyr is described.

The coupling of peptoid monomers during peptoid or peptomer synthesis is rather more difficult due to steric hindrance around secondary amine nitrogen. In such case, the use of PyBroP [5], PyBOP [5, 39], or HATU [40] as coupling reagents is advised.

1. Prepare all solvents, reagents and prepare the workspace under fume hood.
2. Place resin in dry reaction vessel.
3. Remove the Fmoc protecting group by agitating the resin with 20 % piperidine in DMF (3 × 6 min) using 4–5 times resin volume (*see* **Note 2**, pre-swelling).
4. Wash the resin (3 min each) with 3 × 10 mL of DMF; and 3 × 10 mL of DCM.
5. Dissolve 4 eq (relative to resin loading) of Fmoc-*N*-substituted glycine building block, 4 eq of coupling reagent (PyBOP or PyBrOP) and 9 eq of DIPEA, in amount of DMF that equals 2–3 volumes of resin; immediately add this solution to the resin.
6. Allow the resin to agitate at room temperature for 1 h.
7. Wash the resin with 3 × 10 mL of DMF for 3 min. each.
8. Perform ninhydrin test when necessary.
9. Keep repeating **steps 3–8** until throughout desired sequence is obtained.
10. Prepare the resin bound peptoid/peptomer for cleavage (*see* Subheading 3.5 below).

3.5 Cleavage and Deprotection

Upon completion of synthesis, peptoids and/or peptide–peptoid hybrids are released from solid support using “cleavage cocktails” similar to ones used in classical Fmoc peptide chemistry. A wide range of TFA based cleavage cocktails are successfully applied for cleavage of peptoids with typical composition of TFA/triisopropylsilane/H_2O (95:2.5:2.5, v/v). Other scavengers such as thioanisole [41], anisole [42], or triethylsilane [43] may also be used. Cleavage is performed at room temperature and under inert atmosphere for 2 h, with additional extension when side-chain protecting groups are present [44].

1. Place resin in dry reaction vessel.
2. Remove the Fmoc protecting group from the last residue (if necessary) by agitating the resin with 20 % piperidine in DMF (3×6 min) using 4–5 times resin volume (*see* **Note 2**).
3. Wash the resin (3 min each) with 2×10 mL of DMF; and 4×10 mL of DCM (*see* **Note 4**).
4. Additionally, to shrink the resin, wash it with MeOH (polystyrene based resins) or diethyl ether (PEG-PS based resins), 4×10 mL with agitation.
5. Dry the peptoid/peptomer-resin under vacuum for 3 h.
6. Prepare cleavage cocktail containing TFA/triisopropylsilane/H_2O (95:2.5:2.5, v/v) in the amount of 15 mL per 1 g of resin.
7. Add cleavage cocktail to the resin and flush the reaction vessel with nitrogen (balloon).
8. Agitate the resin for 2 h.
9. Prepare 2–3 (depending on the scale of synthesis) conical centrifuge tubes of 50 mL, each filled with 30 mL of cold methyl *tert*-butyl ether (MTBE) (*see* **Note 5**).
10. Remove the resin by filtration, allowing solution to drip directly into cold MTBE containing tubes. Distribute the solution evenly between tubes.
11. Centrifuge for 10 min (1000–3500×*g*) (*see* **Note 6**).
12. Decant the ether layer (*see* **Note** 7) and wash the precipitate again with MTBE. Centrifuge using **step 12**. Repeat washings two times.
13. Dissolve the precipitate in aqueous buffer or 10 % acetic acid solution and lyophilize.

    Remarks: When working with peptoids/peptomers bearing many different protecting groups other cleavage cocktails may be more suitable [44].

The progress of all reactions (acylation/deprotection/amination) can be conveniently monitored by performing “mini-cleavage” of the product from the resin and then analyzing it by means of MALDI-ToF and RP-HPLC.

1. Place several beads of resin in 1.5 mL microcentrifugation tube.
2. Prepare a mixture of 2.5 μL of H_2O, 2.5 μL triisopropylsilane, and 95 μL of TFA and add it to the resin.

3. Agitate the resin in the tube for 1 h using laboratory shaker.
4. Add 200 μL of H_2O and 200 μL of MTBE.
5. Vortex the mixture for 30 s and then centrifuge for 30 s to separate the water and ether layers.
6. Discard the top (ether) layer using pipette.
7. Transfer 100 μL of water layer to new microcentrifugation tube.
8. Dilute the sample with water and perform MALDI-ToF/ HPLC analysis.

4 Notes

1. Other solvents may also be used such as DCM, NMP, THF, or mixture of DMF and DCM
2. When coupling the first peptoid residue, pre-swell the resin in DCM for 10 min (four times the resin volume), then wash with DMF 2 × 10 mL.
3. Amines that are purchased as hydrochloride salts should be neutralized. A protocol developed by Zuckermann and coworkers can be followed [45]. Briefly, amine solution in DMSO is treated with 0.95 eq of KOH, and the precipitated NaCl is then removed by centrifugation.
4. Before cleavage from the resin it is important to remove all traces of DMF since residual basic DMF can have inhibitory effect on TFA based acidolysis.
5. Prepare the tubes by filling it with desired volumes of MTBE ether and by precooling them in dry ice for 1–2 h.
6. Make sure that spark-proof centrifuge is used since highly flammable MTBE is present.
7. Ether washings should be retained until the yield of product has been established.

References

1. Kirshenbaum K, Barron AE, Goldsmith RA, Armand P, Bradley EK, Truong KT, Dill KA, Cohen FE, Zuckermann RN (1998) Sequence-specific polypeptoids: a diverse family of heteropolymers with stable secondary structure. Proc Natl Acad Sci USA 95:4303–4308
2. Butterfoss GL, Renfrew PD, Kuhlman B, Kirshenbaum K, Bonneau R (2009) A preliminary survey of the peptoid folding landscape. J Am Chem Soc 131:16798–16807
3. Choudhary A, Gandla D, Krow GR, Raines RT (2009) Nature of amide carbonyl–carbonyl interactions in proteins. J Am Chem Soc 131:7244–7246
4. Shah NH, Butterfoss GL, Nguyen K, Yoo B, Bonneau R, Rabenstein DL, Kirshenbaum K (2008) Oligo(N-aryl glycines): a new twist on structured peptoids. J Am Chem Soc 130:16622–16632
5. Simon RJ, Kania RS, Zuckermann RN, Huebner VD, Jewell DA, Banville S, Ng S, Wang L, Rosenberg S, Marlowe CK et al (1992) Peptoids: a modular approach to drug discovery. Proc Natl Acad Sci USA 89:9367–9371

6. Gellman SH (1998) Foldamers: a manifesto. Acc Chem Res 31:173–180
7. Armand P, Kirshenbaum K, Goldsmith RA, Farr-Jones S, Barron AE, Truong KT, Dill KA, Mierke DF, Cohen FE, Zuckermann RN, Bradley EK (1998) NMR determination of the major solution conformation of a peptoid pentamer with chiral side chains. Proc Natl Acad Sci USA 95:4309–4314
8. Wu CW, Sanborn TJ, Huang K, Zuckermann RN, Barron AE (2001) Peptoid oligomers with α-chiral, aromatic side chains: sequence requirements for the formation of stable peptoid helices. J Am Chem Soc 123:6778–6784
9. Shin SB, Yoo B, Todaro LJ, Kirshenbaum K (2007) Cyclic peptoids. J Am Chem Soc 129:3218–3225
10. Nam KT, Shelby SA, Choi PH, Marciel AB, Chen R, Tan L, Chu TK, Mesch RA, Lee BC, Connolly MD, Kisielowski C, Zuckermann RN (2010) Free-floating ultrathin two-dimensional crystals from sequence-specific peptoid polymers. Nat Mater 9:454–460
11. Culf AS, Ouellette RJ (2010) Solid-phase synthesis of N-substituted glycine oligomers (alpha-peptoids) and derivatives. Molecules 15:5282–5335
12. Peptoid Summits website. http://www.peptoids.org
13. Olsen CA (2010) Peptoid-peptide hybrid backbone architectures. Chembiochem 11:152–160
14. Goodman M, Wang LL, Feng Y (1994) Synthesis and characterization of sequential peptide-peptoid copolymers. Polym Prepr 35:767–768
15. Hamy F, Felder ER, Heizmann G, Lazdins J, Aboul-ela F, Varani G, Karn J, Klimkait T (1997) An inhibitor of the Tat/TAR RNA interaction that effectively suppresses HIV-1 replication. Proc Natl Acad Sci USA 94:3548–3553
16. Tran TA, Mattern RH, Afargan M, Amitay O, Ziv O, Morgan BA, Taylor JE, Hoyer D, Goodman M (1998) Design, synthesis, and biological activities of potent and selective somatostatin analogues incorporating novel peptoid residues. J Med Chem 41:2679–2685
17. Mattern RH, Tran TA, Goodman M (1998) Conformational analyses of somatostatin-related cyclic hexapeptides containing peptoid residues. J Med Chem 41:2686–2692
18. Holder JR, Bauzo RM, Xiang Z, Scott J, Haskell-Luevano C (2003) Design and pharmacology of peptoids and peptide–peptoid hybrids based on the melanocortin agonists core tetrapeptide sequence. Bioorg Med Chem Lett 13:4505–4509
19. Kruijtzer JAW, Nijenhuis WAJ, Wanders N, Gispen WH, Liskamp RMJ, Adan RAH (2005) Peptoid-peptide hybrids as potent novel melanocortin receptor ligands. J Med Chem 48:4224–4230
20. Masip I, Perez-Paya E, Messeguer A (2005) Peptoids as source of compounds eliciting antibacterial activity. Comb Chem High Throughput Screen 8:235–239
21. Miller SM, Simon RJ, Ng S, Zuckermann RN, Kerr JM, Moos WH (1995) Comparison of the proteolytic susceptibilities of homologous L-amino acid, D-amino acid, and N-substituted glycine peptide and peptoid oligomers. Drug Dev Res 35:20–32
22. Patch JA, Barron AE (2003) Helical peptoid mimics of magainin-2 amide. J Am Chem Soc 125:12092–12093
23. Chongsiriwatana NP, Wetzler M, Barron AE (2011) Functional synergy between antimicrobial peptoids and peptides against Gram-negative bacteria. Antimicrob Agents Chemother 55:5399–5402
24. Shankaramma SC, Moehle K, James S, Vrijbloed JW, Obrecht D, Robinson JA (2003) A family of macrocyclic antibiotics with a mixed peptide-peptoid beta-hairpin backbone conformation. Chem Commun 1842–1843
25. Zhu WL, Hahm KS, Shin SY (2007) Cathelicidin-derived Trp/Pro-rich antimicrobial peptides with lysine peptoid residue (Nlys): therapeutic index and plausible mode of action. J Pept Sci 13:529–535
26. Zhu WL, Song YM, Park Y, Park KH, Yang ST, Kim JI, Park IS, Hahm KS, Shin SY (2007) Substitution of the leucine zipper sequence in melittin with peptoid residues affects self-association, cell selectivity, and mode of action. Biochim Biophys Acta Biomembr 1768:1506–1517
27. Gobbo M, Benincasa M, Bertoloni G, Biondi B, Dosselli R, Papini E, Reddi E, Rocchi R, Tavano R, Gennaro R (2009) Substitution of the arginine/leucine residues in apidaecin Ib with peptoid residues: effect on antimicrobial activity, cellular uptake, and proteolytic degradation. J Med Chem 52:5197–5206
28. Fowler SA, Stacy DM, Blackwell HE (2008) Design and synthesis of macrocyclic peptomers as mimics of a quorum sensing signal from *Staphylococcus aureus*. Org Lett 10:2329–2332
29. Zuckermann RN, Kerr JM, Kent SBH, Moos WH (1992) Efficient method for the preparation of peptoids [oligo(N-substituted glycines)] by submonomer solid-phase synthesis. J Am Chem Soc 114:10646–10647
30. Ostergaard S, Holm A (1997) Peptomers: a versatile approach for the preparation of diverse combinatorial peptidomimetic bead libraries. Mol Divers 3:17–27

31. Robey FA (1994) Bromoacetylated synthetic peptides. Starting materials for cyclic peptides, peptomers, and peptide conjugates. Methods Mol Biol 35:73–90
32. Frey A, Neutra MR, Robey FA (1997) Peptomer aluminum oxide nanoparticle conjugates as systemic and mucosal vaccine candidates: synthesis and characterization of a conjugate derived from the C4 domain of HIV-1MN Gp120. Bioconjugate Chem 8:424–433
33. Huang K, Wu CW, Sanborn TJ, Patch JA, Kirshenbaum K, Zuckermann RN, Barron AE, Radhakrishnan I (2006) A threaded loop conformation adopted by a family of peptoid nonamers. J Am Chem Soc 128:1733–1738
34. Heine N, Ast T, Schneider-Mergener J, Reineke U, Germeroth L, Wenschuh H (2003) Synthesis and screening of peptoid arrays on cellulose membranes. Tetrahedron 59:9919–9930
35. Li S, Bowerman D, Marthandan N, Klyza S, Luebke KJ, Garner HR, Kodadek T (2004) Photolithographic synthesis of peptoids. J Am Chem Soc 126:4088–4089
36. Reddy MM, Kodadek T (2005) Protein "fingerprinting" in complex mixtures with peptoid microarrays. Proc Natl Acad Sci USA 102: 12672–12677
37. Burkoth TS, Fafarman AT, Charych DH, Connolly MD, Zuckermann RN (2003) Incorporation of unprotected heterocyclic side chains into peptoid oligomers via solid-phase submonomer synthesis. J Am Chem Soc 125: 8841–8845
38. Stawikowski M, Stawikowska R, Jaskiewicz A, Zablotna E, Rolka K (2005) Examples of peptide–peptoid hybrid serine protease inhibitors based on the trypsin inhibitor SFTI-1 with complete protease resistance at the P1-P1' reactive site. Chembiochem 6:1057–1061
39. Kruijtzer JAW, Hofmeyer LJF, Heerma W, Versluis C, Liskamp RMJ (1998) Solid-phase syntheses of peptoids using Fmoc-protected N-substituted glycines: the synthesis of (retro) peptoids of Leu-enkephalin and substance P. Chem Eur J 4:1570–1580
40. de Haan EC, Wauben MHM, Grosfeld-Stulemeyer MC, Kruijtzer JAW, Liskamp RMJ, Moret EE (2002) Major histocompatibility complex class II binding characteristics of peptoid–peptide hybrids. Bioorg Med Chem 10:1939–1945
41. Simpson LS, Burdine L, Dutta AK, Feranchak AP, Kodadek T (2009) Selective toxin sequestrants for the treatment of bacterial infections. J Am Chem Soc 131:5760–5762
42. Alluri PG, Reddy MM, Bachhawat-Sikder K, Olivos HJ, Kodadek T (2003) Isolation of protein ligands from large peptoid libraries. J Am Chem Soc 125:13995–14004
43. Lee BC, Chu TK, Dill KA, Zuckermann RN (2008) Biomimetic nanostructures: creating a high-affinity zinc-binding site in a folded nonbiological polymer. J Am Chem Soc 130:8847–8855
44. Dick F (1994) Acid cleavage/deprotection in Fmoc/tBiu solid-phase peptide synthesis. Methods Mol Biol 35:63–72
45. Figliozzi GM, Goldsmith R, Ng SC, Banville SC, Zuckermann RN (1996) Synthesis of N-substituted glycine peptoid libraries. Methods Enzymol 267:437–447

Chapter 5

Synthesis of Side Chain *N,N'*-Diaminoalkylated Derivatives of Basic Amino Acids for Application in Solid-Phase Peptide Synthesis

Jean-Philippe Pitteloud, Nina Bionda, and Predrag Cudic

Abstract

Despite the enormous therapeutic potential, the clinical use of peptides has been limited by their poor bioavailability and low stability under physiological conditions. Hence, efforts have been undertaken to alter peptide structure in ways to improve their pharmacological properties. Inspired by the importance of basic amino acids in biological systems and the remarkable versatility displayed by lysine during the synthesis of complex peptide scaffolds, this chapter describes a simple procedure that enables rapid access to protected *N,N'*-diaminoalkylated basic amino acid building blocks suitable for standard solid-phase peptide synthesis. This procedure allows preparation of symmetrical, as well as unsymmetrical, dialkylated amino acid derivatives that can be further modified, enhancing their synthetic utility. The suitability of the synthesized branched basic amino acid building blocks for use in standard solid-phase peptide synthesis has been demonstrated by synthesis of an indolicidin analog in which the lysine residue was substituted with its synthetic polyamino derivate. The substitution provided indolicidin analog with increase net positive charge, more ordered secondary structure in biological membranes mimicking conditions, and enhanced antibacterial activity without altering hemolytic activity. Taking into consideration the increasing interest for peptides with unusual structural features due to their improved biological properties, the described synthesis of polyfunctional amino acid building blocks is of particular practical value.

Key words *N*-alkylation, Reductive amination, One-pot procedure, Polyfunctional amino acid building blocks, Solid-phase peptide synthesis

1 Introduction

Peptides and proteins play essential roles in many biological systems and are currently used as treatment for a wide variety of human diseases [1]. Despite their great potential, the development of peptides as efficient therapeutic agents has been limited by their high susceptibility to enzymatic proteolysis (low oral bioavailability and short in vivo half-lives) and poor penetration of biological barriers. In order to overcome these limitations, attention has been focused on the synthesis of peptide mimics with enhanced biological and

Predrag Cudic (ed.), *Peptide Modifications to Increase Metabolic Stability and Activity*, Methods in Molecular Biology, vol. 1081, DOI 10.1007/978-1-62703-652-8_5,

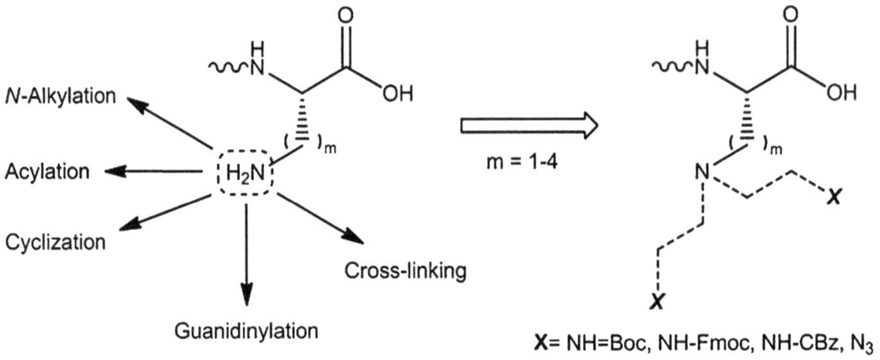

Fig. 1 Common applications of a basic amino acid's side chain and design of novel *N*,*N*′-diaminoalkylated amino acid building blocks

pharmacological properties [2]. For example, the replacement of a proteolytically cleavable peptide bond with bioisosteric (e.g., peptoids [3–5], azapeptides [6, 7], urea-peptidomimetics [8], β-peptidosulfonamides [9], depsipeptides [10]), but less labile chemical entities afforded peptide analogs (pseudopeptides) with increased stability and consistent biological activity. Also, the introduction of small molecules that can imitate specific secondary structural features has resulted in peptides with enhanced biological properties [11]. Other successful approaches comprising the cyclization of biologically active peptides [12, 13], the incorporation of non-proteinogenic amino acids (e.g., D-amino acids, *N*-alkylated amino acids, β-amino acids) [14], as well as peptide glycosylation [15], have also met success in improving the physicochemical and pharmacological properties of peptides.

Due to the reactivity of their side-chain primary amines, basic amino acids such as lysine, ornithine, diaminobutyric acid, and diaminopropionic acid have been extensively used for the synthesis of complex structures in solution and solid-phase (*see* Fig. 1). The availability of orthogonally protected lysine has been pivotal for the synthesis of densely branched multi-antigenic peptide systems (MAPs) [16, 17], constrained cyclic peptides [12, 18–20], α-helical bundle proteins [21, 22], and collagen-like helical structures [23–25]. Also, the high incidence of basic amino acids has been determinant for the unique properties exhibited by many cationic antimicrobial peptides (CAMPs) [26, 27] and cell-penetrating peptides (CPPs) [28–30]. In the case of CAMPs, the protonation of side-chain amino groups provide the cationic character necessary for a selective interaction of the peptide with the negatively charged phospholipid membranes of bacteria and other pathogens. In addition, The presence of properly spaced positive charges has proved crucial for molecules that bind and neutralize bacterial lipopolysaccharides (LPS), the primary cause of the Gram-negative septic shock [31].

Similarly, a net positive charge at physiological pH has proved critical for the membrane translocation properties displayed by CPPs [32–34].

With a special interest in expanding the possible applications of basic amino acids in the synthesis of complex molecular scaffolds, we have developed [35] a simple and efficient one-pot procedure that enables rapid access to orthogonally protected *N,N'*-diaminoalkylated amino acid building blocks suitable for standard Boc and Fmoc-solid-phase peptide synthesis (*see* Fig. 1). Our synthetic approach comprises the double reductive alkylation [36, 37] of commercially available N^{α}-protected diamino acids (lysine, ornithine, 2,4-diaminobutyric acid, and 2,3-diaminopropionic acid) with *N*-protected amino aldehydes in the presence of sodium cyanoborohydride [38]. Some advantages of this approach include mild reaction conditions, use of readily available starting materials, and direct access to symmetrically and unsymmetrically branched basic amino acid building blocks. Moreover, the possibility of further modification of these derivatives in solution or on solid support made them well suited for potential combinatorial chemistry diversity. The suitability of our orthogonally protected *N,N'*-diaminoalkylated amino acid building blocks for utilization in SPPS has been demonstrated [35] through the substitution of the sole lysine residue in the naturally occurring antimicrobial peptide indolicidin (ILPWKWPWWPWRR-NH_2) [39]. The results are summarized in Subheading 3.2.

Taking into consideration the impact that peptides with unusual structural features have on the design of novel agents with enhanced biological and pharmacological properties, we describe in this chapter the protocols for the synthesis of versatile *N,N'*-diaminoalkylated basic amino acids.

2 Materials

1. All commercially available materials and reagents are used as received.
2. Solvents such as methanol (MeOH), dichloromethane (DCM), ethyl acetate (EtOAc), tetrahydrofuran (THF), methyl *tert*-butyl ether, and toluene were ACS grade and can be purchased from Sigma-Aldrich, Fisher, VWR, or other commercial sources.
3. Glacial acetic acid (AcOH) and trifluoroacetic acid (TFA) are available commercially from various vendors.
4. The N^{α}-protected amino acid precursors are commercially available from ChemImpex, Sigma-Aldrich, Bachem, and Novabiochem.
5. Long chain amino acids can be purchased from Sigma-Aldrich.

6. 2-(*N*-*tert*-butoxycarbonyl) aminoacetaldehyde is available from Sigma-Aldrich and 3-(*N*-benzyloxycarbonyl) aminopropionaldehyde can be purchased from Santa Cruz Biotechnology.
7. 2-(*N*-9-fluorenylmethoxycarbonyl) aminoacetaldehyde [40] and 3-(*N*-9-fluorenylmethoxy-carbonyl) aminopropionaldehyde [41] can be obtained by mild oxidation of commercially available Fmoc-*N*-amino alcohols (Sigma-Aldrich, ChemImpex, Santa Cruz Biotechnology) employing Dess-Martin Periodinane (DMP) in excellent yields.
8. 3-Azipropionaldehyde can be readily synthesized from commercially available acrolein (Sigma-Aldrich, Acros) following a reported procedure [42, 43] and stored at –20 °C for several weeks.
9. Sodium cyanoborohydride ($NaBH_3CN$) is available from Sigma-Aldrich and Acros.
10. Thin layer chromatography (TLC), used for monitoring reactions, is performed on Whatman precoated silica gel F-254 plates and visualized by ultraviolet light and/or staining with ninhydrin solution.
11. Silica gel 60 (EMD Chemicals, Germany; particle size 0.040–0.063 mm, 230–400 Mesh ASTM) is used for normal phase flash chromatography.
12. All reagents employed during the synthesis and cleavages of the peptides are readily available from several sources.

3 Methods

3.1 One-Pot Double Reductive Alkylation of N^{α}-Protected Diamino Acids and Long Chain Aliphatic Amino Acids

3.1.1 Synthesis of Symmetrically Dialkylated N^{α}-Protected Diamino Acids and Long Chain Aliphatic Amino Acids

1. Dissolve the (di)amino acid in MeOH (90 mL/mmol of diamino acid) containing AcOH (1–6 %) (*see* **Note 1**).
2. Add the corresponding amino aldehyde (3.0 eq.) as a solution in MeOH (*see* **Note 2**).
3. Stir this solution for 1 h at room temperature.
4. Add dropwise solution of $NaBH_3CN$ (1.5 eq.) in MeOH over 2 h (*see* **Note 3**).
5. Stir reaction mixture overnight at room temperature.
6. Check for completion by thin-layer chromatography (TLC) and MALDI-TOF MS analyses of a representative aliquot.
7. The reaction mixture workup is described in Subheading 3.1.3.

3.1.2 Synthesis of Unsymmetrically Dialkylated N^{α}-Protected Diamino Acids and Long Chain Aliphatic Amino Acids

1. Dissolve the (di)amino acid in MeOH (90 mL/mmol of diamino acid) containing AcOH (1–6 %) (*see* **Note 1**).
2. Load the first protected amino aldehyde (1.2 eq.) as a solution in MeOH on a syringe (*see* **Note 2**).

3. In a similar fashion, load $NaBH_3CN$ (0.8 eq.) as a solution in MeOH on a syringe (*see* **Note 3**).
4. Add simultaneously the aldehyde and $NaBH_3CN$ into the diamino acid solution over 4 h at room temperature (*see* **Note 4**).
5. Stir reaction mixture for 4 h at room temperature.
6. Monitor the progress of the reaction by thin-layer chromatography (TLC) of a representative aliquot (*see* **Note 5**).
7. Repeat **steps 2–4** using the second protected amino aldehyde.
8. Stir reaction mixture overnight at room temperature.
9. Check for completion by TLC and MALDI-TOF MS analyses of a representative aliquot.
10. The reaction mixture workup is described in Subheading 3.1.3.

3.1.3 Aqueous Workup

1. Concentrate the reaction mixture by rotary evaporation and dilute the residue with EtOAc (*see* **Note 6**).
2. Transfer the liquid to a separatory funnel and wash with water.
3. Separate the organic layer.
4. Extract the aqueous layer with fresh portions of the solvent used in **step 1**. Repeat this step three times.
5. Dry the combined organic layers over anhydrous sodium sulfate (Na_2SO_4), filter, and evaporate to dryness (*see* **Note 7**).
6. Purification of the crude product is described in Subheading 3.1.4.

3.1.4 Purification

A. Normal phase flash column chromatography (*see* **Note 8**).
   1. Pack a standard glass chromatographic column with slurry of silica gel in hexane [44].
   2. Transfer the crude mixture of modified amino acid to the column in the minimum amount of DCM.
   3. Elute the less polar components of the mixture using a mixture of hexanes and EtOAc (1:1 v/v).
   4. Upon complete elution of less polar impurities, use a mixture of EtOAc/MeOH/AcOH (8:1:1 v/v) to elute the desired amino acid as a pure compound (*see* **Note 9**).
   5. The collected fractions analyze for purity by TLC.
   6. Remove solvent by rotary evaporation (*see* **Note 7**).

B. Reverse-phase flash column chromatography using an automatic flash chromatography system
   1. Purification by reverse-phase flash chromatography can be performed on a CombiFlash® Companion® flash chromatography system using RediSep Rf Gold® C_{18} columns (Teledyne Isco, USA) of appropriate sizes (*see* **Note 10**).

2. Load the crude mixture of modified amino acid(s) applying a dry loading technique on appropriate C-18 silica gel cartridge.
3. Elute the components of the mixture using 5–85 % gradient of ACN in H_2O containing 0.1 % formic acid. Use UV detection at 214 and 267 nm.
4. Combine and lyophilize fractions containing pure product.

3.1.5 Characterization

A. Nuclear Magnetic Resonance
The structure of the modified building blocks can be confirmed by ^1H and ^{13}C NMR spectroscopy. Use standard borosilicate NMR tubes with a diameter of 5 mm and 0.5 mL of a convenient deuterated solvent to prepare samples (*see* **Note 11**). The NMR spectra can also offer a good estimation of the purity of the synthetic derivatives.

B. Matrix-assisted laser desorption ionization time-of-flight mass spectrometry (MALDI-TOF MS)
Mass spectrometric analyses of the synthetic derivatives can be performed on MALDI-TOF MS employing commonly used matrices such as 3,5-dimethoxy-4-hydroxycinnamic acid (sinapinic acid), 2,5-dihydroxybenzoic acid, and α-cyano-4-hydroxycinnamic acid. The use of any other mass spectrometric technique is encourage and should also provide desired results.

C. Purity assessment by high-performance liquid chromatography (HPLC)
The purity of the protected building blocks (>95 %) can be assessed by analytical reverse-phase high-performance liquid chromatography (RP-HPLC) system equipped with a UV/Vis detector. For this purpose a reverse-phase C_{18} column (e.g., Grace Vydac, 250×4.6 mm, 5 μm, 120 Å), 1 mL/min flow rate, and elution method of 98 % A for 0.5 min followed by linear gradient of 2–85 % B over 32 min. can be used (A is 0.1 % TFA in H_2O and B is 0.08 % TFA in ACN). The eluting products are detected by UV at 214 and 267 nm.

The synthetic protocols described above (*see* Subheading 3.1) rely on a double reductive alkylation of commercially available amino acid precursors with *N*-protected amino aldehydes in the presence of sodium cyanoborohydride. The combination of orthogonal amine protecting groups commonly employed in the synthesis of peptides in solution and on solid support (Fmoc, Boc, Cbz carbamates, and azide [45]), makes compounds **5–15** (*see* Table 1) valuable tools for the construction of more complex structural motifs.

In an attempt to synthesize unsymmetrically *N,N'*-diaminoalkylated amino acid building blocks in one-pot, the N^{α}-*protected* diamino acid **1a** was treated consecutively with two different *N*-protected amino aldehydes according to the protocol described in

Table 1
Double reductive alkylation of N^{α}-protected diamino acids with selected *N*-protected amino aldehydes[a]

R_2—NH—$(CH_2)_{n-1}$—CHO
$NaBH_3CN$
MeOH/AcOH

1 m=1 (Dap)
2 m=2 (Dab)
3 m=3 (Orn)
4 m=4 (Lys)

Series: **a** R_1 = Fmoc
b R_1 = Boc
c R_1 = Cbz

Entry	Substrate	R_1	*n*	R_2	Product	Yield (%)[b]
1	**1a**	Fmoc	2	Boc	**5**	85
2	**2a**	Fmoc	2	Boc	**6**	89
3	**3a**	Fmoc	2	Boc	**7**	87
4	**4a**	Fmoc	2	Boc	**8**	88
5	**1b**	Boc	2	Fmoc	**9**	80
6	**4b**	Boc	3	Fmoc	**10**	89
7	**1a**	Fmoc	3	Cbz	**11**	86
8	**4a**	Fmoc	3	Cbz	**12**	89
9	**1b**	Boc	3	Cbz	**13**	82
10	**1c**	Cbz	2	Boc	**14**	81
11	**1a**	Fmoc	3	N_3	**15**	65[c]

[a]The protected diamino acid dissolved in MeOH/AcOH was treated with the corresponding aldehyde (3.0 eq.) followed by a solution of $NaBH_3CN$ in MeOH and stirring at room temperature
[b]Starting diamino acid as limiting reagent
[c]Purified by RP-flash chromatography with a gradient 5–100 % B, where A was 0.1 % FA in H_2O and B was 0.1 % FA in ACN

Subheading 3.1.2. Initially, the selection of 3-(*N*-benzyloxycarbonyl) aminopropionaldehyde and 2-(*N-tert-butoxycarbonyl*)aminoacetaldehyde as aldehydes, afforded a mixture of homo- and heterodialkylated products **5**, **11**, and **16** (*see* Fig. 2). Although the desired unsymmetrical building block **16** was obtained as the major product, we were unable to isolate **16** in pure form. In an effort to overcome this problem, the tandem reaction of **1a** with 2-(*N-tert*-butoxycarbonyl)aminoacetaldehyde and 3-azidopropionaldehyde according to the protocol described in Subheading 3.1.2, afforded a fully separable mixture of dialkylated products **5**, **15**, and **17**. Besides yielding more separable mixtures in this particular case, the presence of the azide group in compounds **15** and **17** permits also a wide

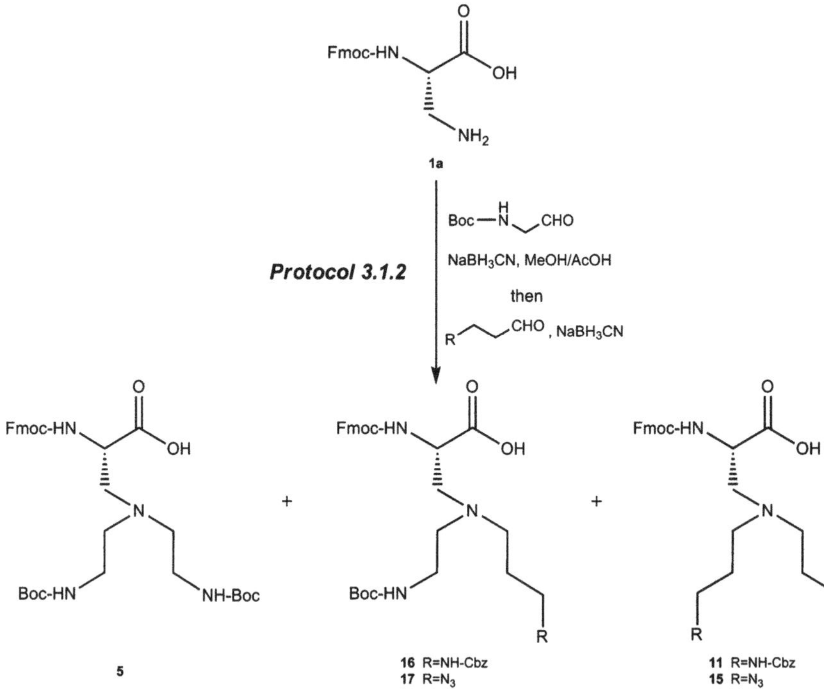

Fig. 2 One-pot synthesis of unsymmetrically *N,N'*-diaminoalkylated amino acid building blocks via tandem double reductive alkylation

range of transformations [46], both in solution and on solid support. For example, the chemoselectivity of the azide group towards aryl/alkyl phosphine allows its transformation into a free amine in the presence of small amounts of water [47]. Efficient protocols for the direct conversion of azides into various carbamates, including allyloxycarbamate (Aloc) have been reported as well [48]. Moreover, organic azides have been popular for their mild and highly chemoselective reaction with terminal alkynes via 1,3-dipolar cycloaddition (Hüisgen reaction) in the presence of copper (I) salts [49, 50]. Copper-free versions of this reaction have been reported to be highly compatible in the study of biological processes in living systems [51–53].

The synthetic protocols described in Subheading 3.1 have been followed for the construction of *N*-protected branched long chain aliphatic amino acids **18–20** (*see* Fig. 3). Treatment of commercially available 6-aminocaproic acid, 8-aminocaprylic acid, or 12-aminolauric acid with 3-(*N*-9-fluorenylmethoxycarbonyl)aminopropionaldehyde according to protocol described in Subheading 3.1.1 afforded products **18–20** respectively. Similarly, the tandem treatment of 12-aminolauric acid with 2-(*N*-Boc)aminoacetaldehyde/3-(*N*-Fmoc)aminopropionaldehyde and 3-(*N*-Fmoc)aminopropionaldehyde/3-azidopropionaldehyde using Subheading 3.1.2 resulted in mixtures of unsymmetrically dialkylated products **22** and **23** along with the corresponding homo-dialkylated byproducts (*see* Fig. 4). The use of

Protocol 3.1.1

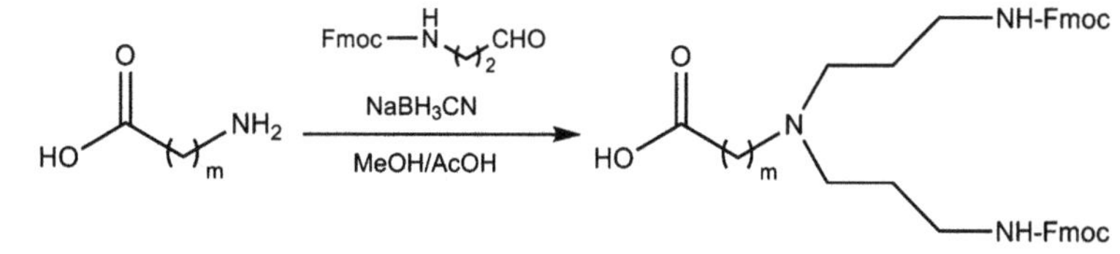

m = 5 (5-aminocaproic acid)
m = 7 (8-aminocaprylic acid)
m = 11 (12-aminolauric acid)

18 m = 5
19 m = 7
20 m = 11

Fig. 3 Double reductive alkylation of long chain amino acids with 3-(*N-9-fluorenylmethoxycarbonyl*) aminopropionaldehyde

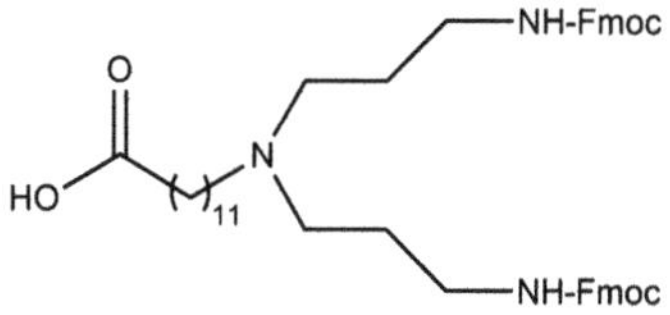

21

Protocol 3.1.2

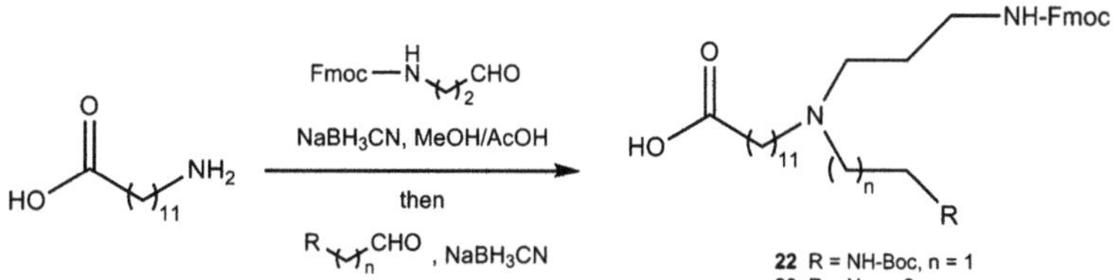

22 R = NH-Boc, n = 1
23 R = N_3, n=2

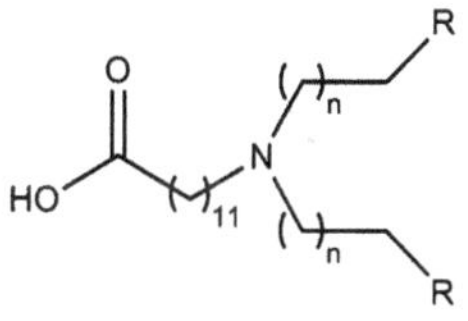

24 R = NH-Boc, n = 1
25 R = N_3, n=2

Fig. 4 One-pot synthesis of unsymmetrically *N,N′*-diaminoalkylated 12-aminolauric acid via tandem double reductive alkylation

unsymmetrically dialkylated peptide building blocks is highly desirable for the construction of more complex scaffolds.

3.2 Synthesis, Structural Characterization, Antimicrobial and Hemolytic Activities of Indolicidin's Analog Indo-5

Indolicidin(H-Ile-Leu-Pro-Trp-Lys-Trp-Pro-Trp-Trp-Pro-Trp-Arg-Arg-NH_2), a 13 amino acid C terminus amidated peptide isolated from the granules of bovine neutrophils, is the smallest known naturally occurring linear cationic antimicrobial peptide. To evaluate the suitability of described *N,N′*-diaminoalkylated building blocks for standard Fmoc SPPS, we synthesized an analog of indolicidin in which the lysine was substituted with a building block **5** [35]. Since the antibacterial activity of cationic peptides can be modulated through alteration of the peptide's hydrophobicity and/or net charge, this substitution also served to evaluate

Table 2
Retention time, charge, antimicrobial and hemolytic activity of indolicidin and its analog Indo-5

			MIC (μg/mL)[b]		
Peptides	**t_R (min)[a]**	**Net charge**	***S. aureus*[c]**	***E. coli*[d]**	**% Hemolysis[e]**
Indolicidin	19.05	+4	64	64	20
Indo-5	18.47	+5	32	32	18

[a]Analytical RP-HPLC conditions described in general methods
[b]*See* **Note 12**
[c]ATCC 33591
[d]ATCC 29181
[e]Determined at MIC of peptides and expressed in percent relative to total hemolysis caused by Triton X-100

the effect of increasing indolicidin's net positive charge without altering its backbone structure on its antibacterial and hemolytic activities. Hence, both indolicidin and its analog **Indo-5** were synthesized using standard solid-phase Fmoc synthetic protocol. The crude peptides were precipitated and washed with cold methyl *tert*-butyl ether, and purified on an Agilent Infinity 1260 HPLC system using a preparative Vydac C_{18} column (Grace Vydac, 250 × 22 mm, 10 μm, 120 Å) with an elution method consisting of a linear gradient of 2–85 % B over 32.5 min, where A was 0.1 % TFA in H_2O and B was 0.08 % TFA in ACN. The final purity of the peptides (>98 %) was assessed by analytical reverse-phase HPLC using conditions described in Subheading 3.1.5, **step C**, and MALDI-TOF MS (indolicidin, found m/z = 1907.0822, calculated m/z = 1906.2845; Indo-5, found m/z = 1951.5239, calculated m/z = 1950.3403).

The antimicrobial activity of both indolicidin and its analog **Indo-5** against multidrug-resistant strains of *Staphylococcus aureus* (ATCC 33591) and *Escherichia coli* (ATCC 29181) were determined and the results are summarized in Table 2. The synthetic analog **Indo-5**, bearing an additional positive charge, showed somewhat improved activity against both bacterial strains in comparison to the natural product indolicidin. These results are in agreement with previous reports in which an increase in net positive charge translates into enhanced indolicidin's antimicrobial activity [54, 55]. However, the obtained MIC values showed that a decrease in overall hydrophobicity, as indicated by the HPLC retention times (*see* Table 2), did not correlate with their antibacterial activities [56, 57], with less hydrophobic **Indo-5** being more active. Not surprisingly, indolicidin and **Indo-5** exhibit comparable hemolytic activities, confirming the dependence of their hemolytic activity with specific tryptophans and the arginines in the sequences [58].

The structural features of indolicidin and **Indo-5** were monitored by CD spectroscopy (*see* Fig. 5) The CD spectra were recorded in water as well as in less polar solvent systems such as 50 and 100 %

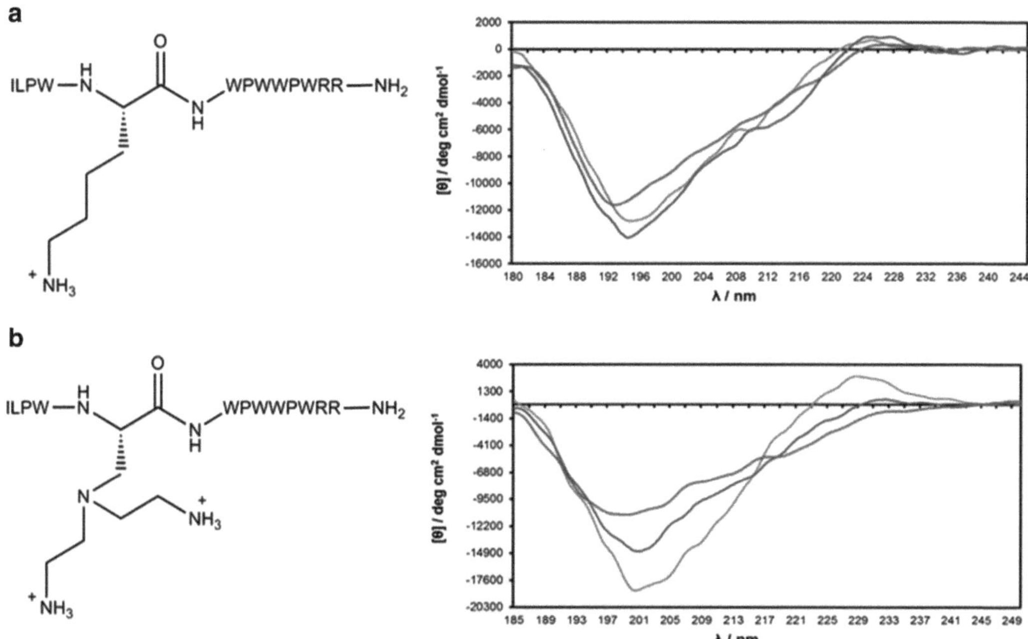

Fig. 5 Structures and CD spectra of (**a**) indolicidin, and (**b**) synthetic analog **Indo-5** in water (*green*), 50 % TFE (*blue*), and 100 % TFE (*red*) (Color figure online)

TFE. Typically, TFE is used as a membrane mimicking solvent, and it is known to induce formation of the stable conformations in peptides which are otherwise unstructured in aqueous solutions [59–61].

The CD spectrum of indolicidin measured in water shows a single negative band at 201 nm, characteristic of an unordered conformation [58, 62]. However, the proposed secondary structure of indolicidin has been somewhat controversial, and poly-L-Pro II helix and β-turn structures were reported as well [63–65]. An increase in the TFE concentration resulted in a slight shift of the minimum from 201 to 198 nm suggesting no major conformational change. In the case of the indolicidin analog **Indo-5**, the CD spectrum recorded in water showed an intense negative band at 201 nm along with a weak positive band at 227 nm, reminiscent of a poly-L-Pro II helix structure. The observed decrease in the intensity of the CD spectrum of **Indo-5** in 50 and 100 % TFE suggest formation of a more ordered secondary structure in TFE [61]. CD experiments show that the substitution of the lysine residue with branched basic amino acids, such as **5**, can induce changes in the peptide's secondary structure.

In summary, substitution of Lys residue in indulacidin peptide sequence with unusual amino acid **5** resulted in an indolicidin analog with increased net positive charge, more ordered secondary structure under biological membranes mimicking conditions, improved antibacterial activity, and with hemolytic activity comparable to the parent natural product.

4 Notes

1. When using N^{α}-Fmoc protected diamino acid, dissolve the material in a minimum amount of glacial AcOH and diluted with MeOH. For others, dissolve the material in MeOH and add glacial AcOH.
2. Other solvents such as DMF or THF can be also employed to dissolve protected amino aldehydes without affecting the reaction outcome.
3. A commercially available solution of $NaBH_3CN$ in THF (1 M) may be used instead of the solid material.
4. To facilitate the simultaneous addition of reducing agent and protected amino aldehydes the use of a syringe pump is highly recommended.
5. Due to the increased reactivity of secondary amines towards further alkylation, treatment of primary amines with less than two equivalents of aldehyde usually leads to mixtures of unreacted primary amine, mono-alkylated, and di-alkylated products.
6. Dichloromethane can be used instead of ethyl acetate.
7. Consecutive co-evaporation with toluene is highly encouraged to remove residual AcOH.
8. The use of automatic flash chromatography instruments significantly reduces the time and solvent required to achieve product purification.
9. Formic acid (b.p. 108 °C) can be used as a more volatile alternative to acetic acid (b.p. 118 °C).
10. Other commercially available instruments meeting the same specifications are expected to give similar results.
11. Although the products are fully soluble in $CHCl_3$-d_1, the use of more polar DMSO-d_6 or MeOH-d_4 gives spectra with considerably better resolution.
12. Minimal inhibitory concentrations (MICs) were determined by twofold serial dilution of peptides in sterile 96-well plates containing Mueller-Hinton broth according to the Clinical and Laboratory Standards Institute (CLSI) guidelines.

References

1. Stevenson CL (2009) Advances in peptide pharmaceuticals. Curr Pharm Biotechnol 10:122–137
2. Liskamp RMJ, Rijkers DTS, Kruijtzer JAW, Kemmink J (2011) Peptides and proteins as a continuing exciting source of inspiration for peptidomimetics. Chembiochem 12: 1626–1653
3. Zuckermann RN, Kerr JM, Kent SBH, Moos WH (1992) Efficient method for the preparation of peptoids [oligo(N-substituted glycines)] by submonomer solid-phase synthesis. J Am Chem Soc 114:10646–10647
4. Fowler SA, Blackwell HE (2009) Structure–function relationships in peptoids: recent advances toward deciphering the structural

requirements for biological function. Org Biomol Chem 7:1508–1524

5. Yoo B, Shin SBY, Huang ML, Kirshenbaum K (2010) Peptoid macrocycles: making the rounds with peptidomimetic oligomers. Chem Eur J 16:5528–5537
6. Gante J, Krug M, Lauterbach G, Weitzel R, Hiller W (1995) Synthesis and properties of the first all-aza analogue of a biologically active peptide. J Pept Sci 1:201–206
7. Zega A (2005) Azapeptides as pharmacological agents. Curr Med Chem 12(5):589–597
8. Boeijen A, Liskamp RMJ (1999) Solid-phase synthesis of oligourea peptidomimetics. Eur J Org Chem 1999:2127–2135
9. de Jong R, Rijkers DTS, Liskamp RMJ (2002) Automated solid-phase synthesis and structural investigation of β-peptidosulfonamides and β-peptidosulfonamide/β-peptide hybrids: β-peptidosulfonamide and β-peptide foldamers are two of a different kind. Helv Chim Acta 85:4230–4243
10. Cudic P, Stawikowski M (2007) Pseudopeptide synthesis via Fmoc solid-phase synthetic methodology. Mini Rev Org Chem 4:268–280
11. Toniolo C (2004) Peptides incorporating secondary structure inducers and mimetics. In: Goodman M (ed) Synthesis of peptides and peptidomimetics, vol E 22c. Houben-Weyl, Stuttgart, pp 693–835
12. Katsara M, Tselios T, Deraos S, Deraos G, Matsoukas M-T, Lazoura E, Matsoukas J, Apostolopoulos V (2006) Round and round we go: cyclic peptides in disease. Curr Med Chem 13:2221–2232
13. Karskela T, Virta P, Lönnberg H (2006) Synthesis of bicyclic peptides. Curr Org Synth 3:283–311
14. Gentilucci L, De Marco R, Cerisoli L (2010) Chemical modifications designed to improve peptide stability: incorporation of non-natural amino acids, pseudo-peptide bonds, and cyclization. Curr Pharm Des 16:3185–3203
15. Solá RJ, Griebenow K (2010) Glycosylation of therapeutic proteins: an effective strategy to optimize efficacy. BioDrugs 24:9–21
16. Sebestik J, Niederhafner P, Jezek J (2011) Peptide and glycopeptide dendrimers and analogous dendrimeric structures and their biomedical applications. Amino Acids 40:301–370
17. Sadler K, Tam JP (2002) Peptide dendrimers: applications and synthesis. Rev Mol Biotechnol 90:195–229
18. Tolle JC, Staples MA, Blout ER (1982) Synthesis of a new type of cyclic peptide: a bicyclic nonapeptide. J Am Chem Soc 104: 6883–6884
19. Zhang W, Taylor JW (1996) Efficient solid-phase synthesis of peptides with tripodal side-chain bridges and optimization of the solvent conditions for solid-phase cyclizations. Tetrahedron Lett 37:2173–2176
20. Yu C, Taylor JW (1996) A new strategy applied to the synthesis of an [alpha]-helical bicyclic peptide constrained by two overlapping i, i+7 side-chain bridges of novel design. Tetrahedron Lett 37:1731–1734
21. Hahn K, Klis W, Stewart J (1990) Design and synthesis of a peptide having chymotrypsin-like esterase activity. Science 248:1544–1547
22. Mutter M, Tuchscherer GG, Miller C, Altmann KH, Carey RI, Wyss DF, Labhardt AM, Rivier JE (1992) Template-assembled synthetic proteins with four-helix-bundle topology. Total chemical synthesis and conformational studies. J Am Chem Soc 114:1463–1470
23. Fields GB, Prockop DJ (1996) Perspectives on the synthesis and application of triple-helical, collagen-model peptides. Biopolymers 40: 345–357
24. Fields CG, Mickelson DJ, Drake SL, McCarthy JB, Fields GB (1993) Melanoma cell adhesion and spreading activities of a synthetic 124-residue triple-helical "mini-collagen". J Biol Chem 268:14153–14160
25. Grab B, Miles AJ, Furcht LT, Fields GB (1996) Promotion of fibroblast adhesion by triple-helical peptide models of type I collagen-derived sequences. J Biol Chem 271: 12234–12240
26. Jenssen H, Hamill P, Hancock REW (2006) Peptide antimicrobial agents. Clin Microbiol Rev 19:491–511
27. Findlay B, Zhanel GG, Schweizer F (2010) Cationic amphiphiles, a new generation of antimicrobials inspired by the natural antimicrobial peptide scaffold. Antimicrob Agents Chemother 54:4049–4058
28. Green M, Loewenstein PM (1988) Autonomous functional domains of chemically synthesized human immunodeficiency virus tat trans-activator protein. Cell 55:1179–1188
29. Derossi D, Joliot AH, Chassaing G, Prochiantz A (1994) The third helix of the Antennapedia homeodomain translocates through biological membranes. J Biol Chem 269:10444–10450
30. Oehlke J, Scheller A, Wiesner B, Krause E, Beyermann M, Klauschenz E, Melzig M, Bienert M (1998) Cellular uptake of an α-helical amphipathic model peptide with the potential to deliver polar compounds into the cell interior non-endocytically. Biochim Biophys Acta Biomembr 1414:127–139

31. David SA (2001) Towards a rational development of anti-endotoxin agents: novel approaches to sequestration of bacterial endotoxins with small molecules. J Mol Recognit 14(6): 370–387

32. Futaki S, Suzuki T, Ohashi W, Yagami T, Tanaka S, Ueda K, Sugiura Y (2001) Arginine-rich peptides. J Biol Chem 276:5836–5840

33. Wender PA, Mitchell DJ, Pattabiraman K, Pelkey ET, Steinman L, Rothbard JB (2000) The design, synthesis, and evaluation of molecules that enable or enhance cellular uptake: peptoid molecular transporters. Proc Natl Acad Sci USA 97:13003–13008

34. Patel L, Zaro J, Shen W-C (2007) Cell penetrating peptides: intracellular pathways and pharmaceutical perspectives. Pharm Res 24:1977–1992

35. Pitteloud J-P, Bionda N, Cudic P (2012) Direct access to side chain N,N′-diaminoalkylated derivatives of basic amino acids suitable for solid-phase peptide synthesis. Amino Acids 44(2): 321–333

36. Levadala MK, Banerjee SR, Maresca KP, Babich JW, Zubieta J (2004) Direct reductive alkylation of amino acids: synthesis of bifunctional chelates for nuclear imaging. Synthesis 2004:1759–1766

37. Bartholoma M, Valliant J, Maresca KP, Babich J, Zubieta J (2009) Single amino acid chelates (SAAC): a strategy for the design of technetium and rhenium radiopharmaceuticals. Chem Commun 493–512

38. Borch RF, Bernstein MD, Durst HD (1971) Cyanohydridoborate anion as a selective reducing agent. J Am Chem Soc 93: 2897–2904

39. Selsted ME, Novotny MJ, Morris WL, Tang YQ, Smith W, Cullor JS (1992) Indolicidin, a novel bactericidal tridecapeptide amide from neutrophils. J Biol Chem 267:4292–4295

40. Diness F, Beyer J, Meldal M (2004) Synthesis of 3-Boc-(1,3)-oxazinane-protected amino aldehydes from amino acids and their conversion into urea precursors. Novel building blocks for combinatorial synthesis. QSAR Comb Sci 23:130–144

41. Reggelin M, Junker B, Heinrich T, Slavik S, Bühle P (2006) Asymmetric synthesis of highly substituted azapolycyclic compounds via 2-alkenyl sulfoximines: potential scaffolds for peptide mimetics. J Am Chem Soc 128: 4023–4034

42. Boyer JH (1951) Addition of hydrazoic acid to conjugated systems. J Am Chem Soc 73: 5248–5252

43. Davies AJ, Donald ASR, Marks RE (1967) The acid-catalysed decomposition of some [small beta]-azido-carbonyl compounds. J Chem Soc C 2109–2112

44. Vogel AI, Tatchell AR, Furnis BS, Hannaford AJ, Smith PWG (1989) Vogel's textbook of practical organic chemistry, 5th edn. Pearson Education Limited, Essex

45. Lundquist JT, Pelletier JC (2001) Improved solid-phase peptide synthesis method utilizing α-azide-protected amino acids. Org Lett 3:781–783

46. Bräse S, Gil C, Knepper K, Zimmermann V (2005) Organic azides: an exploding diversity of a unique class of compounds. Angew Chem Int Ed 44:5188–5240

47. Lundquist JT, Pelletier JC (2002) A new tri-orthogonal strategy for peptide cyclization. Org Lett 4:3219–3221

48. Ariza X, Urpí F, Vilarrasa J (1999) A practical procedure for the preparation of carbamates from azides. Tetrahedron Lett 40:7515–7517

49. Rostovtsev VV, Green LG, Fokin VV, Sharpless KB (2002) A stepwise huisgen cycloaddition process: copper(I)-catalyzed regioselective "ligation" of azides and terminal alkynes. Angew Chem Int Edit 41:2596–2599

50. Tornøe CW, Christensen C, Meldal M (2002) Peptidotriazoles on solid phase: [1,2,3]-triazoles by regiospecific copper(I)-catalyzed 1,3-dipolar cycloadditions of terminal alkynes to azides. J Org Chem 67:3057–3064

51. Chang PV, Prescher JA, Sletten EM, Baskin JM, Miller IA, Agard NJ, Lo A, Bertozzi CR (2010) Copper-free click chemistry in living animals. Proc Natl Acad Sci USA 107: 1821–1826

52. Ning X, Guo J, Wolfert MA, Boons G-J (2008) Visualizing metabolically labeled glycoconjugates of living cells by copper-free and fast Huisgen cycloadditions. Angew Chem Int Edit 47:2253–2255

53. Lutz J-F (2008) Copper-free azide–alkyne cycloadditions: new insights and perspectives. Angew Chem Int Edit 47:2182–2184

54. Nan YH, Park KH, Park Y, Jeon YJ, Kim Y, Park I-S, Hahm K-S, Shin SY (2009) Investigating the effects of positive charge and hydrophobicity on the cell selectivity, mechanism of action and anti-inflammatory activity of a Trp-rich antimicrobial peptide indolicidin. FEMS Microbiol Lett 292:134–140

55. Nan YH, Bang J-K, Shin SY (2009) Design of novel indolicidin-derived antimicrobial peptides with enhanced cell specificity and potent anti-inflammatory activity. Peptides 30: 832–838

56. Lee DL, Powers JPS, Pflegerl K, Vasil ML, Hancock REW, Hodges RS (2004) Effects of single d-amino acid substitutions on disruption of β-sheet structure and hydrophobicity in cyclic 14-residue antimicrobial peptide analogs related to gramicidin S. J Pept Res 63:69–84
57. Chen Y, Guarnieri MT, Vasil AI, Vasil ML, Mant CT, Hodges RS (2007) Role of peptide hydrophobicity in the mechanism of action of α-helical antimicrobial peptides. Antimicrob Agents Chemother 51:1398–1406
58. Ando S, Mitsuyasu K, Soeda Y, Hidaka M, Ito Y, Matsubara K, Shindo M, Uchida Y, Aoyagi H (2010) Structure-activity relationship of indolicidin, a Trp-rich antibacterial peptide. J Pept Sci 16:171–177
59. Sonnichsen FD, Van Eyk JE, Hodges RS, Sykes BD (1992) Effect of trifluoroethanol on protein secondary structure: an NMR and CD study using a synthetic actin peptide. Biochemistry 31:8790–8798
60. Roccatano D, Colombo G, Fioroni M, Mark AE (2002) Mechanism by which 2,2,2-trifluoroethanol/water mixtures stabilize secondary-structure formation in peptides: a molecular dynamics study. Proc Natl Acad Sci USA 99:12179–12184
61. Otvos L (1997) Use of circular dichroism to determine secondary structure of neuropeptides neuropeptide protocols. In: Irvine GB, Williams CH (eds) Methods in molecular biology, vol 73. Humana Press, Clifton, NJ, pp 153–161
62. Falla TJ, Karunaratne DN, Hancock REW (1996) Mode of action of the antimicrobial peptide indolicidin. J Biol Chem 271: 19298–19303
63. Hsu C-H, Chen C, Jou M-L, Lee AY-L, Lin Y-C, Yu Y-P, Huang W-T, Wu S-H (2005) Structural and DNA-binding studies on the bovine antimicrobial peptide, indolicidin: evidence for multiple conformations involved in binding to membranes and DNA. Nucleic Acids Res 33:4053–4064
64. Ladokhin AS, Selsted ME, White SH (1999) CD spectra of indolicidin antimicrobial peptides suggest turns, not polyproline helix. Biochemistry 38:12313–12319
65. Andrushchenko VV, Vogel HJ, Prenner EJ (2006) Solvent-dependent structure of two tryptophan-rich antimicrobial peptides and their analogs studied by FTIR and CD spectroscopy. Biochim Biophys Acta Biomembr 1758:1596–1608

Chapter 6

Study Protein Folding and Aggregation Using Nonnatural Amino Acid *p*-Cyanophenylalanine as a Sensitive Optical Probe

Deguo Du, Haiyang Liu, and Bimlesh Ojha

Abstract

Incorporation of nonnatural amino acids with a variety of special side groups into protein sequences has substantially expanded the experimental means of exploring protein structures and functions. Recently, *p*-cyanophenylalanine (Phe_{CN}), the nitrile analogue of phenylalanine, has been used as a novel optical probe for protein binding and folding studies. The fluorescence emission of Phe_{CN} is sensitive to solvent and local environment of the residue, making it a useful fluorescent probe of protein structural change at residue-specific resolution. Moreover, the utility of Phe_{CN} is increased by its ability to excite tryptophan fluorescence via the mechanism of fluorescence resonance energy transfer. Phe_{CN} could be applied to study a variety of biological problems, e.g., protein folding/unfolding and protein aggregation.

Key words Protein folding, Nonnatural amino acid, *p*-Cyanophenylalanine, Fluorescence, Protein aggregation, Fluorescence resonance energy transfer

1 Introduction

The functions of proteins are determined by their three-dimensional structures. How individual protein sequences efficiently and reliably fold to their native states following synthesis on the ribosome is one of the most fundamental questions in structural biology [1]. Moreover, failure of protein proper folding and formation of aggregated amyloids have been implicated in a number of amyloid diseases including Alzheimer's disease, Parkinson's disease, and Prion disorders [2, 3]. Understanding the mechanisms of protein folding/misfolding and its relation to disease pathology will substantially improve our knowledge of amyloidosis-associated pathological mechanisms, which may lead to the development of effective therapeutic strategies to treat amyloid diseases.

Tremendous advances have been made to shed light on the underlying mechanism of protein folding/misfolding and aggregation

Predrag Cudic (ed.), *Peptide Modifications to Increase Metabolic Stability and Activity*, Methods in Molecular Biology, vol. 1081, DOI 10.1007/978-1-62703-652-8_6, © Springer Science+Business Media New York 2013

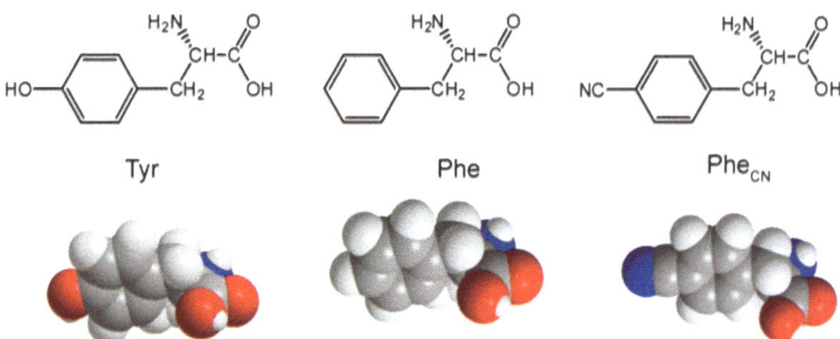

Fig. 1 Molecular structure of Phe_{CN}, and comparison of the size of Phe_{CN} to that of Tyr and Phe

in the past decade due to technological improvement and communication between theoreticians and experimentalists [4–14]. Among the various powerful techniques that have been used in the attempt to study the structural dynamics in protein folding and misfolding, sensitive and fast spectroscopic approaches are of the primary choices. In particular, fluorescence spectroscopy is utilized widely due to its high sensitivity and selectivity, and hence can be used in studies at micromolar concentration regime. Tryptophan (Trp) is a commonly used intrinsic fluorescent probe in studying protein structural dynamics and folding kinetics due to its sensitivity to the polarity of its environment and high quantum yield. Extrinsic dyes have also been employed extensively to characterize the conformational dynamics in protein folding, misfolding, and aggregation, taking the potential uncertainty of the effect of giant dye residues on folding and aggregation properties of target molecules [15–17].

Recently, a novel nonnatural amino acid, *p*-cyanophenylalanine (Phe_{CN}), has been reported as a versatile spectroscopic probe owing to its unique infrared and fluorescence properties [18]. The nitrile group in this nonnatural amino acid (Fig. 1), as highly environment-sensitive fluorescence and/or IR probe [18], can report the solvent accessibility of specific residues of the wild-type conformation due to its sensitivity to the local electrostatic environment. For instance, the band position and half maximum of the CN stretching vibration at 2,210–2,240 cm^{-1} region can be used to estimate the solvent accessibility of the nitrile-labeled side chains [18]. Furthermore, the fluorescence quantum yield of Phe_{CN} strongly depends on the local environment, and thus can be used individually as a novel fluorescent probe to monitor protein structural dynamics at specific sequence region. Importantly, Phe_{CN} has a blue-shifted absorption band with a maximum at approximately 235 nm [19], making selective excitation of Phe_{CN} possible even in the presence of Trp or Tyr. Another significant advantage of using Phe_{CN} is its minimal perturbation fashion. Owing to its small size and high mimic of Phe and Tyr (Fig. 1), Phe_{CN} only minimally perturbs the physical properties of the native polypeptide, especially when it replaces either Phe or Tyr residue in the native sequence.

The utility of Phe_{CN} as a spectroscopic probe is further increased by its ability to excite Trp fluorescence via the mechanism of fluorescence resonance energy transfer (FRET) [19]. FRET is widely used as a molecular ruler to probe the conformational dynamics of proteins as the FRET efficiency highly depends on the distance between donor and acceptor fluorophores [16, 20]. The Föster distance of the Phe_{CN}/Trp FRET pair is ~16 Å [19], making it well suited for probing relatively short separation distance along a polypeptide sequence and the conformational changes during protein folding, protein–protein binding, or self-association. Over the past few years, Phe_{CN} has been used as a sensitive intrinsic probe to study protein folding [21–24], protein–membrane interaction [25, 26], and aggregation process of islet amyloid polypeptide [27].

Here, we briefly discuss the studies of using Phe_{CN} as a unique fluorescence probe to monitor the thermally and chaotropically induced unfolding and the residual packing of a β-hairpin structural motif, and track the aggregation kinetics of Aβ fragment peptides.

2 Materials

1. All the materials and reagents are commercially available and used as received.
2. Peptide synthesis and purification solvents, such as dimethylformamide (DMF), *N*-methylmorpholine (NMM), ethyl ether anhydrous, and acetonitrile, are high-performance liquid chromatography (HPLC)- or peptide synthesis-grade and purchased from Sigma-Aldrich or Fisher Scientific.
3. All Fmoc-group protected natural amino acids and Rink amide resin are purchased from Novabiochem (Gibbstown, NJ). Fmoc-*p*-cyano-L-phenylalanine (Fmoc-Phe_{CN}) is purchased from Synthetech (Albany, OR, USA) and Chem-Impex International Inc. (Wood Dale, IL).
4. Peptide coupling reagent 1-hydroxybenzotriazole monohydrate ($HOBt{\cdot}H_2O$) is purchased from Chem-Impex International Inc. (Wood Dale, IL). *O-Benzotriazole-N,N,N',N'*-tetramethyl-uronium-hexafluoro-phosphate (HBTU) is purchased from Advanced ChemTech (Louisville, KY).
5. Other reagents such as piperidine, trifluoroacetic acid (TFA), and triisopropylsilane (TIS) are purchased from Sigma-Aldrich.
6. Thioflavin T (ThT) dye is purchased from EMD Calbiochem (La Jolla, CA). Millex-GV 0.22 μM filter is purchased from Fisher Scientific (Suwanee, GA).

3 Methods

3.1 Study the Unfolding of a 16-Residue β-Hairpin Peptide Using a Phe_{CN}/Trp FRET Pair

The β-hairpin structural motif is a basic protein secondary structural element and proving to be critical for understanding the thermodynamics and kinetics of β-sheet formation [28–31]. While most β-hairpins are small, some of them show cooperative folding behaviors that are characteristic of proteins. Therefore, the mechanism of β-hairpin folding has been a subject of great interest in recent years [32–48]. To better understand the nature of the unfolded states of β-hairpins, we studied the compactness of the thermally and chaotropically denatured states of a 16-residue β-hairpin, gb1-m3p (KKWTYNPATGK-Phe_{CN}-TVQE), using a Phe_{CN}/Trp FRET pair [49].

The gb1-m3 peptide (KKWTYNPATGKFTVQE) was designed based on the sequence of the gb1 β-hairpin by Andersen and coworkers and has been shown to have a much higher thermal stability than the parent [50]. To use the method of FRET, we replaced the single Phe residue in gb1-m3 with Phe_{CN} nonnatural amino acid, and the resulting peptide was named as gb1-m3p. A pentapeptide (GK-Phe_{CN}-TV) was used as a reference peptide in the study.

3.1.1 Peptide Synthesis and Purification

This protocol was adopted from ref. 49. The peptides were synthesized via solid-phase synthetic strategy on a PS3 peptide synthesizer (Protein Technologies, Inc., Woburn, MA). The Fmoc-group protected natural amino acids and Fmoc-Phe_{CN} were activated by HOBt·H_2O and HBTU using DMF with addition of 4.4 % (v/v) NMM as solvent.

1. Couple the activated amino acid residue onto the Rink amide resin from C terminus, with a subsequent Fmoc deprotection process using 20 % (v/v) piperidine in DMF. The coupling time is 40 min for each amino acid residue.
2. Cleave the peptides from the resin using mixed reagent of 95 % (v/v) TFA, 2.5 % (v/v) TIS, and 2.5 % (v/v) Milli-Q H_2O in stir condition for 1.5 h after synthesis.
3. Blow the mixture with nitrogen gas for ~20 min to volatilize the cleaving reagents.
4. Add 20 mL cold ethyl ether to precipitate the crude peptides, transfer the solution to a 45 mL test tube, and centrifuge the test tube (2,465 × *g*) for 10 min at 4 °C.
5. Discard the upper solvent, wash the precipitated sample using cold ethyl ether for three times, and dry the crude peptide sample with slight air blow to remove ethyl ether.
6. Purify the crude peptides via reverse-phase HPLC with a C18 semipreparative column (Phenomenex) in a slightly acidic

environment using 0.1 % TFA in Milli-Q H_2O as eluent A and 0.1 % TFA in acetonitrile as eluent B, with a gradient of eluent B from 5 to 90 % over 60 min.

7. Identify the purified peptides by matrix-assisted laser desorption/ionization time-of-flight mass spectroscopy; freeze the purified HPLC solution in liquid nitrogen for 10 min and lyophilize for 2 days to obtain homogenized solid peptide sample.

3.1.2 Fluorescence Measurement

This protocol was adopted from ref. 49.

1. Prepare the peptide sample solution by directly dissolving lyophilized peptide solid into 20 mM phosphate buffer (pH 7), with the final concentration of around 25 μM, determined optically using $\varepsilon_{280} = 850\ M^{-1}\ cm^{-1}$ for the reference peptide (GK-Phe_{CN}-TV) and $\varepsilon_{280} = 6{,}540\ M^{-1}\ cm^{-1}$ for gb1-m3p.
2. Transfer 1.4 mL peptide solution into a 1 cm quartz sample cuvette, and load the cuvette onto a Fluorolog 3.10 spectrofluorometer (Jobin Yvon Horiba).
3. Measure the temperature-dependent fluorescence spectra (250–460 nm, 2 nm spectral resolution) of gb1-m3p and the reference peptide solution (25 μM) from 5 to 95 °C, in a step of 10 °C, with an excitation wavelength of 240 nm. Temperature is regulated using a TLC 50 Peltier temperature controller (Quantum Northwest, WA).
4. Measure the fluorescence spectra of gb1-m3p and the reference peptide solution (25 μM) containing various amount of urea. The urea concentration of the peptide solution is adjusted using an equimolar peptide (~25 μM) stock solution containing urea (7 M). Each time, an aliquot of the peptide solution is removed from the fluorescence cuvette, followed by the addition of an equal amount of the urea-containing peptide stock solution, resulting in an increase of the urea concentration by 0.75 M.

3.1.3 FRET Efficiency Calculation

This protocol was adopted from ref. [49] (*see* **Note 1**).

The FRET efficiency, E, is calculated according to the following equation:

$$E = 1 - \frac{I_{DA}}{I_D},$$

where I_{DA} and I_D are the integrated fluorescence intensities of the donor, with and without the presence of the acceptor, respectively. For the current study, I_{DA} corresponds to the integrated area of the Phe_{CN} fluorescence spectrum obtained with gb1-m3p peptide, and I_D corresponds to that obtained with the reference pentapeptide,

GK-Phe_{CN}-TV, under the same conditions. To accurately determine the integrated area of the Phe_{CN} fluorescence, which overlaps the emission spectrum of Trp, we fit the fluorescence spectrum of these peptides by using a linear combination of two profiles generated from fitting the fluorescence spectra of Trp and Phe_{CN} plus a linear background [49].

3.1.4 Investigation of the Thermally and Chaotropically Induced FRET Efficiency Change

The fluorescence emission spectrum of the gb1-m3p peptide at 20 °C, resulting from a selective excitation of the Phe_{CN} residue at 240 nm, shows typical features of fluorescence resonance energy transfer (Fig. 2) [49]. This is further supported by the fluorescence spectra of gb1-m3p obtained in solutions of different urea concentrations. As indicated (Fig. 2), the fluorescence intensity of the donor (i.e., Phe_{CN}) increases with the increase of the urea concentration, suggesting that as the β-hairpin conformation unfolds at higher concentrations of denaturant the FRET efficiency decreases, as a result of the increased separation distance between the FRET donor (Phe_{CN}) and acceptor (Trp).

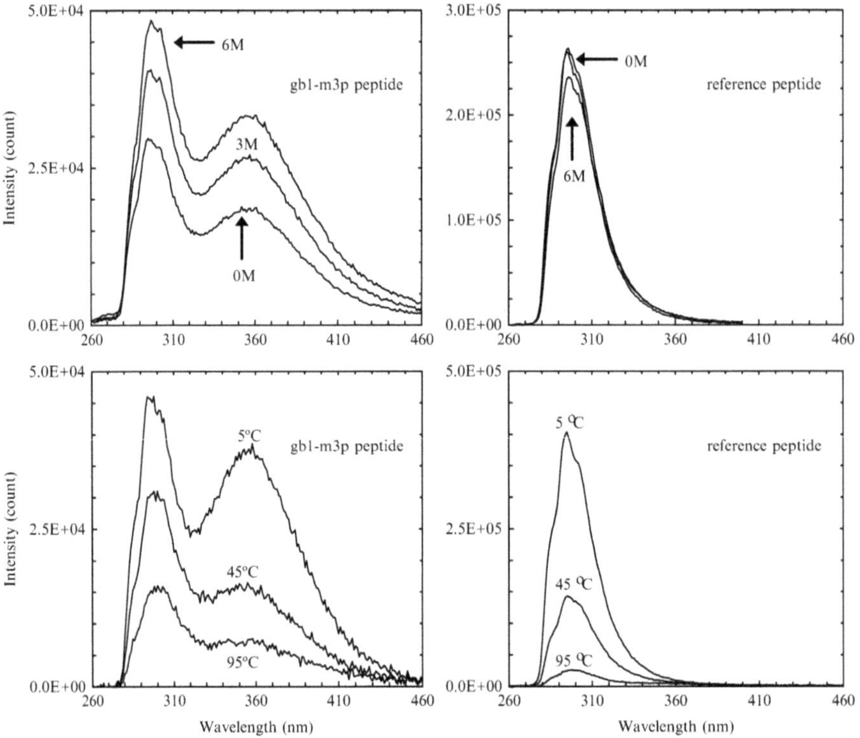

Fig. 2 *Upper panel*: Fluorescence spectra of gb1-m3p (25 μM) and the reference peptide (25 μM) at 20.0 °C as a function of the urea concentration, as indicated. *Lower panel*: Fluorescence spectra of gb1-m3p (25 μM) and the reference peptide (25 μM) collected at different temperatures, as indicated. $\lambda_{ex} = 240$ nm (reproduced with permission from ref. 49)

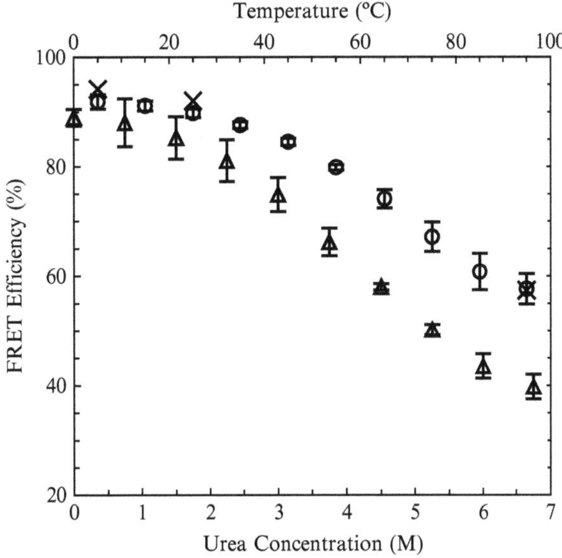

Fig. 3 FRET efficiency of gb1-m3p vs. temperature (*open circle*) and urea concentration (*open triangle*). The peptide concentration was 25 μM. Also shown are FRET efficiencies (*x*) obtained with a peptide concentration of 250 μM. The urea data were collected at 20.0 °C (reproduced with permission from ref. 49)

Thermally induced and chaotropically induced unfolding transitions of gb1-m3p were monitored using this FRET pair. As shown (Fig. 3), the apparent FRET efficiency, which was calculated using the method described, decreases with the increase of either the temperature or the concentration of urea, indicating that the average distance between Phe_{CN} and Trp increases as the β-hairpin conformation unfolds. Interestingly, these results suggest that the thermally denatured state of gb1-m3p is different from the urea-denatured state. For example, the FRET efficiency only changes from ~90 to ~60 % when the temperature is increased from 5 to 95 °C, where the β-hairpin population is very small according to thermodynamic circular dichroism studies [49], whereas in 6.75 M urea the FRET efficiency reduces to ~40 %.

These results suggest that the thermally unfolded state of gb1-m3p is more compact than its urea-denatured state. When folding proceeds from an extended conformation the very first step very possibly corresponds to a collapse process, probably hydrophobic in nature. Since the process of hydrophobic collapse has been shown to occur on the nanosecond timescale [51], an initial hydrophobic collapse process in β-hairpin formation may speed up the rate of folding because it can narrow the conformational space that must be searched associated with the formation of the turn (or loop).

3.2 Study the Aggregation Kinetics of a Short Aβ Fragment Peptide Using Phe_{CN} Fluorescence

Amyloid diseases, including neurodegenerative diseases such as Alzheimer's and Parkinson's, are linked to a poorly understood progression of protein misfolding and aggregation events that culminate in tissue-selective deposition and human pathology. Amyloid formation is usually viewed as a nucleated polymerization mechanism in vitro [52–56], manifested by an apparent initial lag phase followed by a rapid conversion of the monomeric or oligomeric protein to the fibrillar forms. While the accumulation of amyloid fibrils, and their deposition as plaques, has long been associated with cell death and disease progression, recent emerging evidence suggests that the soluble oligomers are the most neurotoxic species [57–63]. Despite the importance of the oligomeric intermediates in pathogenesis of amyloid diseases, there are few suitable experimental methods that can be used to reveal the high-resolution structural characteristics of the liable oligomers, assess the critical interactions that drive the formation of oligomeric intermediates, and monitor the kinetics of protein early aggregation process.

Owing to its high sensitivity to environmental change, Phe_{CN} fluorescence is expected to be employed as a sensitive indicator for all phases of protein aggregation, particularly the heretofore elusive early aggregation stage. Here, the aggregation kinetics of a short peptide $A\beta_{1-23}$ was studied using Phe_{CN} fluorescence. $A\beta_{1-23}$ is the fragment of $A\beta_{1-40}$ peptide, containing the first 23 residues (DAEFRHDSGY^{10} EVHHQKLVFF^{20} AED). A nonnatural amino acid Phe_{CN} residue was incorporated into $A\beta_{1-23}$ to substitute residue Tyr10 ($A\beta_{1-23}$M) in the study.

3.2.1 Peptide Synthesis and Purification

The peptides used in the study were synthesized based on standard Fmoc-protocols on a PS3 automated peptide synthesizer (Protein Technologies, Inc., Woburn, MA). The protocol is the same as described in Subheading 3.1.1, except using a longer coupling time of 1 h for each residue to maximize the coupling efficiency (*see* **Note 2**). The peptides were purified to homogeneity by reverse-phase chromatography and identified by matrix-assisted laser desorption/ionization mass spectroscopy, as described in Subheading 3.1.1.

3.2.2 Preparation of $A\beta_{1-23}$ Wild-Type and Mutant Peptide Stock Solution

1. Dissolve the purified peptide (1.0 mg) in 1 mL Milli-Q H_2O, sonicate the solution in ice bath for 30 min, and then filter the solution through 0.22 μM filter with 1.0 mL syringe. Solubility of the peptide should be examined before filtration since a poor solubility may result in high pressure in the closed system (*see* **Note 3**).
2. Transfer 100 μL peptide solution into a 0.1 mL cuvette and measure the UV–Vis spectra on an Agilent 8453 UV–Vis spectrometer. The result is presented after subtraction of the blank trial and baseline correction, and as average of three parallel tests.

3. Calculate the concentration of the peptide solutions using the absorbance at 280 nm fixed wavelength ($\varepsilon_{280nm} = 1{,}280\ M^{-1}cm^{-1}$) for the wild-type $A\beta_{1-23}$ peptide, and 240 nm fixed wavelength ($\varepsilon_{240nm} = 13{,}000\ M^{-1}cm^{-1}$) for the Phe_{CN}-substituted mutant peptide $A\beta_{1-23}M$.
4. Utilize the fresh peptide stock solution for further experiments immediately.

3.2.3 Phe_{CN} Fluorescence Assay

1. Dilute the peptide stock solution to the target concentration (50 μM) at physiological condition of pH 7.4, 50 mM phosphorous buffer saline solution (150 mM NaCl, 0.025 % NaN_3), with a final concentration of 20 μM ThT dye.
2. Prepare a blank reference in the same condition without peptide.
3. Incubate the sample and reference vials quiescently at 37 °C in the incubator (VWR™) for 7–10 days in order to obtain a steady incubatory process at a relatively slow manner.
4. Take out 100 μL of the incubated sample at desired time points, briefly vortex, sonicate in ice bath for 20 min, and transfer into a 0.1 mL quartz sample cuvette.
5. Collect the Phe_{CN} emission fluorescence spectra on a FluoroMax-4 spectrofluorometer (Jobin Yvon Horiba) from 260 to 500 nm (slit 5 nm) with an excitation wavelength of 240 nm (slit 5 nm). The results are presented using the average of three tests after subtraction of the blank reference.

3.2.4 ThT Fluorescence Measurement

At desired time points, the ThT fluorescence spectra of the same peptide solution are also measured in parallel using the FluoroMax-4 spectrofluorometer (Jobin Yvon Horiba) from 460 to 700 nm (slit 5 nm) with an excitation wavelength of 440 nm (slit 5 nm). The results are presented using the average of three tests after subtraction of the blank reference.

3.2.5 Aggregation Kinetics of $A\beta_{1-23}M$ Followed by ThT Fluorescence and Phe_{CN} Fluorescence

$A\beta_{1-23}$ peptide aggregates when incubated in 50 mM phosphate buffer (pH 7.4). The formed aggregates bind to ThT dye and give rise to a strong fluorescence emission with the maximum at 485 nm, suggesting the formation of a typical β-sheet structure in $A\beta_{1-23}$ amyloid. The Phe_{CN} mutant peptide $A\beta_{1-23}M$ also steadily aggregates to form amyloid fibrils, resulting in an enhanced intensity of ThT fluorescence after binding (*see* **Note 4**). The aggregation kinetics of the $A\beta_{1-23}$ mutant was followed using ThT binding assay and the intrinsic fluorescence of Phe_{CN} in parallel. The Phe_{CN} fluorescence intensity of $A\beta_{1-23}M$ is found to decrease with time, indicating the local environmental change of residue 10, while the ThT binding results do not show obvious fluorescence signal until day 2 of incubation. The results imply that the Phe_{CN} residue at position 10 may be involved in the early aggregation stage and the subsequent conformational conversion process to form fibrillar structure.

The conformational change and packing of the monomeric peptide might gradually put the residue from a polar and solvent accessible environment to a more hydrophobic environment, leading to the decrease of Phe_{CN} fluorescence intensity [64].

The results suggest that the nonnatural amino acid fluorophore Phe_{CN} could be applied as a useful probe to identify the local environmental change of specific residue regions in the early stage of protein aggregation. Mutation studies at other residue positions (e.g., Phe4, Phe19, Phe20) will provide more information of the conformational changes of local regions during aggregation of the peptide. The rational selection of the mutation sites in the target protein could allow investigating the aggregation process at a residue-specific level and provide insights into the mechanisms of protein oligomerization and fibrillization.

4 Notes

1. The intensity of the Trp emission should not be used as a quantitative measure of the FRET efficiency in studying unfolding of gb1-m3p because several amino acids in this peptide, such as Lys and Gln, can quench Trp fluorescence [65].
2. Coupling time of the Fmoc-Phe_{CN} residue in peptide synthesis can be longer than 1 h (e.g., 2 h) to maximize the coupling efficiency.
3. The $A\beta_{1-23}$ and $A\beta_{1-23}M$ stock solutions can be stored at −80 °C and used within 3 weeks. Frequent thaw of the frozen peptide aliquot should be avoided.
4. The effect of Phe_{CN} mutation on peptide aggregation properties should still be considered. The current work shows that the halftime of the aggregation kinetics of $A\beta_{1-23}$ and the mutant $A\beta_{1-23}M$ has no significant difference, suggesting that the replacement of Tyr by Phe_{CN} is tolerated well in the mutant and does not change the aggregation property of the peptide dramatically.

References

1. Dobson CM (2003) Protein folding and misfolding. Nature 426:884–890
2. Kelly JW (1996) Alternative conformations of amyloidogenic proteins govern their behavior. Curr Opin Struct Biol 6:11–17
3. Selkoe DJ (2003) Folding proteins in fatal ways. Nature 426:900–904
4. Fersht AR (2008) From the first protein structures to our current knowledge of protein folding: delights and scepticisms. Nat Rev Mol Cell Biol 9:650–654
5. Kubelka J, Hofrichter J, Eaton WA (2004) The protein folding 'speed limit'. Curr Opin Struct Biol 14:76–88
6. Campos LA, Liu J, Wang X, Ramanathan R, English DS, Munoz V (2011) A photoprotection strategy for microsecond-resolution single-molecule fluorescence spectroscopy. Nat Methods 8:143–146
7. Onuchic JN, Wolynes PG (2004) Theory of protein folding. Curr Opin Struct Biol 14:70–75

8. Luheshi LM, Dobson CM (2009) Bridging the gap: from protein misfolding to protein misfolding diseases. FEBS Lett 583:2581–2586
9. Bartlett AI, Radford SE (2009) An expanding arsenal of experimental methods yields an explosion of insights into protein folding mechanisms. Nat Struct Mol Biol 16:582–588
10. Nelson R, Sawaya MR, Balbirnie M, Madsen AO, Riekel C, Grothe R, Eisenberg D (2005) Structure of the cross-beta spine of amyloid-like fibrils. Nature 435:773–778
11. Luhrs T, Ritter C, Adrian M, Riek-Loher D, Bohrmann B, Dobeli H, Schubert D, Riek R (2005) 3D structure of Alzheimer's amyloid-beta(1-42) fibrils. Proc Natl Acad Sci USA 102:17342–17347
12. Mustata M, Capone R, Jang H, Arce FT, Ramachandran S, Lal R, Nussinov R (2009) K3 fragment of amyloidogenic beta(2)-microglobulin forms ion channels: implication for dialysis related amyloidosis. J Am Chem Soc 131:14938–14945
13. Fersht AR, Daggett V (2002) Protein folding and unfolding at atomic resolution. Cell 108: 573–582
14. Guo M, Xu Y, Gruebele M (2012) Temperature dependence of protein folding kinetics in living cells. Proc Natl Acad Sci USA 109:17863–17867
15. Ferreon AC, Deniz AA (2011) Protein folding at single-molecule resolution. Biochim Biophys Acta 1814:1021–1029
16. Schuler B, Eaton WA (2008) Protein folding studied by single-molecule FRET. Curr Opin Struct Biol 18:16–26
17. Hawe A, Sutter M, Jiskoot W (2008) Extrinsic fluorescent dyes as tools for protein characterization. Pharm Res 25:1487–1499
18. Getahun Z, Huang CY, Wang T, De Leon B, DeGrado WF, Gai F (2003) Using nitrile-derivatized amino acids as infrared probes of local environment. J Am Chem Soc 125:405–411
19. Tucker MJ, Oyola R, Gai F (2005) Conformational distribution of a 14-residue peptide in solution: a fluorescence resonance energy transfer study. J Phys Chem B 109:4788–4795
20. Palmer AE, Tsien RY (2006) Measuring calcium signaling using genetically targetable fluorescent indicators. Nat Protoc 1:1057–1065
21. Aprilakis KN, Taskent H, Raleigh DP (2007) Use of the novel fluorescent amino acid p-cyanophenylalanine offers a direct probe of hydrophobic core formation during the folding of the N-terminal domain of the ribosomal protein L9 and provides evidence for two-state folding. Biochemistry 46:12308–12313
22. Miyake-Stoner SJ, Miller AM, Hammill JT, Peeler JC, Hess KR, Mehl RA, Brewer SH (2009) Probing protein folding using site-specifically encoded unnatural amino acids as FRET donors with tryptophan. Biochemistry 48:5953–5962
23. Taskent-Sezgin H, Marek P, Thomas R, Goldberg D, Chung J, Carrico I, Raleigh DP (2010) Modulation of p-cyanophenylalanine fluorescence by amino acid side chains and rational design of fluorescence probes of alpha-helix formation. Biochemistry 49: 6290–6295
24. Glasscock JM, Zhu Y, Chowdhury P, Tang J, Gai F (2008) Using an amino acid fluorescence resonance energy transfer pair to probe protein unfolding: application to the villin headpiece subdomain and the LysM domain. Biochemistry 47:11070–11076
25. Tang J, Yin H, Qiu J, Tucker MJ, DeGrado WF, Gai F (2009) Using two fluorescent probes to dissect the binding, insertion, and dimerization kinetics of a model membrane peptide. J Am Chem Soc 131:3816–3817
26. Tucker MJ, Tang J, Gai F (2006) Probing the kinetics of membrane-mediated helix folding. J Phys Chem B 110:8105–8109
27. Marek P, Gupta R, Raleigh DP (2008) The fluorescent amino acid p-cyanophenylalanine provides an intrinsic probe of amyloid formation. Chembiochem 9:1372–1374
28. Searle MS, Ciani B (2004) Design of beta-sheet systems for understanding the thermodynamics and kinetics of protein folding. Curr Opin Struct Biol 14:458–464
29. Gellman SH (1998) Minimal model systems for b sheet secondary structure in proteins. Curr Opin Chem Biol 2:717–725
30. Serrano L (2000) The relationship between sequence and structure in elementary folding units. Adv Protein Chem 53:49–85
31. Searle MS (2001) Peptide models of protein b-sheets: design, folding and insights into stabilising weak interactions. J Chem Soc Perkin Trans 2:1011–1020
32. Pande VSP, Rokhsar DS (1999) Molecular dynamics simulations of unfolding and refolding of a b-hairpin fragment of protein G. Proc Natl Acad Sci USA 96:9062–9067
33. Munoz V, Henry ER, Hofrichter J, Eaton WA (1998) A statistical mechanical model for beta-hairpin kinetics. Proc Natl Acad Sci USA 95: 5872–5879
34. Dinner AR, Lazaridis T, Karplus M (1999) Understanding beta-hairpin formation. Proc Natl Acad Sci USA 96:9068–9073

35. Klimov DK, Thirumalai D (2000) Mechanisms and kinetics of beta-hairpin formation. Proc Natl Acad Sci USA 97:2544–2549
36. Bonvin AMJJ, van Gunsteren WF (2000) Beta-hairpin stability and folding: molecular dynamics studies of the first beta-hairpin of tendamistat. J Mol Biol 296:255–268
37. Zhou R, Berne BJ, Germain R (2001) The free energy landscape for beta hairpin folding in explicit water. Proc Natl Acad Sci USA 98:14931–14936
38. Garcia AE, Sanbonmatsu KY (2001) Exploring the energy landscape of a beta hairpin in explicit solvent. Proteins 42:345–354
39. Zhou Y, Linhananta A (2002) Role of hydrophilic and hydrophobic contacts in folding of the second beta-hairpin fragment of protein G: molecular dynamics simulation studies of an all-atom model. Proteins 47:154–162
40. Tsai J, Levitt M (2002) Evidence of turn and salt bridge contributions to beta-hairpin stability: MD simulations of C-terminal fragment from the B1 domain of protein G. Biophys Chem 101–102:187–201
41. Wu X, Wang S, Brooks BR (2002) Direct observation of the folding and unfolding of a beta-hairpin in explicit water through computer simulation. J Am Chem Soc 124: 5282–5283
42. Ma B, Nussinov R (2003) Energy landscape and dynamics of the beta-hairpin G peptide and its isomers: topology and sequences. Protein Sci 12:1882–1893
43. Bolhuis PG (2003) Transition-path sampling of beta-hairpin folding. Proc Natl Acad Sci USA 100:12129–12134
44. Snow CD, Qiu L, Du D, Gai F, Hagen SJ, Pande VS (2004) Trp zipper folding kinetics by molecular dynamics and temperature-jump spectroscopy. Proc Natl Acad Sci USA 101: 4077–4082
45. Paschek D, Garcia AE (2004) Reversible temperature and pressure denaturation of a protein fragment: a replica exchange molecular dynamics simulation study. Phys Rev Lett 93:238105/238101–238105/238104
46. Krivov SV, Karplus M (2004) Hidden complexity of free energy surfaces for peptide (protein) folding. Proc Natl Acad Sci USA 101:14766–14770
47. Kuo NNW, Huang JJT, Miksovska J, Chen RPY, Larsen RW, Chan SI (2005) Effects of turn stability on the kinetics of refolding of a hairpin in a beta-sheet. J Am Chem Soc 127:16945–16954
48. Olsen KA, Fesinmeyer RM, Stewart JM, Andersen NH (2005) Hairpin folding rates reflect mutations within and remote from the turn region. Proc Natl Acad Sci USA 102: 15483–15487
49. Du D, Tucker MJ, Gai F (2006) Understanding the mechanism of beta-hairpin folding via phi-value analysis. Biochemistry 45:2668–2678
50. Fesinmeyer RM, Hudson FM, Andersen NH (2004) Enhanced hairpin stability through loop design: the case of the protein G B1 domain hairpin. J Am Chem Soc 126: 7238–7243
51. Sadqi M, Lapidus LJ, Munoz V (2003) How fast is protein hydrophobic collapse? Proc Natl Acad Sci USA 100:12117–12122
52. Naiki H, Hasegawa K, Yamaguchi I, Nakamura H, Gejyo F, Nakakuki K (1998) Apolipoprotein E and antioxidants have different mechanisms of inhibiting Alzheimer's beta-amyloid fibril formation in vitro. Biochemistry 37: 17882–17889
53. Teplow DB (1998) Structural and kinetic features of amyloid beta-protein fibrillogenesis. Amyloid 5:121–142
54. Harper JD, Lansbury PT (1997) Models of amyloid seeding in Alzheimier's disease and scrapie: mechanistic truths and physiological consequences of the time-dependent solubility of amyloid proteins. Annu Rev Biochem 66:385–407
55. Ferrone FA (2006) Nucleation: the connections between equilibrium and kinetic behavior. Methods Enzymol 412:285–299
56. Powers ET, Powers DL (2008) Mechanisms of protein fibril formation: nucleated polymerization with competing off-pathway aggregation. Biophys J 94:379–391
57. Billings LM, Oddo S, Green KN, McGaugh JL, LaFerla FM (2005) Intraneuronal Abeta causes the onset of early Alzheimer's disease-related cognitive deficits in transgenic mice. Neuron 45:675–688
58. Glabe CG (2008) Structural classification of toxic amyloid oligomers. J Biol Chem 283: 29639–29643
59. Haass C, Selkoe DJ (2007) Soluble protein oligomers in neurodegeneration: lessons from the Alzheimer's amyloid beta-peptide. Nat Rev Mol Cell Biol 8:101–112
60. Lue LF, Kuo YM, Roher AE, Brachova L, Shen Y, Sue L, Beach T, Kurth JH, Rydel RE, Rogers J (1999) Soluble amyloid beta peptide concentration as a predictor of synaptic change in Alzheimer's disease. Am J Pathol 155: 853–862
61. McLean CA, Cherny RA, Fraser FW, Fuller SJ, Smith MJ, Beyreuther K, Bush AI, Masters CL (1999) Soluble pool of Abeta amyloid as a

determinant of severity of neurodegeneration in Alzheimer's disease. Ann Neurol 46: 860–866

62. Terry RD (1996) The pathogenesis of Alzheimer disease: an alternative to the amyloid hypothesis. J Neuropathol Exp Neurol 55:1023–1025
63. Westerman MA, Cooper-Blacketer D, Mariash A, Kotilinek L, Kawarabayashi T, Younkin LH, Carlson GA, Younkin SG, Ashe KH (2002) The relationship between Abeta and memory in the Tg2576 mouse model of Alzheimer's disease. J Neurosci 22:1858–1867
64. Tucker MJ, Oyola R, Gai F (2006) A novel fluorescent probe for protein binding and folding studies: p-cyano-phenylalanine. Biopolymers 83:571–576
65. Chen Y, Barkley MD (1998) Toward understanding tryptophan fluorescence in proteins. Biochemistry 37:9976–9982

Chapter 7

Adamantoylated Biologically Active Small Peptides and Glycopeptides Structurally Related to the Bacterial Peptidoglycan

Ruža Frkanec, Branka Vranešić, and Srdjanka Tomić

Abstract

A large number of novel synthetic compounds representing smaller parts of original peptidoglycan molecules have been synthesized and found to possess versatile biological activity, particularly immunomodulating properties. A series of compounds containing the adamantyl residues coupled to peptides and glycopeptides characteristic for bacterial peptidoglycan was described. The new adamantylpeptides and adamantylglycopeptides were prepared starting from *N*-protected racemic adamantylglycine and dipeptide L-Ala-D-*iso*glutamine. The adamantyl glycopeptides were obtained by coupling the adamantyltripeptides with alpha-D-mannose moiety through spacer molecule of fixed chirality. Since the starting material was D,L-(adamantyl-glycine) the condensation products with the dipeptide were mixtures of diastereoisomers. The obtained diastereoisomers were separated, characterized, and tested for immunostimulating activity. An HPLC method for purity testing was developed and adapted for the particular compounds.

Key words Adamantane, Mannose, Adamantylglycine, Peptidoglycan monomer, Adamantyltripeptide, Bacterial peptidoglycan, Adjuvant

1 Introduction

Numerous nontoxic peptidoglycan fragments of low molecular weight represent an important group of immunomodulators, antitumor and antiviral drugs [1, 2]. They were isolated from natural sources or prepared by chemical synthesis. *N*-acetylmuramyl-L-alanyl-D-*iso*glutamine (muramyl dipeptide, MDP) was recognized as the minimal structure responsible for the biological activity [3]. Chemical modifications of small bioactive peptides containing bacterial peptidoglycan motif in their structure have been the subject of many investigations during the last decades. A number of derivatives and analogues of MDP has been synthesized by modifications of the muramyl moiety or by changing the composition and configuration of the peptide portion [4]. The peptidoglycan monomer, disaccharide pentapeptide

Predrag Cudic (ed.), *Peptide Modifications to Increase Metabolic Stability and Activity*, Methods in Molecular Biology, vol. 1081, DOI 10.1007/978-1-62703-652-8_7,

GlcNAc-MurNAc-L-Ala-D-*iso*Gln-*meso*-DAP(ωNH$_2$)-D-Ala-D-Ala (PGM) isolated from *Brevibacterium divaricatum* is a natural compound with well-defined chemical structure. It is nontoxic, non-pyrogenic and thus suitable for potential use in humans. The peptidoglycan monomer and its derivatives are recognized as potent immunomodulators (*see* Fig. 1). Their adjuvant activity in different experimental models has been demonstrated [5–8]. Structurally related adamantyltripeptides belong to a class of immunomodulating compounds comprising the elements of the bacterial peptidoglycan structures. Thus, the dipeptide, L-Ala-D-*iso*Gln, characteristic for the bacterial cell wall peptidoglycans was bound to the adamantyl moiety through the glycine molecule. Two diastereoisomers (D- and L-(adamant-2-yl)-Gly-L-Ala-D-*iso*-Gln, Ad$_2$TP1 and Ad$_2$TP2) formed due to the different absolute configurations of adamantylglycine. They were characterized and biologically evaluated in several model systems [9, 10]. Among the various chemical groups, the adamantyl moiety is particularly interesting because of its unique chemical structure and properties. It is well known that incorporation of an adamantyl moiety into substances with known biological activity improves their pharmacological properties and enhances their biological activity [11]. The multidimensional value of the adamantyl group in drug design has been stressed in several recent review papers [12, 13].

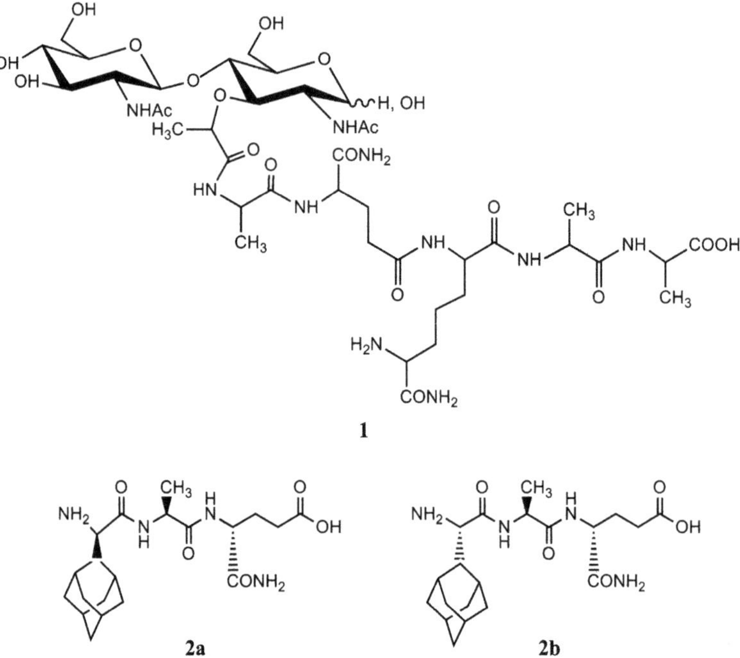

Fig. 1 Chemical structures of the peptidoglycan monomer (PGM) **1** and (adamant-2-yl)tripeptides **2a** (D-L-D, Ad$_2$TP1), **2b** (L-L-D, Ad$_2$TP2). D and L designate the stereochemistry of α-C-atom

As mentioned previously, natural and synthetic peptidoglycans and their fragments of different molecular size exibit remarkable biological activities. In particular, these molecules affect the immune system of mammalian hosts [1, 2], leading to extensive exploatation of peptidoglycan fragments of well defined structures, such as muramyl peptides, as possible adjuvants (immune potentiators or immunomodulators) for human and veterinary vaccines [18, 19]. In addition, it was demonstrated that synthetic adamantyltripeptides containing the adamantylglycyl moiety linked to the L-Ala-D-*iso*Gln sequence (the characteristic sequence of the peptide portion of natural peptidoglycans and synthetic muramyl dipeptides) exibit immunomodulating activity similar to that of the peptidoglycan monomer, natural peptidoglycan [20].

Adjuvants have been used for decades to improve the immune response to vaccine antigens. Although numerous substances have been tested and shown to be potent adjuvants for antibody and cellular (Type 1) immune responses in animal models, very few have proved to be suitable for use in humans due to unacceptable levels of reactogenicity and/or disappointing immuno-enhancing abilities. Selectively, adjuvants can also be employed to optimize a desired immune response, e.g., with respect to immunoglobulin classes. In this case, a well-defined experimental in vivo model suitable for comparison of immunomodulating activity has been established [7]. Adjuvant activity can be evaluated by the immunostimulatory effect on secondary humoral response to ovalbumin (OVA) used as the model antigen in mice. For this purpose, the in-house enzyme-linked immunosorbent assays (ELISA) for qualitative and quantitative determination of OVA-specific IgG (anti-OVA IgG) in mice sera was established. In order to evaluate the influence of potential adjuvants on production of IgG subclass antibodies, qualitative and quantitative determination of OVA-specific IgG subtypes anti-OVA IgG1 and anti-OVA IgG2a as specific indicators of Th2 or Th1 type of immune response can be carried out.

The investigations regarding the interactions of adamantyltripeptides and lipids in the liposome bilayers using electron paramagnetic spectroscopy showed the ability of adamantyl moiety to incorporate into the lipid bilayer despite the hydrophilic, peptide portion of the molecule [21, 22]. The atomic force microscopy study of liposome morphology with encapsulated mannosylated adamantyltripeptides and agglutination assay with concavalin A (ConA) revealed the exposure of the mannose molecule on the surface of liposomes. Based on these studies, the adamantyl moiety may be considered as a potential membrane anchor for different molecules which would be bound on it and thus exposed on liposome surfaces. This opens new possibilities for adamantyl conjugated peptid in targeted drug delivery or as a model for investigations of specific protein interactions and membrane receptors.

In conclusion the importance of adamantoylated peptides and glycopeptides structurally related to the bacterial peptidoglycan should be stressed because:

- of their remarkable biological activities especially adjuvanticity since there is a need for potent adjuvants that are safe in humans and capable of inducing protective systemic humoral and cellular immune responses.
- they are non-toxic, apyrogenic and suitable for use in humans.
- of the unique lypophylic properties of adamantane molecule and its ability to incorporate into the lipid bilayer open up the possibility of their potential applications in investigations of cell membranes as well as studies of the interactions of a specific protein with membrane receptors.

Taking into consideration the importance of the structure–activity relationship study to manipulate the potency of biologically active peptides, we describe herein step-by-step protocols for the synthesis of adamantyltripeptides and glycoconjugates of admantyltripeptides (*see* Fig. 2) [14–16]. More precisely, syntheses of novel adamantyltripeptides in which the peptide portion is coupled to the adamantane moiety through its C-C bond at the position 1 or 2 is described, as well as mannosylation of adamantyltripeptides through a chiral spacer of defined configuration (*R* or *S*). Although incorporation of the adamantyl moiety into a peptide chain improves its pharmacological properties, additional modification of adamantyl conjugated peptide by coupling of a mannosyl moiety may also lead to significant immunostimulating activity. Mannose receptors (MR) present on the cell surface of different immunocompetent cells are considered to be essential for binding

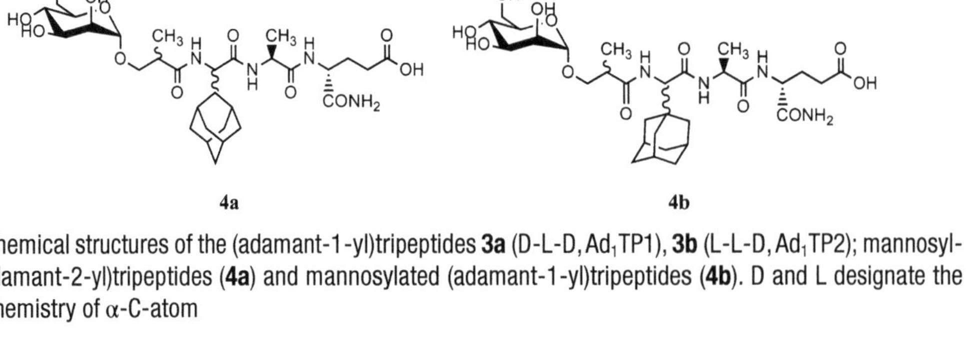

3a

3b

4a

4b

Fig. 2 Chemical structures of the (adamant-1-yl)tripeptides **3a** (D-L-D, Ad_1TP1), **3b** (L-L-D, Ad_1TP2); mannosylated (adamant-2-yl)tripeptides (**4a**) and mannosylated (adamant-1-yl)tripeptides (**4b**). D and L designate the stereochemistry of α-C-atom

Table 1
The condition of RP-HPLC analysis of peptidoglycan monomers, their derivatives and adamantyltripeptides

Compounds	Column	Eluent system	Type of elution	Methods
MDP	Spherisorb ODS	5 mM ammonium acetate (pH 2.5) : MeCN 199:1 v/v	Isocratic	
PGM	LiChrosorb C18	MeCN with 0.035 % TFA: water with 0.05 % TFA	Linear gradient	A
Boc-Tyr-PGM	LiChrosorb C18	MeCN with 0.035 % TFA: water with 0.05 % TFA	Linear gradient	B
Ad-PGM	LiChrosorb C18	MeCN with 0.035 % TFA: water with 0.05 % TFA	Linear gradient	C
Products of enzymatic hydrolysis of PGM	LiChrosorb C18	Water with 0.05 % TFA or water with 0.05 % TFA: MeCN with 0.035 % TFA	Isocratic	D
AdTPs	LiChrosorb C18	MeCN with 0.035 % TFA: water with 0.05 % TFA	Linear gradient	C
Man-AdTPs	LiChrosorb C18	MeCN with 0.035 % TFA: water with 0.05 % TFA	Linear gradient	C

of mannosylated antigens or relevant biologically active molecules containing mannose, thus affecting the immune response.

The synthetized compounds were separated using silica-gel column chromatography and their structures were investigated and confirmed by NMR spectroscopy and mass spectrometry. Absolute configuration of amino acids in the peptide portion was confirmed using the enzyme L-amino acid oxidase [9] (*see* **Note 1**). We have developed a reversed-phase high performance liquid chromatography (RP-HPLC) method based on acetonitrile/water/trifluoroacetic acid to determine purity of the prepared compounds (*see* Table 1) [17]. Different gradient systems within the same solvent mixture had to be applied in order to achieve satisfactory separation and retention time of particular compounds (*see* Table 2).

2 Materials

1. All materials and reagents are commercially available and used as received.
2. Solvents, such as dichloromethane (DCM), ethyl acetate (EtOAc), *tert*-butanol (*t*-BuOH), tetrahydrofurane (THF), methanol (MeOH), ethanol (EtOH), hexane, dioxane, toluene,

Table 2
The composition of linear gradient systems of RP-HPLC methods for analysis of peptidoglycan monomer, its derivatives and adamantyltripeptides

Linear gradient system	*T*/min	Acetonitrile with 0.035 % TFA/%	Water with 0.05 % TFA/%
A	0	3	97
	20	17	83
	25	3	97
B	0	10	90
	25	30	70
	30	10	90
C	0	10	90
	15	30	70
	20	10	90

chloroform ($CHCl_3$), and diethyl ether (Et_2O) were high-performance liquid chromatography (HPLC) or ACS-grade, and can be obtained from Sigma-Aldrich, Fisher, Merck, or other commercial sources. Acetonitrile (MeCN) and trifluoroacetic acid (TFA) were of HPLC grade and can be purchased from Merck or other suppliers. Specific reagents including di-(*tert*-butyl)dicarbonate, 9-fluorenylmethoxycarbonyl chloride Fmoc-AdaGly, 1-octanthiol, and α-mannosyloxy-2-methylpropionic acid, *N*-ethylmorpholine, L-alanyl-D-isoglutamine *tert*-butyl ester, L-alanyl-D-*iso*glutamine benzyl ester hydrochloride salt, 10 % Pd/C, adamantanone, *N*(*tert-butyloxy*)carbonyl-(D,L-adamantyl)glycine can be purchased from Sigma-Aldrich, Novabiochem, or other suppliers.

3. Peptide coupling and deprotection reagents such as *N*-(3-dimethylaminopropyl)-*N*′-ethylcarbodiimide hydrochloride (EDC × HCl) 1-hydroxybenzotriazole (HOBt), *N*-ethyl-5-phenylisoxazolium 3′-sulfonate (Woodward's reagent "K") and 1,8-diazabicyclo[5.4.0]undec-7-ene (DBU) may be obtained from Sigma-Aldrich, ChemImpex, or Novabiochem.
4. HPLC columns LiChrosorb C18 are purchased from Merck, Darmstadt, Germany. A Merck Lichrosorb RP-18 column, 244 mm × 4 mm, 5 μm, and a Merck guard column LiChrospher 100 RP-18 (5 μm) were used. Spherisorb ODS, 250 mm × 3 mm, 5 μm, Spectra-Physics, Santa Clara, CA, USA.
5. A daily supply of water may be obtained from Millipore Simplicity–Personal ultra pure water system (Bedford, MA, USA).
6. Ovalbumin (OVA) was from Serva (Germany), GlcNAc-MurNAc-L-Ala-D-*iso*Gln-*meso*-DAP(ωNH$_2$)-D-Ala-D-Ala

(PGM) was produced in PLIVA Inc. (Croatia) and was obtained by lysozyme digestions of uncross-linked peptidoglycan chains isolated from culture fluid of penicillin treated *Brevibacterium divaricatum* [23], Bovine serum albumin (BSA), Tween-20, monoclonal anti-chicken egg albumin, avidin peroxidase and *o*-phenylenediamine dihydrochloride (OPD) were from Sigma-Aldrich. Biotin conjugated rat anti-mouse IgG1 and anti-mouse IgG2a monoclonal antibodies and streptavidin peroxidase were purchased from PharMingen, Becton Dickinson. Peroxidase-conjugated goat IgG fraction to mouse IgG (whole molecule) was from Cappel (USA).

7. Silica gel used for column chromatography and for TLC were from Merck, silica gel 60 (particle size 0.063–0.2 mm; 70–230 mesh ASTM and 0.04–0.063 mm; 230–400 mesh ASTM) and silica gel 60 F 254, 0.25 mm TLC plates.

3 Methods

3.1 Synthesis of Adamantyltripeptides: D- and L-(Adamant-2-yl)Gly-L-Ala-D-isoGln and D- and L-(adamant-1-yl) Gly-L-Ala-D-isoGln

The adamantyltripeptides can be conveniently prepared by techniques used in the preparation of peptides, preferably by reacting a *N*-protected L-alanyl-D-*iso*glutamine activated ester with D,L-(adamantyl)glycine. Thus, D,L-(adamant-2-yl)glycine may be reacted with a dipeptide derivative bearing a protecting group that can be easily removed by acidic/basic hydrolysis or by hydrogenation (hydrogenolysis). Suitable protecting groups for this purpose include *tert*-butyloxycarbonyl (Boc), 9-fluorenylmethoxycarbonyl (Fmoc), and benzyloxycarbonyl (Cbz). The peptide-bond was formed using *N*-ethyl-5-phenylisoxazolium 3′-sulfonate (Woodward's reagent "K") and the condensation product was easily isolated (*see* **Note 2**) [24]. The synthesis can be also performed using the EDC/HOBt coupling method. Both methods will be described in more detail in this chapter.

The D,L-(adamant-2-yl)glycine and D,L-(adamant-1-yl)glycine can be prepared according to procedures previously described in the literature [25, 26]. (*S*)-(adamant-1-yl)glycine is commercially available (*see* **Note 3**). Boc-carbonyl-L-alanyl-D-*iso*glutamine benzyl ester hydrochloride is prepared from benzyl D-*iso*glutamine hydrochloride and pentachlorphenyl-*N*(Boc)-L-alanine according to protocol described in ref. 27.

3.1.1 Boc-Protection of the Amino Group of D,L-(Adamantyl-glycine)

This protocol was adopted from ref. 9.

1. Add 5 mL mixture of 1 M NaOH and di-(*tert*-butyl)dicarbonate (1.66 and 0.46 mmol) to 9 mL solution of D,L-(adamantyl) glycine (1.55 mmol) in dioxane–H_2O (2:1 v/v) (*see* **Note 4**).
2. Stir the reaction mixture for 24 h at room temperature.

3. Concentrate the reaction mixture by rotary evaporation, acidified to pH 3 with saturated $KHSO_4$ solution and extract with AcOEt three times (*see* **Note 5**).
4. The combined extracts wash twice with H_2O, dried over Na_2SO_4 and evaporate under the reduced pressure to dryness.
5. Recrystallize the crude product from AcOEt–hexane solvent mixture.

3.1.2 Fmoc-Protection of the Amino Group of D,L-(Adamantyl)glycine

This protocol was adopted from ref. 14.

1. Add (2 mmol, 1.2 equiv) of 9-fluorenylmethoxycarbonyl chloride (Fmoc-Cl) and Na_2CO_3 (3 equiv) in 12 mL solution of D,L-(adamantyl)glycine (1.67 mmol) in dioxane–H_2O (2:1 v/v).
2. Stirr the reaction mixture for 1 h at 0 °C and then overnight at room temperature.
3. Acidify the reaction mixture to pH 3 with saturated $KHSO_4$ solution and extract with AcOEt three times. The combined extracts wash twice with H_2O, dry over Na_2SO_4 and evaporate under the reduced pressure to dryness to obtain the crude product.

3.1.3 Selective Deprotection of Boc-L-alanyl-D-isoglutamine Benzyl Ester Hydrochloride: Removal of Boc-Protecting Group

This protocol was adopted from ref. 9.

1. Dissolve (2.5 mmol) Boc-L-alanyl-D-*iso*glutamine benzyl ester hydrochloride in 50 mL AcOH saturated with HCl.
2. Stirr the reaction mixture at room temperature for 2 h.
3. Evaporate the AcOH under the reduced pressure and the residual AcOH co-evaporate several time with benzene.

3.1.4 The Synthesis of D- and L-(Adamantyl) Gly-L-Ala-D-isoGln Using Woodward's Reagent "K"

This protocol was adopted from ref. 9.

1. Dissolve Woodward's reagent "K" (0.38 mmol) in MeCN–DMF (2:1 v/v) and cool the mixture to 0 °C.
2. Add the solution of Boc-(D,L-adamantyl)glycine (0.38 mmol) in MeCN–DMF (2:1 v/v) and Et_3N (0.38 mmol).
3. Stirr the reaction mixture for 1.5 h at 0–5 °C (*see* **Note 6**).
4. Add L-alanyl-D-*iso*glutamine benzyl ester hydrochloride (0.38 mmol) and Et_3N (0.38 mmol) in MeCN–DMF (2:1 v/v) and stir the reaction mixture overnight at room temperature.
5. Evaporate the organic solvent under the reduced pressure to dryness and add 20 mL of H_2O to residual solid and leave the mixture at 0–5 °C for 2 h.
6. Separate the crude material by filtration and purify it by column chromatography on silica gel using 100 % AcOEt as an eluent.

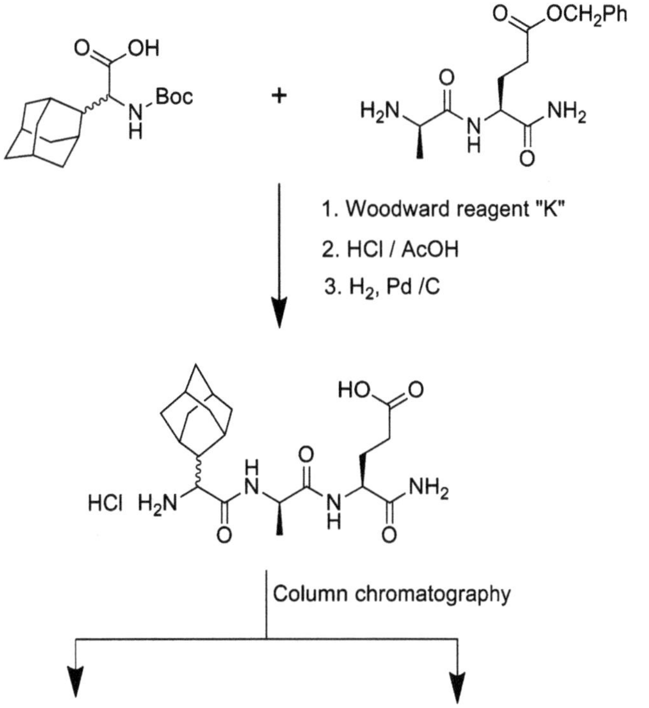

1. Woodward reagent "K"
2. HCl / AcOH
3. H_2, Pd /C

Column chromatography

L-(adamant-2-yl)Gly-L-Ala-D-*iso*Gln D-(adamant-2-yl)Gly-L-Ala-D-*iso*Gln

Scheme 1 Synthesis of adamantyltripeptides (D- *and* L*-(adamant-2-yl)Gly-*L*-Ala-*D*-isoGln*) using Woodward's reagent "K"

7. Recrystallize the crude product from AcOEt–hexane solvent mixture (*see* Scheme 1).

3.1.5 N- and C-Deprotection of Boc[D,L-(adamantyl)] Gly-L-Ala-D-isoGln Benzyl Ester

This protocol was adopted from ref. 9.

1. Dissolve the Boc-(D,L-adamantyl)glycyl-L-alanyl-D-*iso*glutamine benzyl ester (0.6 mmol) in AcOH (25 mL) saturated with HCl and leave the solution at room temperature for 4 h.
2. Evaporate most of the AcOH under the reduced pressure and precipitate the product with absolute Et_2O.
3. Dissolve the obtained Boc-deprotected adamantyltripeptide benzyl ester hydrochloride in 20 mL of 90 % aqueous EtOH and hydrogenate at 4 atm H_2 pressure over 10 % Pd/C (80 mg) at room temperature for 4 h.
4. Filter the catalyst off and evaporate the solvent under the reduced pressure.
5. Purify the mixture of diastereoisomers by the column chromatography on silica-gel (*see* **Note 7**).

3.1.6 Synthesis of D- and L-(Adamantyl)Gly-L-Ala-D-isoGln Using EDC/HOBt Coupling Protocol

This protocol was adopted from ref. 14.

1. Dissolve the Fmoc-(adamantyl)Gly (1.3 mmol) in dry DCM (10 mL) and cool the reaction mixture to 0 °C (*see* **Note 6**).
2. Add EDC×HCl (1.56 mmol, 1.2 equiv.) and HOBt (1.3 mmol, 1 equiv.)
3. Stirr the mixture at 0 °C for 30 min.
4. Add a solution of L-alanyl-D-*iso*glutamine *tert*-butyl ester (1.3 mmol, 1 equiv.) in dioxane (5 mL) into the reaction mixture.
5. Add subsequently *N*-ethylmorpholine (2.6 mmol, 2 equiv.) and stir the mixture for 1 h at 0 °C and then for 48 h at room temperature.
6. Extract the solution with AcOEt three times, combine the organic extracts, wash them with H_2O twice, and dry over $MgSO_4$.
7. Evaporate the solvent under the reduced pressure to dryness and crude diastereoisomeric mixture of Fmoc-adamantyltripeptide *tert*-butyl ester purify by column chromatography on silica gel using 100 % AcOEt as an eluent.

3.1.7 Selective Deprotection of Fmoc-Adamantyltripeptide tert-Butyl Ester: Removal of Fmoc-Protecting Group

This protocol was adopted from ref. 14.

1. Dissolve the Fmoc-adamantyltripeptide *tert*-butyl ester in dry THF (5 mL) under N_2.
2. Add the 1-octanthiol (7.3 mmol, 10 equiv) and DBU (0.24 mmol, 0.33 equiv) and stir the reaction mixture at room temperature over night.
3. Triturate the mixture with Et_2O and purify the solid residue by column chromatography on silica gel using $CHCl_3$–MeOH mixture as an eluent (gradient elution from 10:1 to 1:1 v/v).

3.2 Synthesis of Mannosylated Adamantyltripeptides [(2R)-N-[3-(α-D-mannopyranosyloxy)-2-methylpropanoyl]-D,L-(adamantyl) glycyl-L-alanyl-D-isoglutamine]

Perbenzylated α-mannosyloxy-2-methylpropionic acid of *R* or *S* configuration can be prepared according to the protocol described in ref. 28.

In brief, commercially available methyl (*R*) or (*S*)-3-hydroxy-2-methylpropionate was *O*-mannosylated using benzyl protected mannosyl trichloroacetimidate as a glycosyl donor and catalytic amounts of Lewis acids. The obtained anomeric mixture, *α,β*-glycosides were separated by column chromatography on silica-gel. The alkaline hydrolysis of methyl esters of *O*-glycosides gives the perbenzylated α-mannosyloxy-2-methylpropionic acid. Prepared perbenzylated α-mannosyloxy-2-methylpropionic acid of *R* or *S* configuration can be than coupled to *tert*-butyl protected adamantyltripeptides using EDC/HOBt coupling method. Acid hydrolysis of *tert*-butyl protecting group and catalytic hydrogenation for *O*-debenzylation can be used to obtained the deprotected desired product (*see* Scheme 2).

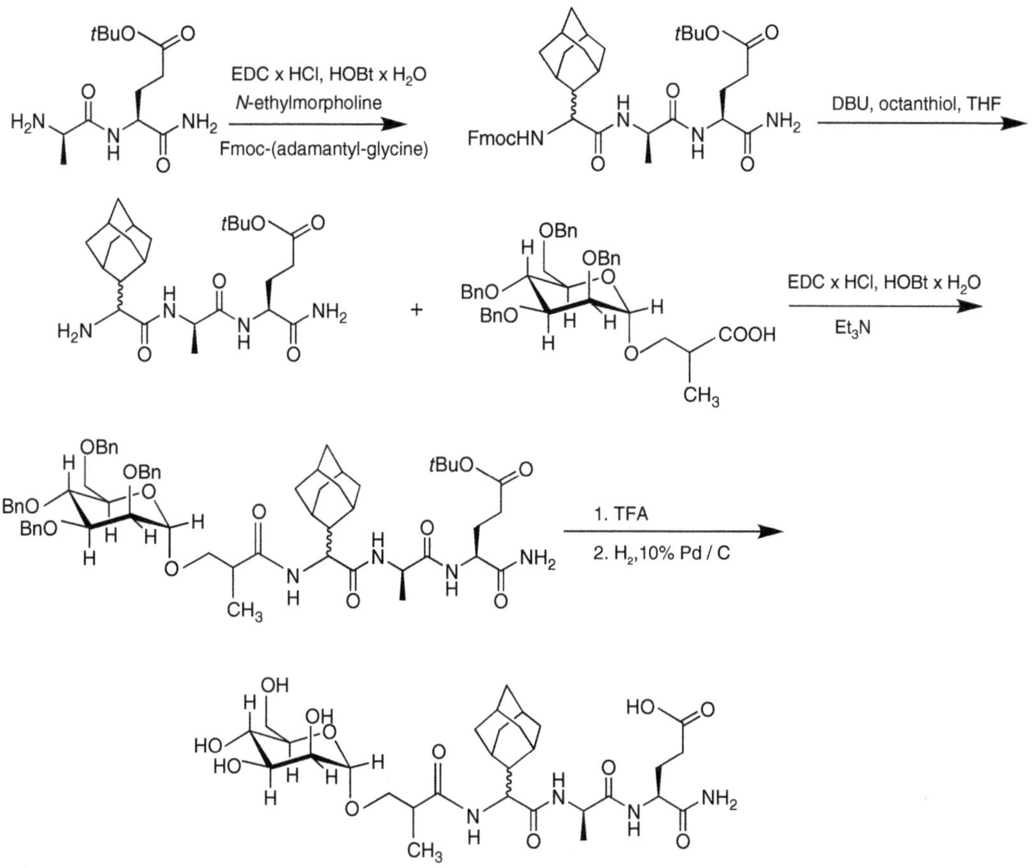

Scheme 2 Synthesis of adamantyltripeptides and mannosylated adamantyltripeptides using EDC/HOBt coupling protocol

3.2.1 Coupling of the Perbenzylated α-Mannosyloxy-2-methylpropionic Acid and Adamantyltripeptides

This protocol was adopted from ref. 14.

1. Dissolve the perbenzylated α-mannosyloxy-2-methylpropionic acid (0.52 mmol) in dry DCM (10 mL) and cool the solution to 0 °C (*see* **Note 6**).
2. Add EDC×HCl (0.63 mmol, 1.2 equiv) and HOBt (0.52 mmol, 1 equiv).
3. Stirr the mixture for 30 min. at 0 °C and then add it to the solution of *tert*-butyl ester of adamantyltripeptides (0.43 mmol) in 5 mL dioxane and Et_3N (1.04 mmol, 2 equiv).
4. Stirr the reaction mixture for 1 h at 0 °C (*see* **Note 6**) and then for 48 h at room temperature.
5. Dilute the reaction mixture with AcOEt and wash it with water. The organic layer dry over Na_2SO_4, filtrate and evaporate the solvent under the reduced pressure to dryness.
6. Purify the solid residue by column chromatography on silica gel using $CHCl_3$–MeOH (10:1 v/v) as an eluent.

3.2.2 Acid Hydrolysis of tert-Butyl Ester of Perbenzylated α-Mannosyloxy-2-methylpropanoyl-D,L-(adamantyltripeptides)

This protocol was adopted from ref. 14.

1. Dissolve the completely protected mannosylated adamantyltripeptide *tert*-butyl ester (0.28 mmol) in $CF_3COOH–H_2O$ (3.3 mL, 95:5, v/v) and stir the mixture at room temperature for 2 h.
2. Concentrate the reaction mixture under the reduced pressure and purify the residue by flash chromatography on silica-gel using $CHCl_3$–MeOH (5:1 v/v) as an eluent.

3.2.3 Catalytic Hydrogenation of Perbenzylated α-Mannosyloxy-2-methylpropanoyl-D,L-(adamantyltripeptides)

This protocol was adopted from ref. 14.

1. Dissolve the perbenzylated α-mannosyloxy-2-methylpropanoyl-D,L-(adamantyltripeptides) (0.132 mmol) in 5 mL $CHCl_3$–MeOH (3:1 v/v), add 50 mg 10 % Pd/C and 20 mL of 50 % EtOH.
2. Stirr the mixture under an atmosphere of hydrogen (4 atm) at room temperature for 24 h.
3. Filtrate the catalyst off, remove the solvent under reduced pressure and purify the crude product by column chromatography on silica-gel using $CHCl_3$–MeOH (1:1 v/v) as an eluent.

3.3 HPLC Analysis of Synthesized Adamantyltripeptides and Mannosylated Adamantyltripeptides

This protocol was adopted from refs. 14, 17.

Chromatographic analyses of synthesized mannosylated adamantyltripeptides were carried out using the Waters HPLC System equipped with 2996 PDA detector and Empower software (Milford, MA, USA). A LiChrosorb RP-18 column, 244 mm × 4 mm, 5 μm, and a LiChrospher guard column 100 RP-18 (5 μm) were used (*see* **Note 8**). Analyses were run at a flow rate of 1.0 mL/min at room temperature and the eluate was monitored at 200 nm. The peptide bonds exhibit absorption in UV spectra near 200 nm and all compounds have been detected by UV absorbance. The gradient solvent system used was made of acetonitrile containing 0.035 % TFA and water containing 0.05 % TFA. The percentage of each peak in the respective chromatograms was calculated by the integration of the UV response (peak area).

3.4 Assessment of Immunostimulatory Activity of Synthesized Mannosylated Adamantyltripeptides

This protocol was adopted from ref. 7.

The ability of antigens to induce protective immune responses in a host can be enhanced by combining the antigen with immunostimulants or adjuvants. Ovalbumin was used as a customary antigen for studying the adjuvant effect [29]. The immunostimulatory effects on secondary humoral response to ovalbumin in mice of two genetically different strains were evaluated. CBA (H-2^k) and NIH/OlaHsd (H-2^q) mice were immunized according to our previously described mouse model in vivo [7]. The comparison of induced anti-OVA levels was carried out quantitatively and the subclasses of IgG, IgG1 as an indicator of Th2 and IgG2a as an indicator of Th1 type of immune response were also determined. For in vivo experiments special batches of endotoxin-free compounds were prepared (*see* **Note 9**). All mice used were females 2–2.5 months old. Commercial food and water were provided ad libitum.

3.4.1 Enzyme-Linked Immunosorbent Assay for Qualitative and Quantitative Determination of OVA-Specific IgG (Anti-OVA IgG) in Mice Sera

ELISA was performed on flat-bottomed high binding microtiter plates according to previously described procedures [7]. Experimental group of five mice was immunized and boosted two times subcutaneously (s.c.) into the tail base at 21-day intervals. Mice were anesthetized prior to blood collection on 7th day after the second booster. Sera were collected, decomplemented at 56 °C for 30 min, and stored at −20 °C until tested. The dose of OVA antigen was 10 μg per mouse. The dose of tested immunostimulators was 200 μg per mouse. OVA and tested substances were dissolved in saline, and the injection volume in all experimental groups was 0.1 mL per mouse.

The relative quantities of anti-OVA IgG were determined by parallel line assay comparing each serum to monoclonal anti-chicken egg albumin, declared as standard preparation, to which 20,000 arbitrary units per milliliter (AU/mL) were voluntarily assigned.

3.4.2 Enzyme-Linked Immunosorbent Assay for Qualitative and Quantitative Determination of OVA-Specific IgG Subtypes Anti-OVA IgG1 and Anti-OVA IgG2a in Mice Sera

ELISA for determination of anti-OVA IgG1 and anti-OVA IgG2a were performed as described in [7]. The relative quantities of antibody subtypes were determined by parallel line assay using appropriate standard preparation. The monoclonal anti-OVA IgG1 was a standard for relative quantification of anti-OVA IgG1 to which 400,000 AU/mL was assigned, while polyclonal mouse serum containing high levels of anti-OVA IgG2a was used as a standard for relative quantification of IgG2a specific antibodies with voluntarily assigned 20,000 AU/mL. The ratio of anti-OVA IgG1 and anti-OVA IgG2a (IgG1/IgG2a) was used as indication of the Th1/Th2-bias of induced immune response. The results were subjected to the statistical analyses (*see* **Note 10**).

4 Notes

1. HCl acid hydrolysis: dissolve 1 mg of compound in 1 mL of 6 M HCl in a sealed tube, evacuate the tube under the reduced pressure and leave at 100 °C at least for 16 h. Evaporate the HCl and dissolve the residue in phosphate buffer (pH 6.5). Add 20 μL of 1 M NaOH to adjust pH to 7. Add 80 μL of aqueous solution of L-amino acid oxidase (7.0 mg protein/mL, 8.9 units/mg protein) and incubate at 37 °C for 24 h.

TLC of the incubation mixture in comparison with TLC of the hydrolysate will reveal the absence of L-Ala and L-Adamantyl-glycine.

2. This isoxazolium salt reagent, *N*-ethyl-5-phenylisoxazolium 3′-sulfonate, a zwitterionic salt, is commonly known and sold as Woodward's Reagent K (see for example: The Merck Index, 9th ed., 1976, entry 9715). The use of the Woodward's reagent "K" in the carboxyl-activating step during the peptide synthesis has been proven practical and very useful [24]. The peptide bond formation and isolation of the product proceed easy and under mild reaction conditions, but the special practical merit of the method is that synthesized peptides are obtained directly in an unusually high degree of purity and almost no racemization.
3. (*S*)-Boc-(adamant-1-yl)glycine is commercially available from the Chem-Impex Int'l Inc.
4. Add slowly dropwise the mixture of NaOH and di-(*tert*-butyl) dicarbonate at room temperature. Repeat this procedure twice. However, second time use less di-(*tert*-butyl)dicarbonate (0.46 mmol).
5. Use universal pH indicator paper to estimate the pH of aqueous solution.
6. To maintain a temperature of the reaction mixture between 0 and 5 °C use a mixture of ice and water as a cooling bath.
7. The fully deprotected diastereoisomers of the adamantyltripeptides are separated by column chromatography on silica gel using the complex mixture of eluent $CHCl_3$–i-PrOH–MeOH–H_2O–AcOH (20:15:6:4:2).
8. For HPLC analyses any of the columns with the octadecyl silica (ODS) adsorbent can be used. The ODS adsorbents are the most popular adsorbents for RP-HPLC separation and several commercial adsorbents (Spherisorb ODS, LiChrosorb, Zorbax ODS and Shodex ODSpak) are used in analysis of peptidoglycans and related compounds according to the previously published paper [30].
9. Endotoxin-free compounds were prepared by passing the aqueous solutions of compounds through a *Detoxy* gel column (Pierce, The Netherlands), followed by liophylization.
10. Statistical analyses were performed with Statistica 6.0 for Windows, StatSoft Inc. The significant difference between experimental groups was evaluated by Kruskal–Wallis ANOVA, followed by multiple Mann–Whitney *U*-nonparametric tests. A probability values less than 0.05 ($p \leq 0.05$) were considered significant.

Acknowledgments

We thank the Ministry of Science, Education and Sport of the Republic of Croatia for its financial support. Projects No. 021-0212432-2431 and 119-1191344-3121 are acknowledged.

References

1. Seidl PH, Schleifer KH (1986) Structure and immunochemistry of peptidoglycan. In: Seidl PH, Schleifer KH (eds) Biological properties of peptidoglycan. Walter de Gruyter, Berlin, pp 1–2
2. Stewart-Tull DES (1988) Immunostimulation with peptidoglycan or its synthetic derivatives. Prog Drug Res 32:305–328
3. Ellouz F, Adam A, Ciorbaru R, Lederer E (1974) Minimal structural requirements for adjuvant activity of bacterial peptidoglycan derivatives. Biochem Biophys Res Commun 59:1317–1324
4. Dzierzbicka K, Kolodziejczyk AM (2003) Muramyl peptides—synthesis and biological activity. Polish J Chem 77:373–395
5. Hršak I, Tomašić J (1986) Immunostimulating and antimalignant activity of peptidoglycan monomer and its metabolites. Period Biol 88: 22–23
6. Hršak I, Ljevaković Đ, Tomašić J, Vranešić B (1994) Preparation properties and biological activities of tert-butyloxycarbonyl-L-tyrosyl peptidoglycan monomer. In: Masihi N (ed) Immunotherapy of infection. Marcel Dekker, New York, pp 249–257
7. Tomašić J, Hanzl-Dujmović I, Špoljar B, Vranešić B, Šantak M, Jovičić A (2000) Comparative study of the effects of peptidoglycan monomer and structurally related adamantyltripeptides on humoral immune response to ovalbumin in the mouse. Vaccine 18:1236–1243
8. Ljevaković Đ, Tomašić J, Šporec V, Halassy Špoljar B, Hanzl Dujmović I (2000) Synthesis of novel adamantylacetyl derivative of peptidoglycan monomer-biological evaluation of immunomodulatory peptidoglycan monomer and respective derivatives with lipophilic substituents on amino group. Bioorg Med Chem 8:2441–2447
9. Vranešić B, Tomašić J, Smerdel S, Kantoci D, Benedetti F (1993) Synthesis and antiviral activity of novel adamantyltripeptides. Helv Chim Acta 76:1752–1758
10. Vranešić B, Tomašić J, Ljevaković Đ, Hršak I (1994) Biological activity of novel adamantyltripeptides with the emphasis on their immunorestorative effect. In: Masihi N (ed) Immunotherapy of infection. Marcel-Deker, New York, pp 241–248
11. Lamoureux G, Artavia G (2010) Use of the adamantane structure in medicinal chemistry. Curr Med Chem 17:2967–2978
12. Gerzon K, Krumkalus EV, Brindle RL, Marshall FJ, Root MA (1963) The adamantyl group in medicinal agents. J Med Chem 6:760–763
13. Liu J, Obando D, Liao V, Lifa T, Codd R (2011) The many faces of the adamantyl group in drug design. Eur J Med Chem 46: 1949–1963
14. Ribić R, Habjanec L, Vranešić B, Frkanec R, Tomić S (2011) Synthesis and biological evaluation of new mannose derived immunomodulating adamantyltripeptides. Croat Chem Acta 84:233–244
15. Ribić R, Habjanec L, Vranešić B, Frkanec R, Tomić S (2012) Influence of mannosylation on immunostimulating activity of adamant-1-yl tripeptide. Chem Biodivers 9:1373–1381
16. Ribić R, Habjanec L, Vranešić B, Frkanec R, Tomić S (2012) Synthesis and immunostimulating properties of novel adamant-1-yl tripeptides. Chem Biodivers 9(4):777–788
17. Krstanović M, Frkanec R, Vranešić B, Ljevaković Đ, Šporec V, Tomašić J (2002) Reversed-phase high-performance liquid chromatographic method for the determination of peptidoglycan monomers and structurally related peptides and admantyltripeptides. J Chromatogr B 773:167–174
18. Azuma I (1992) Synthetic immunoadjuvants: application to non-specific host stimulation and potentiation of vaccine immunogenicity. Vaccine 10:1000–1006
19. Yoo YC, Yoshimatsu K, Koike Y, Hatsuse R, Yamanishi K, Tanishita O, Arikawa J, Azuma I (1998) Adjuvant activity of muramyl dipeptide derivatives to enhance immunogenicity of a hantavirus-inactivated vaccine. Vaccine 16:216–224
20. Vranešić B, Tomašić J, Ljevaković D, Hršak I (1994) Biological activity of novel adamantyltripeptides with the emphasis on their immunorestorative effect. In: Masihi N (ed) Immunotherapy of infection. Marcel Dekker, New York, pp 241–248

21. Frkanec R, Noethig Laslo V, Vranešić B, Mirosavljević K, Tomašić J (2003) A spin labelling study of immunomodulating peptidoglycan monomer and adamantyltripeptides entrapped into liposomes. Biochim Biophys Acta 1611:187–196
22. Štimac A, Šegota S, Dutour Sikirić M, Ribić R, Frkanec L, Svetličić V, Tomić S, Vranešić B, Frkanec R (2012) Surface modified liposomes by mannosylated conjugates anchored via the adamantyl moiety in the lipid bilayer. BBA-Biomembr 1818:2252–2259
23. Keglević D, Ladešić B, Tomašić J, Valinger Z, Naumski R (1979) Isolation procedure and properties of monomer unit from lysozyme digest of peptidoglycan complex excreted into the medium by penicillin-treated Brevibacterium divaricatum mutant. Biochem Biophys Acta 585:281–373
24. Woodward RB, Olofson RA, Mayer H (1961) A new synthesis of peptides. J Am Chem Soc 83:1010–1012
25. Clariana J, Garcia-Granda S, Gotor V, Gutierrez-Fernandez A, Luna A, Moreno-Manas M, Wallribera A (2000) Preparation of (*R*)-(1-adamantyl)glycine and (*R*)-2-(1-adamantyl)-2-aminoethanol: a combination of cobalt-mediated-ketoester alkylation and enzyme-based aminoalcohol resolution. Tetrahedron Asymmetry 11:4549–4557
26. Gašpert B, Hromadko S, Vranešić B (1976) Synthesis of α-amino-1-adamantylacetic and α-amino-2-adamantylacetic acid. Croat Chem Acta 48:169–178
27. Lefrancier P, Bricas E (1967) Synthesis of the peptidic subunit of cell wall peptidoglycan of 3 g positive bacteria and of peptides of analogous structure. Bull Soc Chim Biol 49:1257–1271
28. Ribić R, Kovačević M, Petrović-Perokovi V, Gruić-Sovul I, Rapić V, Tomić S (2010) Synthesis and biological activity of mannose conjugates with 1-adamantamine and ferrocene amines. Croat Chem Acta 83:421–431
29. Stewart-Tull DES (1989) Recommendations for the assessment of adjuvants (immunopotentiators). In: Gregoriadis G, Allison AC, Poste G (eds) Immunological adjuvants and vaccines. Plenum Press, New York, pp 213–226
30. Frkanec R, Tomašić J (2008) High performance liquid chromatography in analysis of compounds comprising the elements of bacterial peptidoglycan structure with immunological activity. J Liq Chromatogr Relat Technol 38:107–133

Chapter 8

Optimization of Physicochemical and Pharmacological Properties of Peptide Drugs by Glycosylation

Maria C. Rodriguez and Mare Cudic

Abstract

Many biological interactions and functions are mediated by glycans, leading to the emerging importance of carbohydrate and glycoconjugate chemistry in the design of novel drug therapeutics. In addition to direct effects on biological activity, sugar addition appears to alter many physicochemical and pharmacological properties of the peptide backbone. Consequently, glycosylation has been often used to improve various less than optimal features of peptide drug leads.

In order to study the effects that naturally occurring and/or nonnatural glycans have on peptide drug solubility, conformation, proteolytic resistance, membrane permeability, and toxicity, it is essential to have convenient synthetic access toward synthesis of glycopeptide analogs. The crucial step in the synthesis of glycopeptides is the introduction of the carbohydrate group. The preformed glycosyl amino acid building block is the most commonly employed approach used in glycopeptide synthesis.

In this review, we will describe various synthetic approaches to prepare *N*- and *O*-glycopeptides bearing simple monosaccharides as a tool to improve peptide therapeutic efficacy by glycosylation.

Key words Peptide drug modifications *via* glycosylation, *N*- and *O*-glycosyl amino acid building blocks, Fmoc solid-phase glycopeptide synthesis

1 Introduction

An increased interest in the development of peptide drugs within the last decade has resulted in the search for new ways to prolong and/or enhance the biological activity of the peptide drug at the target site [1–3]. However, this has been a persistent challenge to pharmaceutical scientists because of several unfavorable physicochemical properties, including susceptibility to enzymatic degradation, low lipophilicity, the lack of specific transport systems to direct such peptide drugs into cells or across the blood–brain barrier, and the high conformational flexibility [4, 5]. Based on the positive effects that glycosylation makes on some properties of synthetic peptides, it is not surprising that incorporation of glycans into key positions of the peptide chain has become one of the most effective

Predrag Cudic (ed.), *Peptide Modifications to Increase Metabolic Stability and Activity*, Methods in Molecular Biology, vol. 1081, DOI 10.1007/978-1-62703-652-8_8,

strategies to overcome the drawbacks of peptide drugs [6, 7]. Glycosylation appears to modulate in vivo efficacy of protein drugs by altering the balance between their potencies (PD-pharmacodynamics) and exposure time (PK-pharmacokinetics). Changes to protein PK parameters induced by glycosylation include improved absorption and distribution, longer circulation lifetimes, and decreased clearance rates. However, many aspects regarding the mechanisms by which glycosylation induces such effects remain unclear. In addition, it is not well understood how the density, type, size, and charge of glycans affect the overall protein physicochemical and pharmacological properties [8, 9].

1.1 Effects of Glycosylation on Peptide Solubility and Aggregation

It is generally accepted that glycosylation can affect biophysical properties such as solubility and aggregation of peptides and proteins [10]. Increased solubility of glycosylated proteins and reduced aggregation propensity have been attributed to glycan hydrophilicity [11]. The linear dependence of solubility with the glycosylation degree was later explained by the increased overall molecular solvent accessible surface area (SASA) of glycoproteins that leads to an increase in the number of possible interactions between the glycoprotein surface and the surrounding solvent molecules [8, 12]. However, recent findings suggest that glycosylation effects are likely to be more complex than previously recognized [8, 13–15]. It was suggested that the amphiphilic properties of some glycans, in particular the *N*-acetylglucosamine (GlcNAc) moiety of *N*-linked glycans, may play a role in the poor packing of protein aggregates [14, 16, 17].

1.2 Effects of Glycosylation on Peptide Conformation

Structural elements in peptides include α-helices, parallel and antiparallel β-sheets, other types of helices, as well as α-, β-, and γ-turns. The effect of glycosylation on the peptide backbone conformation has been extensively studied [8, 16, 18, 19]. Synthetic glycopeptides are considered good models for detailed structural analysis since the type and site of glycan attachment are well defined compared to the sugar heterogeneity on native glycoproteins. When *N*-glycosylation modifies the conformation, the resulting structures are more ordered than the peptide chain without sugar addition [18, 20]. In the case of *O*-glycopeptides, the final conformations can be either more ordered [21] or less ordered [22]; but in any event, only the innermost carbohydrates make contact with the peptide backbone [16, 18]. Multiple *O*-glycosylations can have a pronounced effect on peptide conformation. The clustering of *O*-GalNAc glycans on the mucin scaffold leads to peptide chain extension and increased rigidity of the backbone [23, 24]. It also appears that the extent of these effects depends on whether glycosylation occurs at a serine or threonine residue. The serine glycosylated peptides show more flexibility as the sugar amide protons in these peptides are not involved in peptide–carbohydrate interactions [18, 25, 26].

1.3 Effects of Glycosylation on Proteolytic Resistance

One of the obstacles of bringing peptide drugs to the market is their inherent instability toward metabolic degradation. Glycosylation has been utilized in the design of protease-resistant peptides [27–31]. For example, sialylation has been used efficiently to increase the half-life of therapeutic proteins [9, 29]. The effect of glycosylation on the stability of peptides against enzymatic degradation has been very often explained by the steric hindrance of the proteolytic site. Thereby, the direct contact between the enzyme and amino acids adjacent to the glycan attachment site is prevented [10]. However, the long-range stabilization by glycosylation is possible even when the glycosylation site is several amino acids away from the predominant site(s) of cleavage. This effect is likely a consequence of altered peptide conformation [32].

1.4 Effects of Glycosylation on Blood–Brain Barrier (BBB) Penetration

The BBB plays a crucial role in the central nervous system (CNS) [33]. Due to its protective role and highly regulated selective transport of molecules that are essential for brain function, delivery of therapeutics across the BBB represents a major challenge in treating most disorders of the CNS [34–36]. Numerous peptides have been investigated to date whereby glycosylation increases metabolic stability and enhances BBB penetration leading to increased pharmacological activity [37–40].

1.5 Effects of Glycosylation on Biological Activity

Glycosylation has also been explored as a tool to modify the biological activity of peptides. It has been observed that attachment of carbohydrate residues to peptides, even to ones that are not glycosylated in nature, can influence their biological functions. This approach has been used in the case of some enzyme inhibitors [41], antimicrobial peptides [42, 43], neuropeptides [39, 40, 44], and hormones [29, 31, 45–47]. The enhanced biological activity of glycosylated analogs can be a consequence of improved bioavailability, or as an indirect result from conformational changes induced in the protein backbone by the carbohydrate moieties. Regardless, the specific contribution of carbohydrate moieties often remains elusive.

2 Materials

1. Extra dry or anhydrous solvents and reagents for glycosylation-related reactions, such as dichloromethane (DCM), acetic acid (AcOH), chloroform ($CHCl_3$), diethyl ether, petroleum ether, acetonitrile (ACN), tetrahydrofuran (THF), pyridine, ethyl acetate (EtOAc), methanol (MeOH), toluene, *N*,*N*′-dimethylformamide (DMF), hydrobromic acid (HBr), hydrochloric acid (HCl), sulfuric acid (H_2SO_4), perchloric acid ($HClO_4$), hydrogen peroxide (H_2O_2), acetic anhydride (Ac_2O), trifluoroacetic acid (TFA), thioacetic acid (CH_3COSH),

trifluoroacetic anhydride (TFAA), ethanolamine, and *N,N*-diethylethanamine (Et_3N) (Sigma-Aldrich, Fisher Scientific, VWR, Acros Organics).

2. Solvents used for flash chromatography, extractions, and crystallizations are HPLC or ACS grade (Sigma-Aldrich, Fisher Scientific, VWR, Acros Organics).
3. Sodium hydrogen carbonate ($NaHCO_3$), sodium carbonate (Na_2CO_3), ammonium hydrogen carbonate (NH_4HCO_3), anhydrous sodium sulfate (Na_2SO_4), sodium chloride (NaCl), potassium carbonate (K_2CO_3), and sodium azide (NaN_3) (Sigma-Aldrich, Fisher Scientific, VWR, and Acros Organics).
4. Thin-layer chromatography on silica gel $60F_{254}$ (Whatman) plates (Fischer Scientific).
5. Silica gel (60 Å, 230–400 Mesh) for flash chromatography (Fisher Scientific).
6. Acid-washed molecular sieves (AW-300), activated molecular sieves (4 Å), and Celite (Celite 521) (Sigma-Aldrich).
7. Special reagents, such as silver trifluoromethanesulfonate (CF_3SO_3Ag), silver carbonate (Ag_2CO_3), silver triflate (AgOTf), silver chlorate ($AgClO_3$), boron trifluoride diethyl etherate ($BF_3{\cdot}Et_2O$), trimethylsilyl trifluoromethanesulfonate (TMSOTf), trichloroacetonitrile (Cl_3CCN), hydrazine acetate ($N_2H_4{\cdot}CH_3CO_2H$), tetrabutylammonium hydrogensulfate (Bu_4NHSO_4), iron(III) chloride ($FeCl_3$), zinc chloride ($ZnCl_2$), zinc powder, sodium hydride (NaH), and 4-dimethylaminopyridine (DMAP) (Alfa-Aesar, and Acros Organics).
8. Palladium on activated carbon (10–20 % Pd) (Pd/C) (Sigma-Aldrich, Fisher Scientific, VWR, and Acros Organics).
9. Peptide synthesis reagents, such as 1,8-diazabicycloundec-7-ene (DBU), *N,N'*-dicyclohexylcarbodiimide(DCC), *N,N'*-diisopropylcarbodiimide (DIC), *N*-hydroxybenzotriazole (HOBt), and *O*-benzotriazole-*N,N,N',N'*-tetramethyluronium-hexafluorophosphate (HBTU) (Sigma-Aldrich, Novabiochem, AnaSpec, and ChemImpex).
10. *N*-(9-Fluorenylmethoxycarbonyloxy)succinimide (Fmoc-OSu) (Sigma-Aldrich and ChemImpex).
11. 2,2,2-Trichloroethyl chloroformate (Troc-Cl) (Alfa-Aesar and Acros Organics).
12. Non-protected monosaccharides (Acros Organics, Sigma-Aldrich, Carbosynth, and Fluka).
13. Peracetylated monosaccharides, such as glucose pentaacetate, galactose pentaacetate, mannose pentaacetate, glucosamine pentaacetate, galactosamine pentaacetate, and their corresponding 1-*O* deacetylated products (CarboSynth).

14. Pentabenzoylated monosaccharides, such as glucopyranose pentabenzoate, galactopyranose pentabenzoate, and mannose pentabenzoate, and their corresponding 1-*O* debenzoylated products (Carbosynth).
15. Peracetylated galactal and glucal (Carbosynth).
16. Tetraacetylated glucosamine and galactosamine hydrochloride (Sigma-Aldrich and Carbosynth).
17. Fmoc-protected amino acids (Novabiochem, ChemImpex, and Advanced ChemTech).

3 Methods

The assembly of glycopeptides can be performed either in solution or on solid-phase, and comprises two options for the introduction of the carbohydrate moiety: the direct glycosylation of a properly protected full-length peptide, or the stepwise synthesis using preformed glycosylated amino acid building blocks [48–54]. Although the direct glycosylation method offers a more convergent approach, it has been less employed. The glycosyl amines often undergo formation of intramolecular aspartimides [55] and direct *O*-glycosylation is even more complex [56–58]. The use of glycosylated building blocks in the stepwise solid-phase peptide assembly is considered the most reliable and efficient technique for preparation of glycopeptides [50–52, 59–61]. Nevertheless, the desired glycosyl amino acid building blocks are either too expensive or not commercially available (*see* **Note 1**).

In the laboratory setting, solid-phase glycopeptide synthesis is usually more convenient and provides better yields in comparison to the solution synthesis [62]. Due to the sensitivity of the glycosidic bond to strong acids, the combination of protecting groups that can be used in solid phase glycopeptide chemistry is often limited, with Fmoc- usually favored over Boc-approach [63].

In this protocol, we will review the synthetic approaches towards *N*- and *O*-glycopeptides bearing simple monosaccharides as a tool to improve peptide therapeutic efficacy by glycosylation. We have chosen to describe the synthesis of the glycosylated Fmoc amino acid building blocks for which, nowadays, efficient synthetic procedures can be employed. In addition, the resulting correctly protected glycosyl amino acid can be coupled in the same way as the simple amino acid in a stepwise glycopeptide chain assembly. The most commonly used coupling procedures are the DCC/HOBt [64], and the pentafluorophenyl (Pfp) active ester method [65].

Synthesis of O-glycosylated amino acids. There are two major types of *O*-glycosides, which are, depending on nomenclature, most commonly defined as α- and β-, or 1,2-*cis* and 1,2-*trans* glycosides. The stereochemical outcome of a glycosylation reaction is

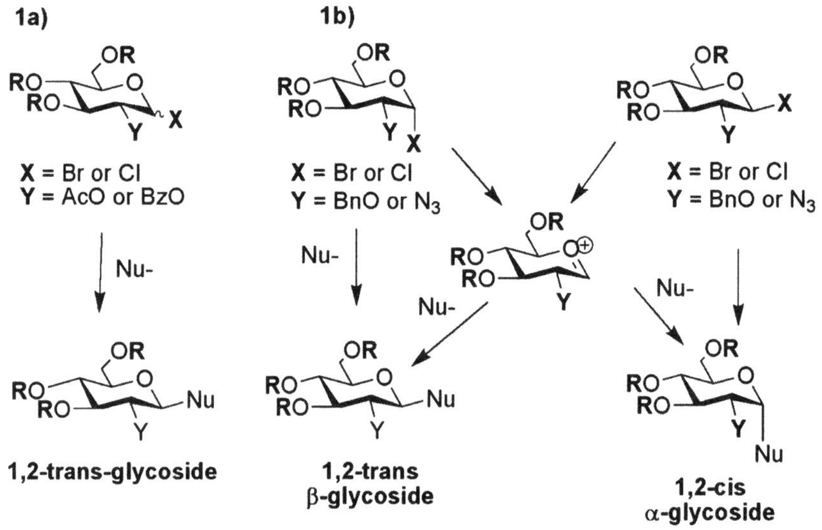

Fig. 1 Glycosylation reaction with the neighboring group participation leading to (**a**) 1,2-*trans* glycosides, and (**b**) without neighboring group participation leading to formation of both, 1,2-*trans* and 1,2-*cis* glycosides

usually controlled by the choice of the neighboring group participation of the substituent at C-2. A neighboring group at C-2 will lead to a 1,2-*trans* glycoside while glycosylation without a neighboring effect will result in the formation of both, 1,2-*trans* as well as 1,2-*cis* glycosides [66] (Fig. 1). The most widely used approaches in the synthesis of the *O*-glycosylated amino acids, Köenigs-Knorr, activation of anomeric acetate, and trichloroacetimidate methods will be explained in detail.

3.1 The Köenigs-Knorr Method

The Köenigs-Knorr method was, for a long time, the only available method for glycosylation and still represents the most widely used procedure for the synthesis of 1,2-*trans* glycosides in the chemistry of carbohydrate derivatives [66]. The α-glycosyl halide generated in the activation step [67], can be readily further activated in the glycosylation step by promoters, i.e., heavy metal salts, resulting in an irreversible glycosyl transfer to the acceptor. Insoluble silver salts, such as Ag_2CO_3, as well as soluble ones, such as AgOTf and $AgClO_3$, have been employed in the synthesis of glycosides.

3.1.1 Preparation of Glycosyl Halides

1. Dissolve peracetylated sugar (5 mmol) in HOAc (5 mL) and cool to 0 °C.
2. Add HBr in HOAc (33 % v/v, 2 mL) dropwise over a period of 10 min.
3. Leave reaction mixture overnight at +4 °C.
4. Add ice into the reaction mixture, extract with $CHCl_3$, and wash the organic extract twice with water.
5. Evaporate, and crystallize the product from diethyl ether (*see* **Note 2**).

3.1.2 Köenigs-Knorr Reaction

1. Dissolve the glycosyl acceptor (1 eq) in DCM and stir with activated molecular sieves (4 Å) under an inert atmosphere at +20 °C (*see* **Note 3**).
2. Add silver salt (2 eq) in the absence of light and cool suspension to −15 °C (*see* **Note 4**).
3. At −15 °C, add glycosyl donor (2 eq), stir for 30 min, and then stir overnight at room temperature (*see* **Note 5**).
4. Filter reaction mixture over Celite and wash the Celite layer well with DCM.
5. Combine filtrates, evaporate, and purify by flash chromatography.

3.2 The Anomeric Acetate Method

The anomeric acetate method uses 1,2-*trans* peracetates in the presence of Lewis acids as a promoter to glycosylate N^{α}-Fmoc amino acids with an unprotected α-carboxyl group [68]. The great advantage of this method is that the starting materials are readily available and the method does not require extensive experience in synthetic carbohydrate chemistry.

3.2.1 Activation of Anomeric Acetate

1. Dissolve the glycosyl acceptor, N^{α}-Fmoc amino acid (1.2 eq), in DCM (or ACN) under an inert atmosphere at room temperature (*see* **Note 3**).
2. Add Lewis acid, $BF_3{\cdot}Et_2O$ (3 eq).
3. Add glycosyl donor, 1,2-*trans*-peracetylated monosaccharide (1 eq), and stir overnight at room temperature (*see* **Note 5**).
4. Dilute the reaction mixture with DCM, and wash with 1 M aqueous HCl and water.
5. Combine organic filtrates, dry over Na_2SO_4, evaporate, and purify by flash chromatography.

3.3 The Trichloroacetimidate Method

The trichloroacetimidate method has been developed into a widely applicable method and is often superior to the Köenigs-Knorr method [69, 70]. The general significance of *O*-glycosyl trichloroacetimidates lies in their ability to act as strong glycosyl donors under relatively mild acid catalysis. In addition, β-glycosyl trichloroacetimidates are more stable than the respective glycosyl bromide.

3.3.1 Synthesis of Trichloroacetimidate

1. Dissolve the reducing sugar derivative (1 eq) in DCM.
2. Add trichloroacetonitrile (2.9 eq) and base (1.1 eq) (*see* **Note 6**).
3. Evaporate the solvent, and purify the reaction mixture by flash chromatography (*see* **Note 7**).

3.3.2 Trichloro-acetimidate Method

1. Dissolve the glycosyl acceptor (1 eq) in DCM (or ACN), and stir with activated molecular sieves (4 Å) under an inert atmosphere at 20 °C (*see* **Note 3**).
2. Add glycosyl donor-trichloroacetimidate (2 eq), and stir for 15 min at −15 °C (*see* **Note 8**).
3. Add TMSOTf (0.1 eq) slowly, and stir reaction mixture at −15 °C for 1 h, followed by at room temperature overnight (*see* **Note 9**).
4. Dilute the reaction mixture with DCM and wash with saturated aqueous $NaHCO_3$.
5. Combine filtrates, evaporate, and purify by flash chromatography.

3.4 Synthesis of β-D-Glc/Gal-(1→O)-Ser/Thr

The building blocks in which glucose or galactose is β-linked to serine or threonine have been used in the synthesis of β-linked glycosides to study the effects of glycosylation on the biological activity, and physicochemical and pharmacological properties of the peptide backbone.

The anomeric acetate activation method offers easy access to these building blocks in one step from commercially available and inexpensive starting materials [68]. The glycosylation of Fmoc-Ser-OH with β-glucose pentaacetate (Fig. 2a) and β-galactose pentaacetate (Fig. 2b) gave the *O*-linked building blocks in 37 and 53 % yields, respectively (*see* Subheading 3.2.1). This building block approach has been applied to the synthesis of two analogs of the antidiuretic drug vasopressin (DDAVP) containing β-D-galactosylated serine at position 4 [47], *O*-β-Gal-Ser^5 analogs of Met-enkephalin to increase the bioavailability and penetration of the BBB [39], model peptides containing one to three β-D-Gal-Ser units to study how glycosylation influences amyloid formation [71],

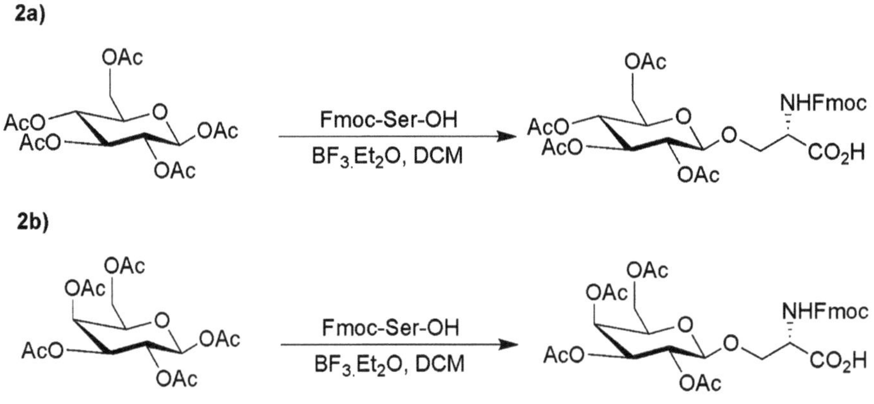

Fig. 2 Synthesis of glycosylated Fmoc-Ser-OH building block containing (**a**) 1-*O*-β-Glc and (**b**) 1-*O*-β-Gal by the anomeric acetate method [68]

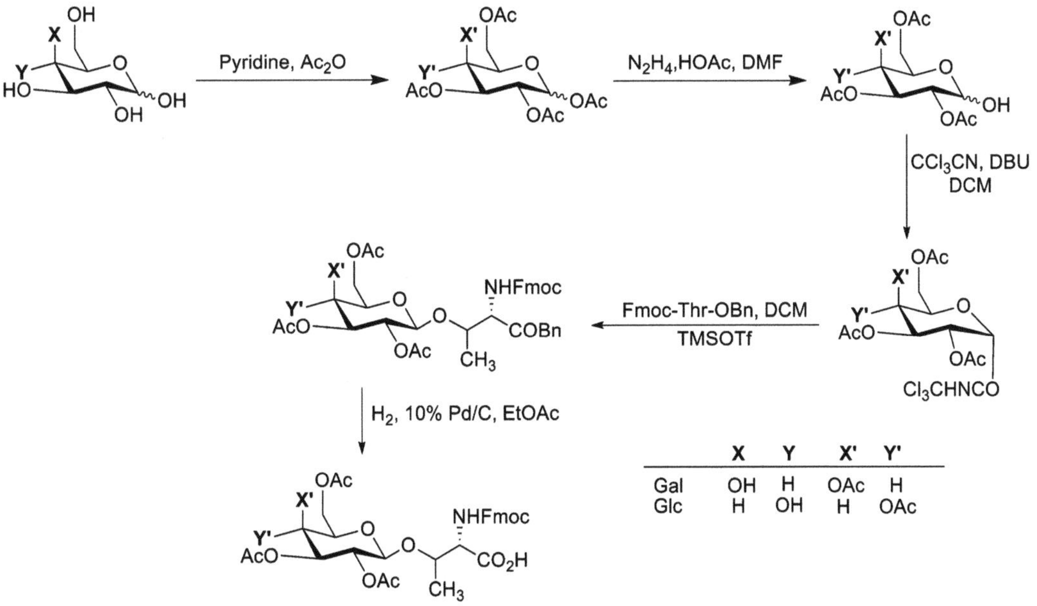

	X	Y	X'	Y'
Gal	OH	H	OAc	H
Glc	H	OH	H	OAc

Fig. 3 Synthesis of glycosylated Fmoc-Thr-OH building block containing 1-*O*-β-Glc/Gal by the trichloroacetimidate method [75]

and antibacterial cyclic glycosylated D,L-α-peptides to explore the consequences of glycosylation on hemolytic and bactericidal activity against multidrug-resistant gram-positive species [42].

Several glycosylated enkephalin analogs have been synthesized and tested in vitro and in vivo for their opiate receptor binding affinity and ability to cross the BBB [40, 72–74]. The Köenigs-Knorr method was chosen to prepare the β-*O*-glucosylated serine building block that was used in the synthesis of Leu-enkephalin analogs [73]. Glycosylation of serine at position 6 led to a compound exhibiting a significant increase in its enzymatic stability in serum, and bioavailability to the brain [73]. Talat et al. [75] applied the trichloroacetimidate method for the synthesis of an Fmoc-protected β-*O*-glycosylated threonine building block that was used in the synthesis of the peptide analogs of formaecin I and drosocin, as highlighted in Fig. 3. The glycosylation of N^{α}-Fmoc-threonine benzyl ester with the per-*O*-acetylated glucose or galactose trichloroacetimidate in the presence of catalytic amounts of TMSOTf afforded a high yield of the β-linked glycosides (*see* Subheading 3.3.2). Subsequent removal of benzyl ester by hydrogenolysis yielded the final product ready for the Fmoc solid-phase peptide synthesis (Fig. 3) [75]. Protocols for synthesis of β-peracetylated sugars, selective 1-*O*-deacetylation and removal of the benzyl ester group by hydrogenolysis are summarized in supplemental protocols (*see* Subheading 3.4.1).

3.4.1 Supplemental Protocols

Synthesis of β-Peracetylated Sugars (with Pyridine as Catalyst) [67]

1. Add Ac_2O (10 eq, 2 eq per OH-group) into dry pyridine (approx. 10 mL pyridine for 1 g of sugar) and cool down the reaction mixture to 0 °C (*see* **Note 10**).
2. Add sugar (1 eq), and stir the solution until it dissolves completely at 0 °C.
3. Let solution stand overnight at room temperature.
4. Pour solution over iced water, and extract the desired peracetylated sugar into CH_3Cl.
5. Wash organic extract with water, $NaHCO_3$, and again with water.
6. Dry the organic layer with Na_2SO_4, evaporate, and isolate the β-anomer by crystallization, or perform flash chromatography to separate the α- from the β-anomer (*see* **Note 11**).

Selective 1-*O*-Deacetylation of Peracetylated Sugars [76]

1. Dissolve the peracetylated sugar (1 eq) in DMF (4 mL).
2. Add $N_2H_4{\cdot}CH_3CO_2H$ (1.5 eq), and stir for 20 min at room temperature.
3. Dilute mixture with EtOAc (25 mL), and wash with water, then with saturated aq. $NaHCO_3$, and again with water, followed by drying with $MgSO_4$.
4. Purify residue by flash chromatography.

Removal of Benzyl Ester by Hydrogenolysis [75]

1. Dissolve the glycosylated amino acid protected benzyl ester (1 eq) in anhydrous EtOAc (20 mL).
2. Add 10 % Pd/C (0.1 eq), and stir the mixture under one atmosphere of hydrogen gas until the reaction is complete (approx. 2 h) (*see* **Notes 12** and **13**).
3. Remove the catalyst by filtration, and after solvent evaporation, purify the residue by flash chromatography (*see* **Note 14**).

3.5 Synthesis of α-D-Glc/Gal-(1→O)-Ser/Thr

The building blocks in which glucose or galactose is α-linked to threonine have been prepared by the trichloroacetimidate method in the presence of TMSOTf, as highlighted in Fig. 4 [77] (*see* Subheading 3.3.2). Protocols for the synthesis of β-perbenzylated sugars and selective 1-*O*-deprotection (*see* Subheading 3.5.1), removal of the benzyl ester group by hydrogenolysis (*see* Subheading 3.4.1), and the *N*-protection of the glycosylated amino acid by acylation with Fmoc-OSu (*see* Subheading 3.5.1) are summarized in the specified subheadings. These building blocks have been used in the synthesis of a series of differentially glycosylated analogs of antimicrobial peptides [43]. The activity of drosocin analogs against several gram-negative bacteria appears to be modulated by the sugar moiety (Gal vs. GalNAc) and the type of glycosidic linkage (α-O-, β-O-, or α-C-), while insertion of a glycosylated threonine residue in the apidaecin sequence reduces the antimicrobial activity [43].

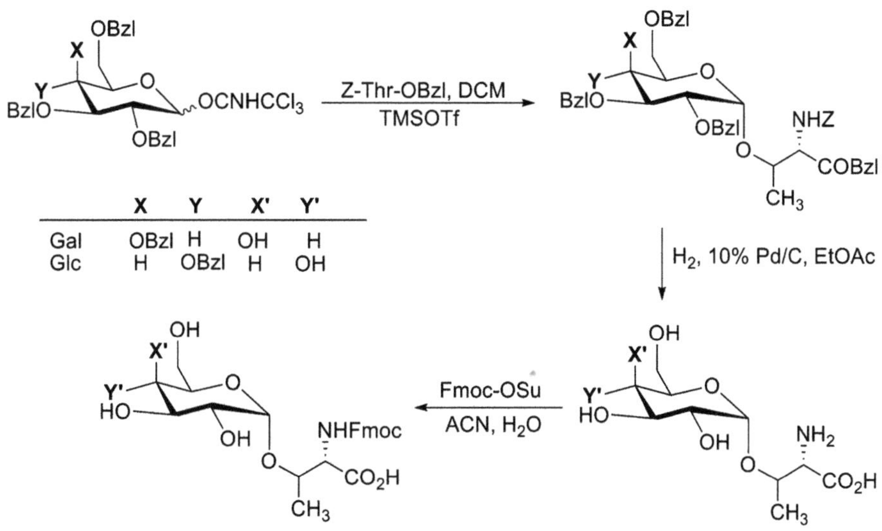

Fig. 4 Synthesis of glycosylated Fmoc-Thr-OH building blocks containing 1-*O*-α-Gal/Glc by the trichloroacetimidate method [77]

3.5.1 Supplemental Protocols

Synthesis of Perbenzoylated Glc/Gal [78]

1. Dissolve the sugar (1 eq) and a catalytic amount of DMAP in pyridine (12 mL), and cool to 0 °C.
2. Add benzoyl chloride (7.5 eq) and stir overnight at room temperature.
3. Once reaction is complete, concentrate the solution in vacuo, and dissolve the residue in DCM (approx. 150 mL).
4. Add water carefully with cooling and vigorous stirring (*see* **Note 15**).
5. Extract the product with DCM, and wash the organic phase with sat. NaCl, $NaHCO_3$, NaCl, and dry with $MgSO_4$, and concentrate in vacuo.
6. Dissolve the crude product in a mixture of acetone–methanol (1:1, 14 mL) and let stand overnight.
7. Dilute the solution with methanol (7 mL) and let stand for 2 h.
8. Dilute the solution again with methanol (14 mL) and let stand for 2 days at +4 °C.
9. Collect crystallized product by filtration.

Selective 1-*O*-Debenzoylation of Perbenzoylated Sugars [78]

1. Dissolve the perbenzoylated sugar (1 eq) in THF (8 mL).
2. Add ethanolamine (1.2 eq) dropwise, and stir overnight at room temperature (*see* **Note 16**).
3. Filter the precipitate, evaporate the solvent in vacuo, and purify the crude mixture by flash chromatography.

Fmoc-Protection by Acylation with Fmoc-OSu [77]

10. Dissolve the glycosylated amino acid (1 eq) in water (3 mL), and add Et_3N (1.2 eq).
11. Add Fmoc-OSu (1.2 eq) in ACN (7 mL), and stir for 45 min at room temperature, maintaining pH at 8.5.
12. Dilute mixture with water (25 mL), and extract with EtOAc.
13. Dilute the aqueous solution with sat. NaCl and acidify to pH 2 with 0.1 N HCl.
14. Extract the oily precipitate with EtOAc, combine the extracts and wash with sat. NaCl solution, then re-extract with 5 % aq. $NaHCO_3$.
15. Acidify the alkaline solution to pH 2 with 0.1 N HCl, and let stand at 4–0 °C for precipitation to occur.
16. Collect the solid by filtration, wash with diethyl ether and dry the product in a desiccator.

3.6 Synthesis of β-D-GlcNAc/GalNAc-(1→O)-Ser/Thr

The synthesis of the building blocks in which *N*-acetyl-glucosamine (GlcNAc) or galactosamine (GalNAc) is β-linked to serine or threonine [79], is often accomplished using a participating group at C-2 of the glycosyl donor. Due to the poor participation of the *N*-acetyl group itself, it is desirable to replace it with other electron-withdrawing groups like *N*-allyloxycarbonyl (Aloc) [80], *N*-trichloroethoxycarbonyl (Troc) [68, 81–83], *N*-trimethylsilylethylcarbonyl (Teoc) [75], or *N*-dithiasuccinoyl (Dts) [84, 85] in order to decrease side product (oxazoline) formation. The *N*-Troc protective group has been used with most success. The Lewis acid-promoted glycosylation with peracetylated *N*-Troc glucosamine provides the β-glycoside in 70 % yield (Fig. 5) (*see* Subheading 3.2.1). The subsequent removal of the Troc group by treatment with zinc in acetic acid and acetic acid anhydride (Fig. 5) provides the final glycosylated building block ready for Fmoc solid-phase peptide synthesis [68] (*see* Subheading 3.6.1).

3.6.1 Supplemental Protocols

N-Troc Protection [86]

1. Dissolve the sugar hydrochloride (1 eq) in dry pyridine (100 mL) and stir vigorously.
2. Add Troc-Cl (2.5 eq) dropwise at room temperature and stir vigorously for 1 h.
3. Add MeOH (3 mL), concentrate, and then purify by flash chromatography.

N-Troc Conversion to Acetamido Group [76]

1. Dissolve the glycosylated amino acid (1 eq) in Ac_2O (2.5 mL).
2. Add Zn powder (500 mg), and stir the mixture at room temperature until reaction is complete (TLC monitoring).
3. Filter the residue and concentrate in vacuo, then purify by flash chromatography.

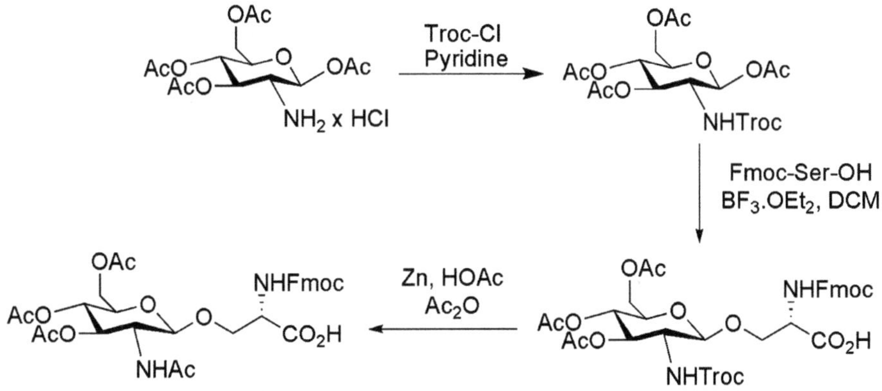
OAc
AcO
AcO
O
OAc
NH2 x HCl
Troc-Cl
Pyridine
NHTroc
Fmoc-Ser-OH
BF3.OEt2, DCM
NHFmoc
CO2H
Zn, HOAc
Ac2O
NHAc

Fig. 5 Synthesis of glycosylated Fmoc-Ser-OH building block containing 1-*O*-*β*-GlcNAc by the anomeric acetate method [68]

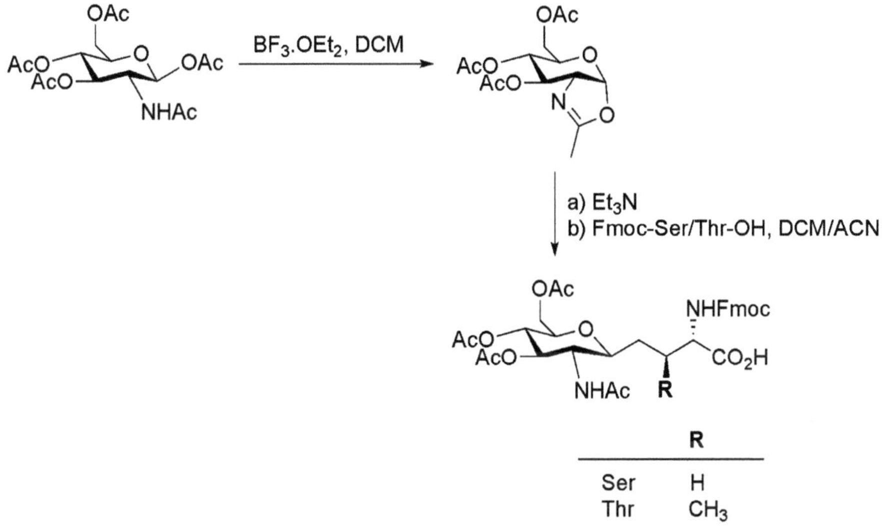
OAc
AcO
AcO
OAc
NHAc
BF3.OEt2, DCM
N
O
a) Et3N
b) Fmoc-Ser/Thr-OH, DCM/ACN
NHFmoc
CO2H
R
Ser H
Thr CH3

Fig. 6 Glycosylation of Fmoc-Ser/Thr-OH with peracetylated β-*N*-acetylglucosamine using oxazoline method [87]

The β-D-GlcNAc-(1→O)-Ser/Thr building blocks have been prepared from β-glucosamine pentaacetate [87] (Fig. 6) using oxazoline procedure in 55 % (serine) and 53 % (threonine) yields, respectively. The excess of glycosyl donor was used in order to avoid formation of stable oxazoline (*see* Subheading 3.6.2).

3.6.2 Supplement Protocol

Oxazoline Derivative Method [87]

1. Dissolve the peracetylated sugar (1 eq) in DCM (4 mL), containing 4 Å molecular sieves (*see* **Note 3**).
2. Cool solution to 0 °C, and add freshly distilled $BF_3{\cdot}Et_2O$ (3.1 eq) dropwise to the suspension.
3. Stir reaction overnight at room temperature until oxazoline formation can be seen on TLC.

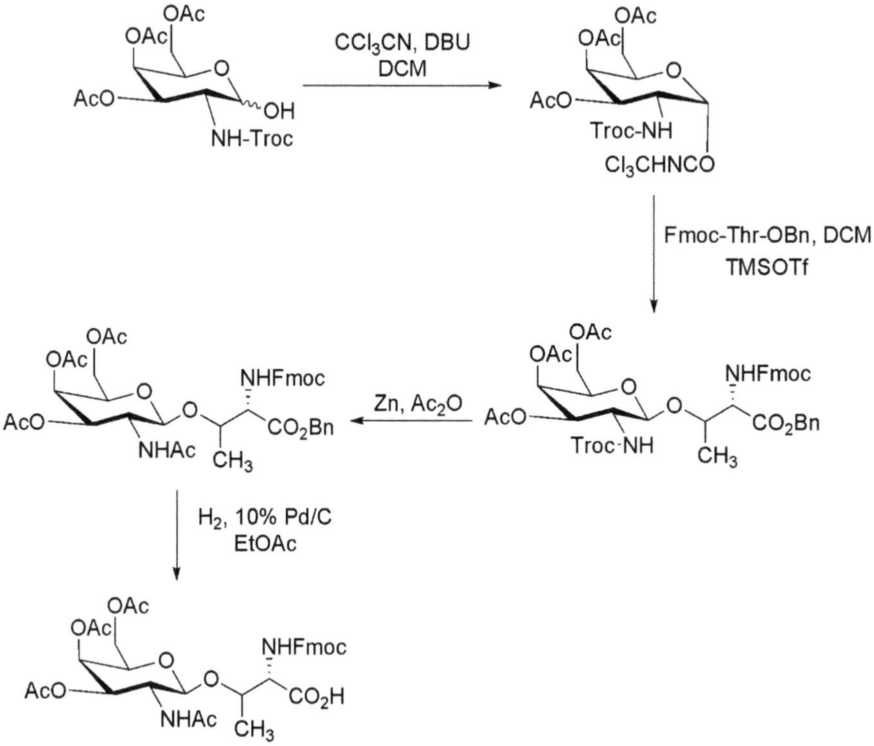

Fig. 7 Synthesis of glycosylated Fmoc-Thr-OH building block containing 1-*O*-β-GalNAc by the trichloroacetimidate method [75]

4. Cool reaction to 0 °C and add Et_3N (1 eq) dropwise.
5. Allow the reaction mixture to stir for 10 min, and add Fmoc-protected amino acid (1 eq) dissolved in a mixture of DCM:ACN (2:1).
6. Let reaction stand at room temperature and monitor by TLC.
7. After reaction is completed, filter through Celite, and concentrate the filtrate.
8. Purify the residue by column chromatography.

The building blocks in which GalNAc or GlcNAc is β-linked to threonine have been used in the synthesis of two antibacterial peptide analogs of formaecin I and drosocin in order to study the impact of differential glycosylation on their antibacterial activities [75]. Reaction of the glycosyl trichloroacetimidate with N^{α}-Fmoc-threonine benzyl ester in the presence of catalytic amounts of TMSOTf provided the β-linked glycoside in good yields. Subsequent replacement of the *N*-Troc group by the *N*-acetyl group with zinc in the presence of acetic anhydride, and finally benzyl ester reduction with Pd/C under H_2 atmosphere provided the desired product (Fig. 7). See the trichloroacetimidate glycosylation protocol (*see* Subheading 3.3.2), and supplemental protocols for the Troc group protection/deprotection strategy (*see* Subheading 3.6.1), and benzyl group deprotection (*see* Subheading 3.4.1) for experimental details.

3.7 Synthesis of α-D-GlcNAc/GalNAc-(1→O)-Ser/Thr

1,2-*cis*-*O*-Linked glycosylated amino acids, α-D-GalNAc-(1→O)-Ser/Thr, have been prepared by utilizing the nonparticipating azido group at C-2 of the GalNAc donor [88]. Often, 1-bromo [23, 89] and 1-chlorosugars [90–93] have been employed in Köenigs-Knorr type glycosylations of Fmoc-protected serine and threonine esters. The highest α-anomer selectivities were obtained when an insoluble promoter, such as silver perchlorate, was used. In order to obtain the desired α-D-GalNAc-(1→O)-Ser/Thr building block, the azido group must be reduced and subsequently acetylated, or reductive acetylation can be performed in a one-step procedure with thioacetic acid, before or after on-resin peptide assembly [23, 89, 90].

For the synthesis of N^{α}-Fmoc-Thr(Ac_3-α-D-GlcNAc)-OH (Fig. 8), per-*O*-acetyl 2-azido-2-deoxy-glucopyranosyl bromide was selected as the glycosyl donor containing the nonparticipating group at C-2. The glycosylation of N^{α}-Fmoc-threonine benzyl ester with the peracetylated 2-azido glycosyl bromide afforded the desired glycoside [75] (*see* Subheading 3.1.2). Reductive acetylation of the crude mixture (*see* Subheading 3.7.1), followed by deprotection of benzyl ester by hydrogenation (*see* Subheading 3.4.1) yielded the final product.

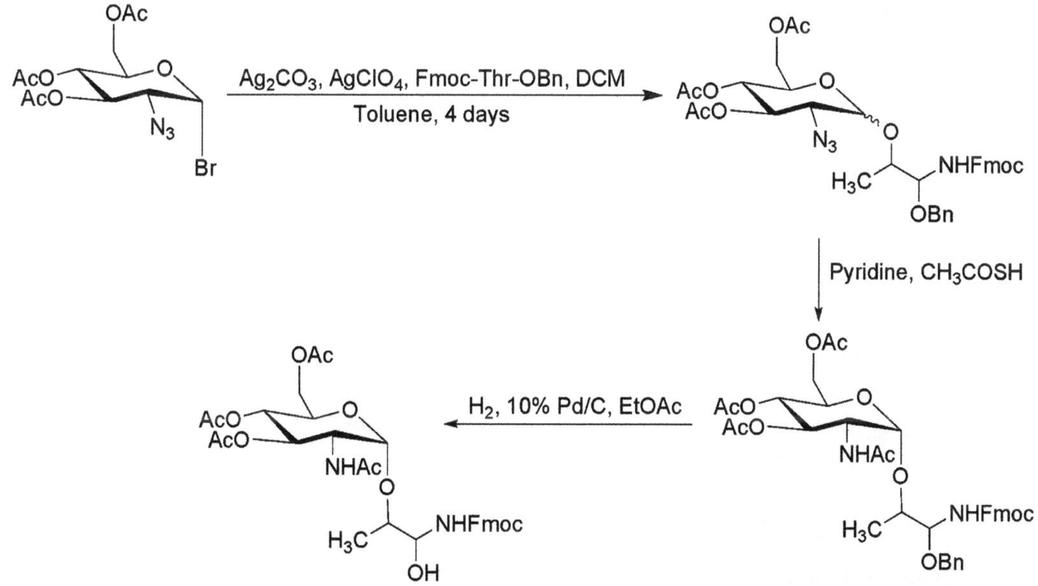

OAc
AcO
AcO
O
N3
Br
Ag_2CO_3, $AgClO_4$, Fmoc-Thr-OBn, DCM
Toluene, 4 days
OAc
AcO
AcO
O
N3
O
H_3C
NHFmoc
OBn
Pyridine, CH_3COSH
OAc
AcO
AcO
O
NHAc
O
H_3C
NHFmoc
OBn
H_2, 10% Pd/C, EtOAc
OAc
AcO
AcO
O
NHAc
O
H_3C
NHFmoc
OH

Fig. 8 Synthesis of glycosylated Fmoc-Thr-OH building block containing 1-*O*-α-GlcNAc by the Köenigs-Knorr method [75]

3.7.1 Supplemental Protocols

Reductive Acetylation [75]

1. Dissolve the glycoside (1 eq) in pyridine (10 mL).
2. Add CH_3COSH (5 mL) and stir the mixture for 4 h at room temperature.
3. Dilute the reaction mixture with toluene and concentrate under reduced pressure.
4. Purify the residue by flash chromatography.

The one-pot azidochlorination procedure, recently described by Plattner et al. [94], can be used to obtain 2-azido substituted glycosyl chlorides (*see* Subheading 3.7.2). This procedure is a significant improvement to the 5-step procedure described by Lemieux and Ratcliffe [63, 95], and offers a simple and efficient approach toward glycosyl donors with the nonparticipating azido group at C-2. Additionally, these 2-azido substituted glycosyl chlorides can be used without a purification step to prepare *O*-glycosylated Thr building blocks, under typical Köenigs-Knorr activation conditions (Fig. 9). The one-pot azidochlorination provides the α-anomer as the major product (85–90 %), with minimal amounts of the β-anomer detected [94]. This building block was incorporated into a mimic-type mucin 2 glycopeptide for conformational studies [96].

3.7.2 Supplemental Protocols

Azidochlorination [94]

1. Dissolve the acetylated sugar (1 eq) in ACN (8 mL).
2. Add $FeCl_3$ (0.8 eq), NaN_3 (1.1 eq), and H_2O_2 (1.1 eq, 33 % aq. solution) subsequently, and stir the mixture at −20 °C for 4–7 h until the starting material is consumed by TLC monitoring.

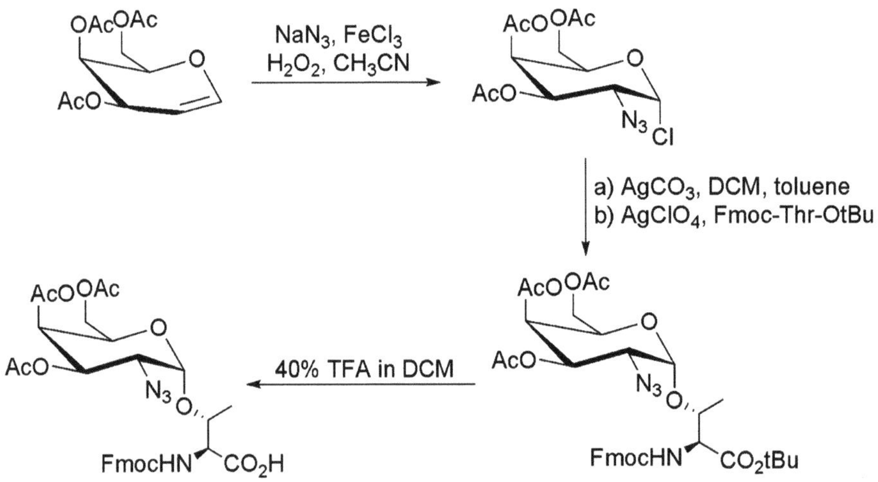

Fig. 9 Synthesis of *O*-linked 2-acetamido-2-deoxy-glycosyl Thr building block using one-pot azidochlorination procedure [94, 97]. (*Note: hydrolysis of the amino ester is an additional step added for the purposes of this chapter*)

3. Dilute the mixture with diethyl ether and wash with water, sat. $NaHCO_3$, and NaCl solution until the organic layer is colorless.
4. The crude product can be used without further purification after the workup.

Hydrolysis of Amino Ester [97]

1. Dissolve the glycosylated Fmoc-protected amino ester (1 eq) with 40 % TFA in DCM (10 mL) at room temperature for 1.5 h.
2. Evaporate the solvent in vacuo and recrystallize the product from cold petroleum ether.

The trichloroacetimidate method has been successfully used in the preparation of the building block in which GalNAc is α-linked to serine or threonine [43, 98]. The building block in which GalNAc is α-linked to threonine has been used in the synthesis of a glycosylated analog of the insect antimicrobial peptide drosocin [43], in glycosylated analogs of salivary mucin (MUC7) to study how *O*-glycosylation affects the peptide backbone [25, 99] and in N/OFQ glycopeptide analogs to study the effect of glycosylation on binding to the nociceptin receptor (NOP) [44].

3.8 Synthesis of α-D-Man-(1→O)-Ser/Thr

The synthesis of the α-D-mannosides is usually straightforward since both the participating acetyl groups and the anomeric effect favor the formation of the desired α-linkage. In Köenigs-Knorr coupling, peracetylated mannosyl bromide was reacted with protected amino acid acceptor, N^{α}-Fmoc-threonine benzyl ester, to yield nearly quantitative yields of the α-*O*-linked mannosyl threonine (Fig. 10) [75] (*see* Subheadings 3.1.2 and 3.4.1). The preparation of *O*-mannosyl serine and threonine building blocks was used in the synthesis of *O*-glycosylated analogs of formaecin I and drosocin containing 1-*O*-α-Man linked to threonine [75].

Alternatively, peracetylated mannose can be used to glycosylate N^{α}-Fmoc amino acids with an unprotected α-carboxyl group in the presence of Lewis acids as a promoter [68] (Fig. 11). Motiei et al. [42] used this method to synthesize cyclic mannosylated D,L-α-PEPTIDES analogs to study the glycosylation effect on membrane activity, biophysical properties, and *in vitro* antibacterial activities [42].

3.9 Synthesis of β-D-Glc/Gal-(1→O)-Tyr

Tyrosine *O*-glycosylation has been explored by Pedersen et al. [31] to study how varying glycosylation in the sequence of the peptide hormone PYY3-36 may affect receptor selectivity. The synthesis of the glycosylated N^{α}-Fmoc-Tyr(Ac_4-β-D-Glc)-OPfp and N^{α}-Fmoc-Tyr(Ac_4-β-D-Glc)-OAll has been achieved by the Köenigs-Knorr method with 64 and 68 % yields, respectively [101] (Fig. 12a) (*see* Subheading 3.1.2). Regardless of the reduced reactivity of the aromatic hydroxyl group of tyrosine compared to the aliphatic hydroxyl

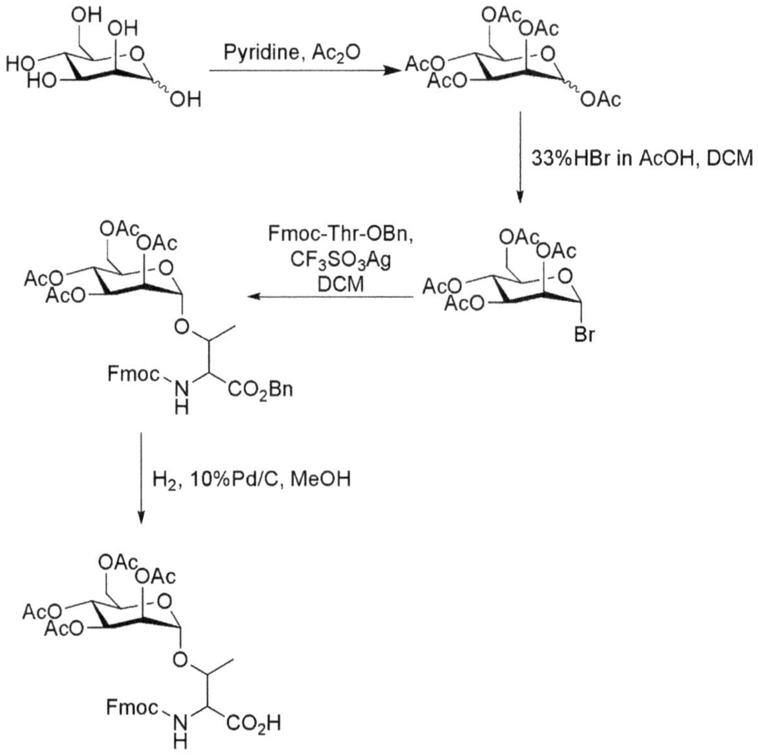

Fig. 10 Synthesis of glycosylated Fmoc-Thr-OH building blocks containing 1-*O*-α-Man by the classical Köenigs-Knorr method [75]

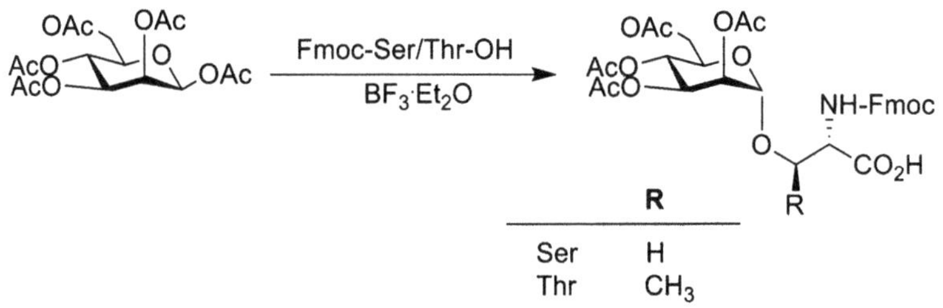

	R
Ser	H
Thr	CH_3

Fig. 11 Synthesis of glycosylated Fmoc-Ser-OH [68] and Fmoc-Thr-OH [100] building blocks containing 1-*O*-α-Man

groups of serine and threonine, C-terminally unprotected N^{α}-Fmoc-Tyr-OH was coupled to β-galactose/glucose pentaacetate in boron trifluoride etherate promoted glycosylation (Fig. 12b) in lower but still satisfying yields [68] (*see* Subheading 3.2.1).

3.10 Synthesis of β-D-Glc/Gal-(1→O)-Hyp

In structure–function studies, proline is sometimes substituted by hydroxyproline (Hyp) in order to examine the effect of glycosylation on physicochemical properties of glycosylated peptide analogs [102–104]. The anomeric acetate method is the most commonly used method in the synthesis of Fmoc-protected glycosylated Hyp

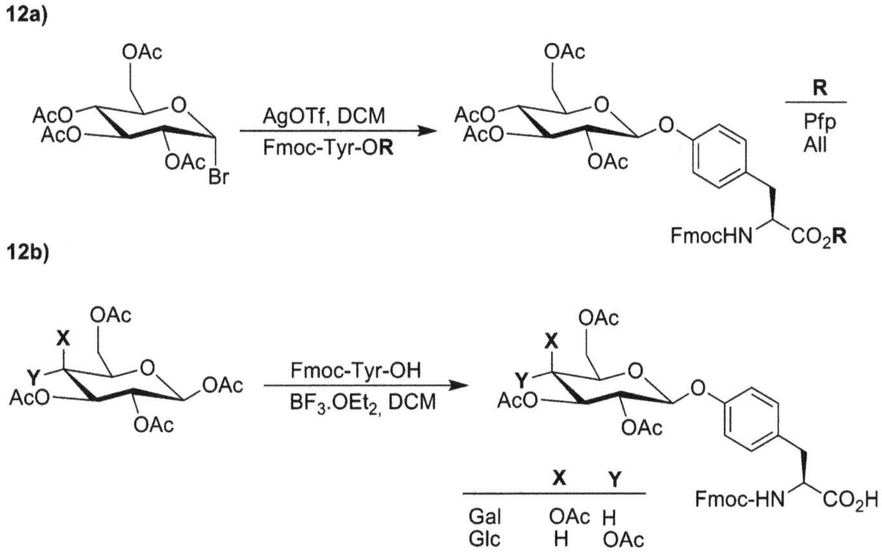

Fig. 12 Synthesis of glycosylated Fmoc-Tyr-OR building blocks containing 1-*O*-β-Glc/Gal by (**a**) the Köenigs-Knorr method (R = Pfp, All) [101], and (**b**) anomeric acetate method (R = OH) [68]

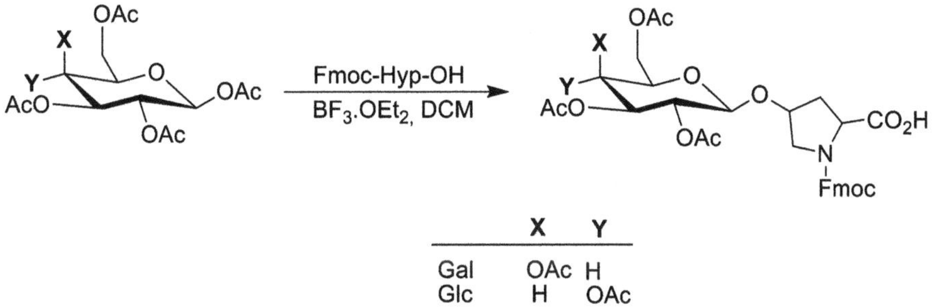

Fig. 13 Synthesis of β-O-linked glycosylated hydroxyproline building block by anomeric acetate method [68, 105]

building blocks for peptide synthesis, since the starting materials are commercially available and the yields of reaction are satisfactory [68, 105]. Glycosylation of Fmoc-Hyp-OH with β-galactose pentaacetate in the presence of Lewis acid gave galactosylated Hyp building block in 51 % yield (Fig. 13) [68, 105]. In a reaction with peracetylated glucose, β-O-linked glycosylated hydroxyproline building block was obtained in similar yield [105].

Synthesis of N-glycosylated amino acids. *N*-glycosidic linkage is commonly formed between a glycosyl amine and an activated aspartic acid derivative instead of glycosylation of asparagine. It is important to note that while this leads to an *N*-glycosidic linkage, the chemistry is an amide bond formation rather than a glycosidic bond formation [106]. Therefore, the first step toward the synthesis of *N*-glycosylated amino acid building blocks is the preparation of glycosyl amines. Two of the most commonly used procedures for the synthesis of glycosyl amines include the reduction of glycosyl azides and the treatment of reducing sugars with saturated ammonium hydrogen carbonate.

3.11 Preparation of Glycosyl Amines

Glycosylamines are accessible from different precursors; however, glycosyl halides are most often converted to their corresponding glycosyl azide [107–109] and then reduced to the glycosyl amines by several available approaches (Fig. 14) [108, 110–116].

3.11.1 Synthesis of Glycosylamine from Glycosyl Azides

1. Prepare glycosyl halide (*see* Subheading 3.1.1).
2. Dissolve glycosyl halide (1 eq) in DMF, add NaN_3 (1.5 eq), and stir overnight at 80 °C.
3. Dilute the reaction mixture with water, extract the glycosyl azide with CH_3Cl, and evaporate the solvent in vacuo.
4. Dissolve formed glycosyl azide in MeOH (containing 1 eq of HCl), add Pd-C (20 mol%) and stir at room temperature under a hydrogen atmosphere until reaction is complete by TLC monitoring (usually overnight) (*see* **Notes 13** and **17**).
5. Filter reaction mixture through Celite, wash with methanol, and concentrate in vacuo.
6. Purify product by flash chromatography.

Alternatively, glycosylamines can be synthesized via the introduction of the amino group at the reducing end of an unprotected glycosylamine in the presence of saturated ammonium bicarbonate solution [117, 118] (Fig. 15). Bejugam and Flitsch have improved this procedure by means of microwave-assisted radiation. The resulting glycosylamines are isolated in 80–90 % yields [119].

3.11.2 Synthesis of Glycosylamines from Reducing Sugars

1. Dissolve the reducing sugar in saturated aqueous NH_4HCO_3 (ca. 20 mL per 0.5 g of sugar) and stir for 6 days at 30 °C (*see* **Note 18**).
2. Dilute reaction mixture with water (10 mL) and concentrate in vacuo to half of its original volume in order to remove excess of NH_4HCO_3. Repeat this **steps 7–8** times (*see* **Note 19**).
3. Evaporate remaining water, and dry the product in a desiccator (*see* **Note 20**).

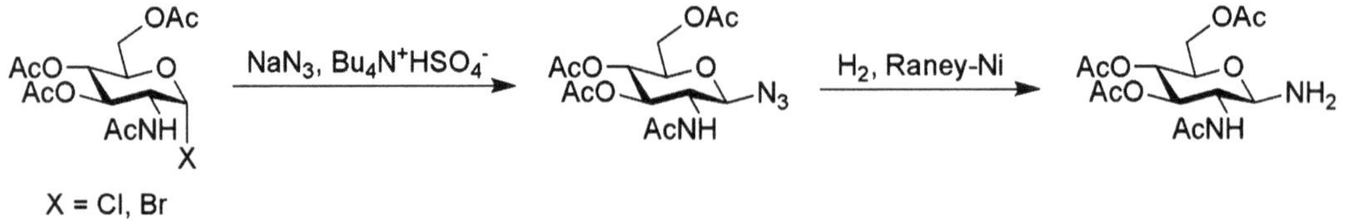

Fig. 14 Synthesis of glycosyl amine by reduction of glycosyl azide [109, 116]

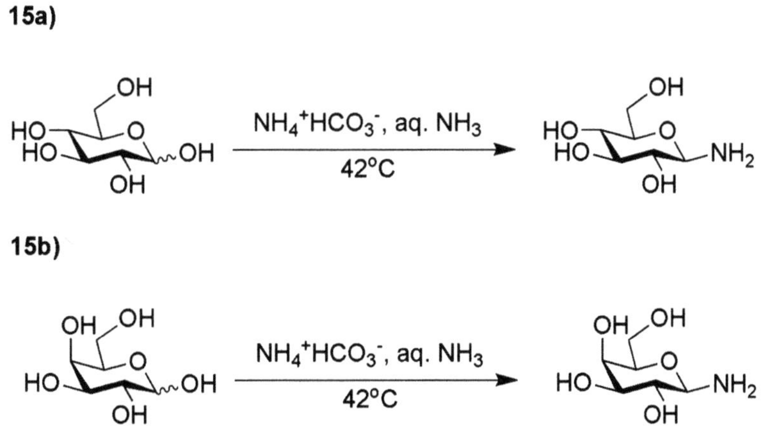

Fig. 15 Synthesis of (**a**) D-glucosylamine, and (**b**) D-galactosylamine by amination of their corresponding reducing sugar [118]

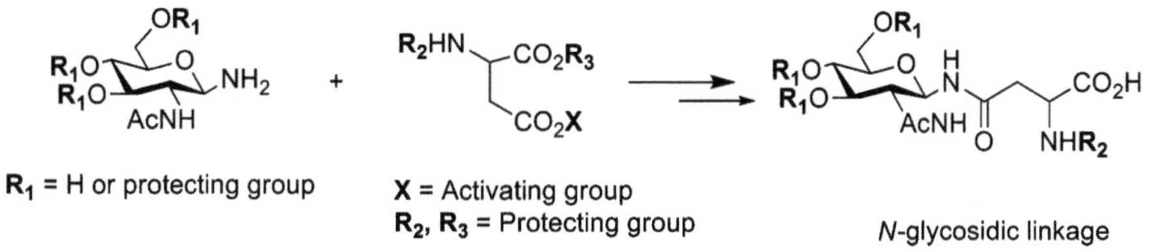

R_1 = H or protecting group

X = Activating group

R_2, R_3 = Protecting group

N-glycosidic linkage

Fig. 16 Formation of the *N*-glycosidic linkage

3.12 Formation of the N-Glycosidic Linkage

The formation of an amide bond between a glycosyl amine in its protected or unprotected form and an activated aspartic acid derivative (Fig. 16) is achieved by methods used for the formation of a peptide bond [120–123].

3.12.1 Formation of N-Glycoside Bond

1. Dissolve Fmoc-Asp-OtBu (1 eq) and pentafluorophenol (1 eq) in DMF (3 mL).
2. Add DCC or DIC (2 eq) and stir at room temperature for 30 min (*see* **Note 21**).
3. Dissolve glycosyl amine (1 eq) in DMF:H_2O (2:1 v/v, 2 mL), and add to the reaction mixture. Stir overnight.
4. Remove solvent in vacuo and wash the residue with cold diethyl ether and cold water.
5. Purify product by flash chromatography.

3.13 Synthesis of β-GlcNAc

A successful approach in the synthesis of *N*-acetyl glucosamine *N*-*β*-linked to Asn side chain involved the use of selectively protected aspartic acid derivatives as outlined in Fig. 17. *N*-*Z*-aspartic acid *tert*-butyl ester was reacted with peracetylated *N*-acetylglucosylamine to provide the fully protected glycosyl asparagine in the presence of

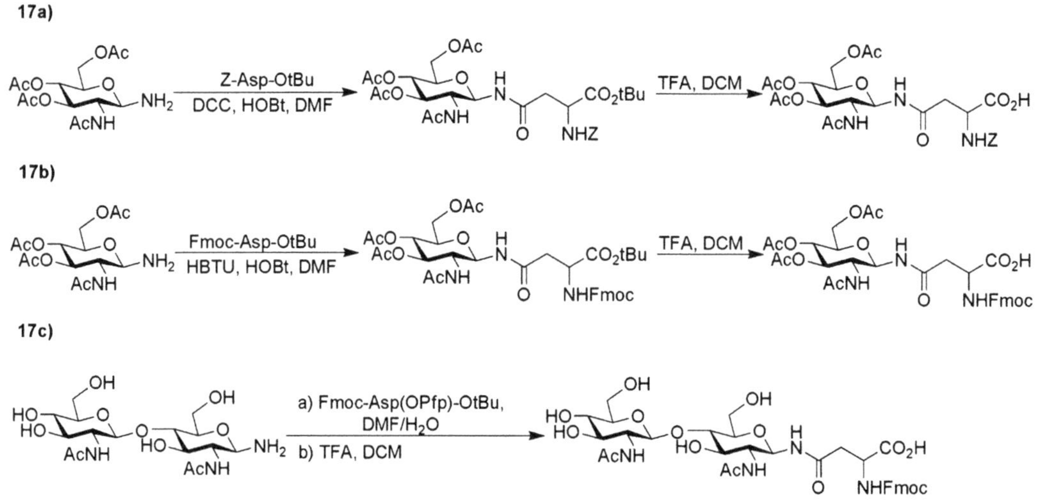

Fig. 17 Synthesis of β-*N*-linked asparagines glycosides: (**a**) DCC and HOBt coupling [120], (**b**) HBTU and HOBT coupling [119], and (**c**) pentafluorophenyl (OPfp) ester activation [121]

DCC/HOBt coupling reagents. Subsequent treatment with TFA gave the desired *N*-linked glycosyl asparagine (Fig. 17a) [120]. Alternative coupling procedures include the HBTU/HOBt method (Fig. 17b) [119] and pentafluorophenyl ester approach (Fig. 17c) [121].

This *N*-glycosylated building block has been used in the synthesis of glycopeptides to study the measles virus binding domains of the CD46 complement cofactor in humans, since structural and functional studies have shown that the core saccharides, in particular, the first two GlcNAc residues interact strongly with the polypeptide backbone, and show that glycosylation is essential for measles virus recognition [124]. However, *N*-glycosylation does not always improve the pharmacokinetics of a peptide, as was seen in a study with the immunogenic peptide OVA(323–339) [125]. The glycopeptides carrying a β-GlcNAc at positions 325 and 335 did not bind to the class II MHC molecule. In addition, the introduction of an *N*-glycosylated Asn residue can stabilize the overall structure of a peptide. Such is the case with the stabilization of the suggested type I1 β-turn conformation of a principal neutralizing determinant (PND) peptide analog [126]. In addition, a range of *N*-glycosylated Asn glycopeptides containing protease cleavage sites were synthesized and studied as substrates for the proteases chymotrypsin and thermolysin. In the case of chymotrypsin, *N*-glycosylation of an Asn residue with β-GlcNAc and β-Glc at the P_2 site appears to reduce hydrolysis whereas glycosylation of the P_1 site does not appear to affect peptide hydrolysis by thermolysin [127]. Also, *N*-glycosylation has been shown to improve and prolong *in vivo* activity of protein by serum half-life extension [128–130]. For example, erythropoietin has been mutated to incorporate two additional *N*-linked glycosylation sites, which resulted in a protein analog with a longer plasma half-life and increased in vivo activity [128, 129].

4 Notes

1. Some of the Fmoc-protected amino acid building blocks are commercially available and can be purchased from a limited number of sources, such as Bachem, Toronto Research Chemicals, V-Labs, and Sussex Research.
2. It is recommended to place the sample in a 4–0 °C environment overnight if crystallization does not occur at room temperature.
3. Reaction must be performed under argon atmosphere and in the presence of molecular sieves (4 Å). Solvent must be extra dry, containing less than 50 ppm of water.
4. The Köenigs-Knorr reaction will yield exclusively 1,2-*trans* glycosides, especially when insoluble silver salts are used as promoters.
5. In the case of GlcNAc or GalNAc, it is desirable to replace the *N*-acetyl with another electron-withdrawing group like *N*-allyloxycarbonyl (Aloc), *N*-trichloroethoxycarbonyl (Troc), or *N*-dithiasuccinoyl (Dts) in order to decrease side product (oxazoline) formation.
6. Both forms of the anomeric glycosyl trichloroacetimidate can be obtained in pure form, depending on the base (NaH, K_2CO_3 or DBU) used for the deprotonation of the reducing sugar.
7. When using NaH, always filter the reaction mixture over Celite before the evaporation step.
8. The high reactivity of the glycosyl trichloroacetimidate can lead to side reactions, or even decomposition of the donor before reacting with the acceptor. To improve yield and stereocontrol of the reaction, the so-called inverse glycosylation procedure is often used. In this case, the glycosyl acceptor and the catalyst are dissolved together, followed by the addition of the glycosyl donor.
9. The amount of Lewis acid employed varies from case to case. In many cases, a few drops of diluted solution of TMSOTf is sufficient to complete the reaction. In other cases, repeated additions of the catalyst are required. This might be particularly necessary when orthoesters, formed as intermediates, are to be isomerized to the respective glycosides.

10. The hydroxyl groups of a carbohydrate will react readily with acetic anhydride in the presence of an acidic or basic catalyst, such as sodium acetate or pyridine [67]. The acidic catalyst may be zinc chloride, hydrogen chloride, sulfuric acid, trifluoroacetic anhydride, perchloric acid, or an acidic ion-exchange resin. The complete acetylation of nonreducing sugars can be effected by any method. In the case of a reducing sugar, the anomeric configuration obtained depends on the catalyst used in acetylation and the temperature. Acetylation of α-D- or β-D-glucopyranose with acetic anhydride and pyridine at 0 °C occurs without appreciable anomerization. The formation of the acetylated β-D-anomer is favored at higher temperatures with sodium acetate as a catalyst.
11. This step is recommended since the α-anomer can be converted to the β-anomer.
12. Cleavage of the Fmoc group may occur during the hydrogenolysis. The use of EtOAc as solvent and adjustment of the amount of Pd/C will lead to selective removal of benzyl ester [131].
13. Caution: Hydrogen gas is highly flammable!
14. Store the used catalyst and filter aid in a properly labeled bottle. Cover the solids with water in order to prevent them from drying out.
15. This step is crucial in order to decompose excess benzoyl chloride [78].
16. During this time, precipitation should occur [78].
17. The use of Pd/C [108, 110], neutral Raney Ni (W-2) [112, 116], and Lyndlar's catalyst [111] had been reported in literature. Conditions used for the reduction of glycosyl azide have to be carefully optimized in order to suppress anomerization and the formation of undesired α-glycosides. Use of basic conditions and procedures that avoid noble metals lead to decreased α-glycosides formation [115].
18. Temperature, time of reaction, and final yields may vary.
19. Glycosyl amines are not very stable; therefore, the water bath temperature should not exceed 30 °C. Always prepare them freshly before further conversion to their respective glycopeptides.
20. The attempts to separate the starting reducing sugars from the glycosyl amines by using Amberlit 15 (H^+) ion-exchange resin was not successful with all sugars [117, 121].
21. DCC and DIC are known allergens and should be handled with care.

References

1. Filice M, Palomo JM (2012) Monosaccharide derivatives as central scaffolds in the synthesis of glycosylated drugs. RSC Adv 2:1729
2. Gracia SR, Gaus K, Sewald N (2009) Synthesis of chemically modified bioactive peptides: recent advances, challenges and developments for medicinal chemistry. Future Med Chem 1:1289–1310
3. Adessi C, Soto C (2002) Converting a peptide into a drug: strategies to improve stability and bioavailability. Curr Med Chem 9:963–978
4. Jensen KJ, Brask J (2005) Carbohydrates in peptide and protein design. Biopolymers 80:747–761
5. Otvos L (ed) (2008) Peptide-based drug design: methods and protocols, vol 494. Methods in molecular biology. Humana Press, Totowa, NJ, USA
6. Simerska P, Moyle PM, Toth I (2011) Modern lipid-, carbohydrate-, and peptide-based delivery systems for peptide, vaccine, and gene products. Med Res Rev 31:520–547
7. Sola RJ, Griebenow K (2010) Glycosylation of therapeutic proteins an effective strategy to optimize efficacy. BioDrugs 24:9–21
8. Solá RJ, Rodríguez-Martínez JA, Griebenow K (2007) Modulation of protein biophysical properties by chemical glycosylation: biochemical insights and biomedical implications. Cell Mol Life Sci 64:2133–2152
9. Byrne B, Donohoe GG, O'Kennedy R (2007) Sialic acids: carbohydrate moieties that influence the biological and physical properties of biopharmaceutical proteins and living cells. Drug Discov Today 12:319–326
10. Solá RJ, Griebenow K (2009) Effects of glycosylation on the stability of protein pharmaceuticals. J Pharm Sci 98:1223–1245
11. Tams JW, Vind J, Welinder KG (1999) Adapting protein solubility by glycosylation. *N*-Glycosylation mutants of Coprinus cinereus peroxidase in salt and organic solutions. Biochim Biophys Acta 1432:214–221
12. Solá RJ, Griebenow K (2006) Influence of modulated structural dynamics on the kinetics of alpha-chymotrypsin catalysis. FEBS J 273:5303–5319
13. Bagger HL, Fuglsang CC, Westh P (2005) Hydration of a glycoprotein: relative water affinity of peptide and glycan moieties. Eur Biophys J 35:367–371
14. Cheng S, Edwards SA, Jiang Y, Gräter F (2010) Glycosylation enhances peptide hydrophobic collapse by impairing solvation. Chem Phys Chem 11:2367–2374
15. Lawson EQ, Hedlund BE, Ericson ME, Mood DA, Litman GW, Middaugh R (1983) Effec of carbohydrate on protein stability. Arch Biochem Biophys 220:572–575
16. Wormald MR, Petrescu AJ, Pao Y, Glithero A, Elliott T, Dwek RA (2002) Conformational studies of oligosaccharides and glycopeptides: complementarity of NMR, X-ray crystallography, and molecular modeling. Chem Rev 102:371–386
17. Høiberg-Nielsen R, Westh P, Arleth L (2009) The effect of glycosylation on interparticle interactions and dimensions of native and denatured phytase. Biophys J 96:153–161
18. Otvos L Jr, Cudic M (2003) Conformation of glycopeptides. Mini-Rev Med Chem 3: 703–711
19. Tagashira M, Kazunori T (2005) Effect of peptide glycosylation on the conformation of the peptide backbone. Trends Glycosci Glyc 17:145–157
20. Bosques CJ, Tschampel SM, Woods RJ, Imperiali B (2004) Effects of glycosylation on peptide conformation: a synergistic experimental and computational study. J Am Chem Soc 126:8421–8425
21. Banna JG, Peyton DH, Bachinger HP (2000) Sweet is stable: glycosylation stabilizes collagen. FEBS Lett 473:237–240
22. Tagashira M, Iijima H, Isogai Y, Hori M, Takamatsu S, Fujibayashi Y, Yoshizawa-Kumagaye K, Isaka S, Nakajima K, Yamamoto T, Teshima T, Toma K (2001) Site-dependent effect of *O*-glycosylation on the conformation and biological activity of calcitonin. Biochemistry 40:11090–11095
23. Cudic M, Hildegund CJE, Otvos LJ (2002) Synthesis, conformation, and T-helper cell stimulation of an *O*-linked glycopeptide epitope containing extended carbohydrate side-chains. Bioorg Med Chem 10:3859–3870
24. Borgert A, Heimburg-Molinaro J, Song X, Lasanajak Y, Ju T, Liu M, Thompson P, Ragupathi G, Barany G, Smith DF, Cummings RD, Live D (2012) Deciphering structural elements of mucin glycoprotein recognition. ACS Chem Biol 7:1031–1039
25. Naganagowda GA, Gururaja TL, Satyanarayana J, Levine MJ (1999) NMR analysis of human salivary mucin (MUC_7) derived *O*-linked model glycopeptides: comparison of structural features and carbohydrate-peptide interactions. J Pept Res 54:290–310
26. Kindahl L, Sandström C, Craig AG, Norberg T, Kenne L (2002) ^1H NMR studies on the

solution conformation of contulakin-G and analogues. Can J Chem 80:1022–1031

27. Sinclair AM, Elliott S (2005) Glycoengineering: the effect of glycosylation on the properties of therapeutic proteins. J Pharm Sci 94:1626–1635
28. Disney MD, Hook DF, Namoto K, Seeberger PH, Seebach D (2005) *N*-linked glycosylated beta-peptides are resistant to degradation by glycoamidase A. Chem Biodiv 2:1624–1634
29. Ueda T, Tomita K, Notsu Y, Ito T, Fumoto M, Takakura T, Nagatome H, Takimoto A, Mihara S, Togame H (2009) Chemoenzymatic synthesis of glycosylated glucagon-like peptide 1: effect of glycosylation on proteolytic resistance and *in vivo* blood glucose-lowering activity. J Am Chem Soc 131:6237–6245
30. Huang W, Groothuys S, Heredia A, Kuijpers BHM, Rutjes FPJT, van Delft FL, Wang LF (2009) Enzymatic glycosylation of triazole-linked GlcNAc/Glc-peptides: synthesis, stability and anti-HIV activity of triazole-linked HIV-1 gp41 glycopeptide C34 analogues. ChemBioChem 10:1234–1242
31. Pedersen SL, Steentoft C, Vrang N, Jensen KJ (2010) Glyco-scan: varying glycosylation in the sequence of the peptide hormone PYY3-36 and its effect on receptor selectivity. ChemBioChem 11:366–374
32. Powell MF, Stewart T, Otvos L Jr, Urge L, Gaeta FCA, Sette A, Arrhenius T, Thomson D, Soda K, Colon SM (1993) Peptide stability in drug development. II. Effect of single amino acid substitution and glycosylation on peptide reactivity in human serum. Pharm Res 10:1268–1273
33. Pardridge WM (1999) Blood–brain barrier biology and methodology. J Neurovirol 5:556–569
34. Chen Y, Liu L (2012) Modern methods for delivery of drugs across the blood–brain barrier. Adv Drug Deliver Rev 64:640–665
35. Pardridge WM (2005) The blood–brain barrier: bottleneck in brain drug development. NeuroRx 2:3–14
36. Mercadante S, Arcuri E (2009) Delivery of opioid analgesics to the brain: the role of blood–brain barrier. Gene Ther Mol Biol 13:82–90
37. Egleton RD, Davis TP (2005) Development of neuropeptide drugs that cross the blood–brain barrier. NeuroRx 2:44–53
38. Witt KA, Gillespie TJ, Huber JD, Egleton RD, Davis TP (2001) Peptide drug modifications to enhance bioavailability and blood–brain barrier permeability. Peptides 22: 2329–2343
39. Masand G, Hanif K, Sen S, Ahsan A, Maiti S, Pasha S (2006) Synthesis, conformational and pharmacological studies of glycosylated chimeric peptides of Met-enkephalin and FMRFa. Brain Res Bull 68:329–334
40. Egleton RD, Mitchell SA, Huber JD, Palian MM, Polt R, Davis TP (2001) Improved blood–brain barrier penetration and enhanced analgesia of an opioid peptide by glycosylation. J Pharmacol Exp Ther 299:967–972
41. Hsieh YSY, Taleski D, Wilkinson BL, Wijeyewickrema LC, Adams TE, Pike RN, Payne RJ (2012) Effect of *O*-glycosylation and tyrosine sulfation of leech-derived peptides on binding and inhibitory activity against thrombin. Chem Commun 48:1547–1549
42. Motiei L, Rahimipour S, Thayer DA, Wong C, Ghadiri MR (2009) Antibacterial cyclic D,L-α-glycopeptides. Chem Commun 25: 3693–3695
43. Gobbo M, Biondi L, Filira F, Gennaro R, Banincasa M, Scolaro B, Rocchi R (2002) Antimicrobial peptides: synthesis and antibacterial activity of linear and cyclic drosocin and apidaecin 1b analogues. J Med Chem 45:4494–4504
44. Arsequell G, Rosa M, Mayato C, Dorta R, Gonzalez-Nunez V, Barreto-Valer K, Marcelo F, Calle LP, Vazquez JT, Rodriguez RE, Jimenez-Barbero J, Valencia G (2011) Synthesis, biological evaluation and structural characterization of novel glycopeptide analogues of nociceptin N/OFQ. Org Biomol Chem 9:6133–6142
45. Kovalszky I, Surmacz E, Scolaro L, Cassone M, Feria R, Sztodola A, Olah J, Hatfield MPD, Lovas S, Otvos L Jr (2010) Leptin-based glycopeptide induces weight loss and simultaneously restores fertility in animal models. Diabetes Obes Metab 12:393–402
46. Ueda T, Ito T, Tomita K, Togame H, Fumoto M, Asakura K, Oshima T, Nihimura S-I, Hanasaki K (2010) Identification of glycosylated exendin-4 analogue with prolonged blood glucose-lowering activity through glycosylation scanning substitution. Bioorg Med Chem Lett 20:4631–4634
47. Kihlberg J, Ahmab J (1995) Glycosylated peptide hormones: pharmacological properties and conformational studies of analogues of [1-desamino,8-D-arginine]vasopressin. J Med Chem 38:161–169
48. Kihlberg J, Elofsson M, Salvador LA (1997) Direct synthesis of glycosylated amino acids from carbohydrate peracetates and Fmoc amino acids: solid-phase synthesis of biomedicinally interesting glycopeptides. Method Enzymol 289:221–245

49. Meldal M, Bock K (1994) A general approach to the synthesis of *O*- and *N*-linked glycopeptides. Glycoconjugate J 11:59–63
50. Kunz H (1987) Synthesis of glycopeptides, partial structures of biological recognition components. Angew Chem 99:297–311
51. Mogemark M, Kihlberg J (2006) Glycopeptides. In: Levy DE, Fügedi P (eds) Organic chemistry of sugars. CRC Press, Taylor and Francis Group, Boca Raton, pp 755–801
52. Seitz O (2000) Glycopeptide synthesis and the effects of glycosylation on protein structure and activity. ChemBioChem 1:214–246
53. Taylor CM (1998) Glycopeptides and glycoproteins: focus on the glycosidic linkage. Tetrahedron Lett 54:11317–11362
54. Conroy T, Jolliffe KA, Payne RJ (2010) Synthesis of *N*-linked glycopeptides via solid-phase aspartylation. Org Biomol Chem 8:3723–3733
55. Bodanszky M, Natarajan S (1975) Side reactions in peptide-synthesis. 2. Formation of succinimide derivatives from aspartyl residues. J Org Chem 40:2495–2499
56. Hollosi M, Kollat E, Laczko I, Medzihradsky KF, Thurin J, Otvos LJ (1991) Solid-phase synthesis of glycopeptides: glycosylation of resin-bound serine-peptides by 3,4,6-tri-*O*-acetyl-D-glucose-oxazoline. Tetrahedron Lett 32:1531–1534
57. Andrews DM, Seale PW (1993) Solid-phase synthesis of *O*-mannosylated peptides: two strategies compared. Int J Pept Protein Res 42:165–170
58. Paulsen H, Schleyer A, Mathieux N, Meldal M, Bock K (1997) New solid-phase oligosaccharide synthesis on glycopeptides bound to a solid phase. J Chem Soc Perkin Trans 1 3:281–293
59. Meldal M (1994) Glycopeptide synthesis. In: Lee YC, Lee RT (eds) Neoglycoconjugates: preparation and applications. Academic, San Diego, pp 145–198
60. Kihlberg J, Elofsson M (1997) Solid-phase synthesis of glycopeptides: immunological studies with T cell simulating glycopeptides. Curr Med Chem 4:79–110
61. Herzner H, Reipen T, Schultz M, Kunz H (2000) Synthesis of glycopeptides containing carbohydrates and peptide recognition motifs. Chem Rev 100:4495–4537
62. Merrifield B (1997) Concept and early development of solid-phase peptide synthesis. Method Enzymol 289:3–13
63. Liu M, Young VG, Lohani S, Live D, Barany G (2005) Syntheses of TN building blocks Nα-(9-fluorenylmethoxycarbonyl)-*O*-(3,4,6-tri-O-acetyl-2-azido-2-deoxy-α-D-galactopyranosyl)-l-serine/l-threonine pentafluorophenyl esters: comparison of protocols and elucidation of side reactions. Carbohydr Res 340: 1273–1285
64. Koenig W, Geiger R (1970) New method for the synthesis of peptides: activation of the carboxyl group with dicyclohexylcarbodiimide by using 1-hydroxybenzotriazoles as additives. Chem Ber 103:788–798
65. Meldal M, Jensen KJ (1990) Pentafluorophenyl esters for the temporary protection of the α-carboxy group in solid phase glycopeptide synthesis. J Chem Soc Chem Commun 6:483–485
66. Lindhorst TK (2003) Essentials of carbohydrate chemistry and biochemistry, 2nd edn. Wiley-VCH Verlag GmbH & Co KGaA, Weinheim
67. Wolfrom ML, Thompson A (1963) Acetylation. Meth Carbohydr Chem 2:211–215
68. Salvador LA, Elofsson M, Kihlberg J (1995) Preparation of building blocks for glycopeptide synthesis by glycosylation of Fmoc amino acids having unprotected carboxyl groups. Tetrahedron 51:5643–5656
69. Schmidt RR, Kinzy W (1994) Anomeric-oxygen activation for glycoside synthesis: the trichloroacetimidate method. Adv Carbohydr Chem Biochem 50:21–123
70. Schmidt RR (1986) New methods for the synthesis of glycosides and oligosaccharides—are there alternatives to the Koenigs-Knorr method? Angew Chem 98:213–236
71. Broncel M, Falenski J, Wagner S, Hackenberger CP, Koksch B (2010) How post-translational modifications influence amyloid formation: a systematic study of phosphorylation and glycosylation in model peptides. Chem Eur J 16:7881–7888
72. Polt R, Szabo L, Treiberg J, Li Y, Hruby V (1992) General methods for alpha- or beta-*O*-Ser/Thr glycosides and glycopeptides. Solid-phase synthesis of O-glycosyl cyclic enkephalin analogues. J Am Chem Soc 114: 10249–10258
73. Bilsky E, Egleton R, Mitchell S, Palian M, Davis P, Huber J, Jones H, Yamamura H, Janders J, Davis T, Porreca F, Hruby V, Polt R (2000) Enkephalin glycopeptide analogues produce analgesia with reduced dependence liability. J Med Chem 43:2586–2590
74. Mitchell SA, Pratt MR, Hruby VJ, Polt R (2001) Solid-phase synthesis of *O*-linked glycopeptide analogues of enkephalins. J Org Chem 66:2327–2342
75. Talat S, Thiruvikraman M, Kumari S, Kaur KJ (2011) Glycosylated analogs of formaecin I

and drosocin exhibit differential pattern of antibacterial activity. Glycoconjugate J 28: 537–555

76. Dullenkopf W, Castro-Palomino JC, Manzoni L, Schmidt RR (1996) *N*-trichloroethoxycarbonyl-glucosamine derivatives as glycosyl donors. Carbohydr Res 296:135–147
77. Filira F, Biondi L, Cavaggion F, Scolaro B, Rocchi R (1990) Synthesis of *O*-glycosylated tuftsins by utilizing threonine derivatives containing an unprotected monosaccharide moiety. Int J Pept Prot Res 36:86–90
78. Brimble MA, Kowalczyk R, Harris PWR, Dunbar PR, Muir VJ (2008) Synthesis of fluorescein-labelled *O*-mannosylated peptides as components for synthetic vaccines: comparison of two synthetic strategies. Org Biomol Chem 6:112–121
79. Jones JKN, Perry MB, Shelton B, Walton DT (1961) Carbohydrate-protein linkage in glycoproteins. I. Syntheses of some model substituted amides and a (2-amino-2-deoxy-D-glucosyl)-L-serine. Can J Chem 39:1005–1016
80. Vargas-Berenguel A, Meldal M, Paulsen H, Bock K (1994) Convenient synthesis of *O*-(2-acetamido-2-deoxy-β-D-glucopyranosyl)-serine and -threonine building blocks for solid-phase glycopeptide assembly. J Am Chem Soc Perkin Trans 1 18:2615–2619
81. Schultz M, Kunz H (1992) Enzymatic glycosylation of *O*-glycopeptides. Tetrahedron Lett 33:5319–5322
82. Meinjohanns E, Meldal M, Bock K (1995) Efficient synthesis of *O*-(2-acetamido-2-deoxy-β-D-glucopyranosyl)-Ser/Thr building blocks for SPPS of *O*-GlcNAc glycopeptides. Tetrahedron Lett 36:9205–9208
83. Saha U, Schmidt R (1997) Efficient synthesis of *O*-(2-acetamido-2-deoxy-β-D-glucopyranosyl) -serine and -threonine building blocks for glycopeptide formation. J Chem Soc Perkin Trans 1 12:1855–1860
84. Meinjohanns E, Vargas-Berenguel A, Meldal M, Paulsen H, Bock K (1995) Comparison of *N*-Dts and *N*-Aloc in the solid-phase synthesis of *O*-GlcNAc glycopeptide fragments of RNA-polymerase II and mammalian neurofilaments. J Chem Soc Perkin Trans 1 17:2165–2175
85. Jensen KJ, Hansen PR, Venugopal D, Barany G (1996) Synthesis of 2-acetamido-2-deoxy-β-D-glucopyranose *O*-glycopeptides from *N-dithiasuccinoyl-protected* derivatives. J Am Chem Soc 118:3148–3155
86. Ellervik U, Magnusson G (1996) Glycosylation with *N*-Troc-protected glycosyl donors. Carbohydr Res 280:251–260
87. Arsequell G, Krippner L, Dwek RA, Wong SYC (1994) Building blocks for solid-phase glycopeptide synthesis: 2-acetamido-2-deoxy-β-D-glycosides of FmocSerOH and FmocThrOH. J Chem Soc Chem Commun 20:2383–2384
88. Norberg T, Luning B, Tejbrant J (1994) Solid-phase synthesis of *O*-glycopeptides. Method Enzymol 247:87–106
89. Kuduk SD, Schwarz JB, Chen X-T, Glunz PW, Sames D, Ragupathi G, Livingston PO, Danishefsky SJ (1998) Synthetic and immunological studies on clustered modes of mucin-related Tn and TF *O*-linked antigens: the preparation of a glycopeptide-based vaccine for clinical trials against prostate cancer. J Am Chem Soc 120:12474–12485
90. Vuljanic T, Bergquist KE, Clausen H, Roy S, Kihlberg J (1996) Piperidine is preferred to morpholine for Fmoc cleavage in solid phase glycopeptide synthesis as exemplified by preparation of glycopeptides related to HIV gp120 and mucins. Tetrahedron 52:7983–8000
91. Kunz H, Birnbach S (1986) Synthesis of *O*-glycopeptides of the tumor-associated Tn- and T-antigen type and their binding to bovine serum albumin. Angew Chem Int Ed 98:360–362
92. Paulsen H, Adermann K (1989) Synthesis of *O*-glycopeptides of the *N*-terminus of interleukin-2. Liebigs Ann Chem 8:751–769
93. Liebe B, Kunz H (1997) Solid-phase synthesis of a tumor-associated sialyl-TN antigen glycopeptide with a partial sequence of the "tandem repeat" of the MUC-1 mucin. Angew Chem Int Ed 36:618–621
94. Plattner C, Hoefener M, Sewald N (2011) One-pot azidochlorination of glycals. Org Lett 13:545–547
95. Lemieux RU, Ratcliffe RM (1979) The azidonitration of tri-*O*-acetyl-d-galactal. Can J Chem 57:1244–1251
96. Liu M, Barany G, Live D (2005) Parallel solid-phase synthesis of mucin-like glycopeptides. Carbohydr Res 340:2111–2122
97. Papini AM, Nardi E, Nuti F, Uziel J, Ginanneschi M, Chelli M, Brandi A (2002) Diastereoselective alkylation of schiff bases for the synthesis of lipidic unnatural Fmoc-protected α-amino acids. Eur J Org Chem 16:2736–2741
98. Biondi L, Filira F, Gobbo M, Scolaro B, Rocchi R, Cavaggion F (1991) Synthesis of glycosylated tuftsins and tuftsin-containing IgG fragment undecapeptide. Int J Pept Prot Res 37:112–121
99. Satyanarayana J, Gururaja TL, Naganagowda GA, Ramasubbu N, Levine MJ (1998) A concise methodology for the stereoselective synthesis of *O*-glycosylated amino acid building blocks: complete ^{1}H NMR assignments and

their application in solid-phase glycopeptide synthesis. J Pept Res 52:165–179

100. Kragol G, Otvos LJ (2001) Orthogonal solid-phase synthesis of tetramannosylated peptide constructs carrying three independent branched epitopes. Tetrahedron 57:957–966
101. Jensen KJ, Meldal M, Bock K (1993) Glycosylation of phenols: preparation of 1,2-cis and 1,2-trans glycosylated tyrosine derivatives to be used in solid-phase glycopeptide synthesis. J Chem Soc Perkin Trans 1 17:2119–2129
102. Rodriguez RE, Rodriguez FD, Sacristan MP, Torres JL, Valencia G, Garcia Anton JM (1989) New glycosylpeptides with high antinociceptive activity. Neurosci Lett 101:89–94
103. Bardaji E, Torres JL, Clapes P, Albericio F, Barany G, Valencia G (1990) Solid-phase synthesis of glycopeptidamides under mild conditions: morphiceptin analogs. Angew Chem 102:311–313
104. Owens NW, Stetefeld J, Lattova′ E, Schweizer F (2010) Contiguous *O*-galactosylation of 4(R)-hydroxy-L-proline residues forms very stable polyproline II helices. J Am Chem Soc 132:5036–5042
105. Arsequell G, Shrries N, Valencia G (1995) Synthesis of glycosylated hydroxyproline building blocks. Tetrahedron Lett 36:7323–7326
106. Arsequell G, Valencia G (1999) Recent advances in the synthesis of complex *N*-glycopeptides. Tetrahedron-Assymetr 10:3045–3094
107. Kunz H, Waldman H, Marz J (1989) Synthesis of partial structures of *N*-glycopeptides representing the linkage regions of the transmembrane neuraminidase of an influenza virus and of factor B of the human complement system. Liebigs Ann Chem 1:45–49
108. Thiem J, Wiemann T (1990) Combined chemoenzymic structure of *N*-glycoprotein synthons. Angew Chem 102:78–80
109. Tropper FD, Andersson FO, Braun S, Roy R (1992) Phase transfer catalysis as a general and stereoselective entry into glycosyl azides from glycosyl halides. Synthesis 7:618–620
110. Marks GS, Neuberger A (1961) Synthetic studies relating to the carbohydrate–protein linkage in egg albumin. J Chem Soc 4872–4879
111. Nakabayashi S, Warren CD, Jeanloz RW (1988) The preparation of a partially protected heptasaccharide-asparagine intermediate for glycopeptide synthesis. Carbohydr Res 174:279–289
112. McDonald FE, Danishefsky SJ (1992) A stereoselective route from glycals to asparagine-linked *N*-protected glycopeptides. J Org Chem 57:7001–7002
113. von dem Bruch K, Kunz H (1994) Synthesis of *N*-glycopeptide clusters with Lewis antigen side chains and their binding of carrier proteins. J Org Chem 33:101–103
114. Saha UK, Roy R (1995) First synthesis of *N*-glycopeptoid as new glycopeptidomimetics. Tetrahedron Lett 36:3635–3638
115. Unverzagt C (1996) Chemoenzymic preparation of a sialylated undecasaccharide—asparagine conjugate. Angew Chem Int Ed Engl 35:2350–2353
116. Kunz H, Unverzagt C (1988) Protecting group dependant stability of intersaccharide bonds. Synthesis of a fucosyl-chitobiose glucopeptides. Angew Chem 100:1763–1765
117. Likhosherstov LM, Novikova OS, Derevitskaja VA, Kochetkov NK (1986) A new simple synthesis of amino sugar β-D-glycosylamines. Carbohydr Res 146:1–5
118. Lubineau A, Auge J, Drouillat B (1995) Improved synthesis of glycosylamines and a straightforward preparation of *N*-acylglycosylamines as carbohydrate-based detergents. Carbohydr Res 266:211–219
119. Bejugam M, Flitsch SL (2004) An efficient synthetic route to glycoamino acid building blocks for glycopeptide synthesis. Org Lett 6:4001–4004
120. Clark RS, Banerjee S, Coward JK (1990) Yeast oligosaccharyltransferase: glycosylation of peptide substrates and chemical characterization of the glycopeptide product. J Org Chem 55:6275–6285
121. Urge L, Kollat E, Hollosi M, Laczko I, Wroblewski K, Thurin J, Otvos LJ (1991) Solid-phase synthesis of glycopeptides: synthesis of *N-α-fluorenylmethoxycarbonyl* L-asparagine *N*-β-glycosides. Tetrahedron Lett 32:3445–3448
122. Anisfeld ST, Lansbury PT Jr (1990) A convergent approach to the chemical synthesis of asparagine-linked glycopeptides. J Org Chem 55:5560–5562
123. Cohen-Anisfeld ST, Lansbury PT Jr (1993) A practical, convergent method for glycopeptide synthesis. J Am Chem Soc 115: 10531–10537
124. Casasnovas JM, Larvie M, Stehle T (1999) Crystal structure of two CD46 domains reveals an extended measles virus-binding surface. EMBO J 18:2911–2922
125. Kihlberg JE, M. (1997) Solid-phase synthesis of glycopeptides: immunological studies with T cell stimulating glycopeptides. Curr Med Chem 4:85–116
126. Laczko I, Hollosi M, Urge L, Ugen K, Weiner DB, Mantsch HH, Thurin J, Otvos L Jr

(1992) Synthesis and conformational studies of *N*-glycosylated analogues of the HIV- 1 principal neutralizing determinant. Biochemistry 31:4282–4288
127. Bejugam M, Maltman BA, Flitsch SL (2005) Synthesis of *N*-linked glycopeptides on solid support and their evaluation as protease substrates. Tetrahedron-Asymmetr 16:21–24
128. Elliott S, Lorenzini T, Asher S, Aoki K, Brankow D, Buck L, Busse L, Chang D, Fuller J, Grant J, Hernday N, Hokum M, Hu S, Knudten A, Levin N, Komorowski R, Martin F, Navarro R, Osslund T, Rogers G, Rogers N, Trail G, Egrie J (2003) Enhancement of therapeutic protein *in vivo* activities through glycoengineering. Nat Biotechnol 21:414–421
129. Bretthauer RK (2003) Genetic engineering of *Pichia pastoris* to humanize *N*-glycosylation of proteins. Trends Biotechnol 21:459–462
130. Ceaglio N, Etcheverrigaray M, Kratje R, Oggero M (2008) Novel long-lasting interferon alpha derivatives designed by glycoengineering. Biochimie 90:437–449
131. Broddefalk J, Backlund J, Almqvist F, Johansson M, Holmdahl R, Kihlberg J (1998) T cells recognize a glycopeptide derived from type II collagen in a model for rheumatoid arthritis. J Am Chem Soc 120: 7676–7683

Chapter 9

The Maillard Reaction Induced Modifications of Endogenous Opioid Peptide Enkephalin

Andreja Jakas

Abstract

Nonenzymatic glycation (Maillard reaction) is a posttranslational modification of peptides and proteins by sugars, which, after a cascade of reactions, leads to the formation of a complex family of irreversibly changed advanced glycation end products (AGE) implicated in the pathogenesis of human diseases. Last reversible intermediates of this reaction are Amadori/Heyns compounds formed in glucose/fructose induced modification of peptides. The stability of these compounds determines the further course of the reaction.

To provide information concerning the preparation of model systems as well as the fate of glycated opioid peptides introduced in the human circulation, the enzymatic (80 % human serum) and chemical (PBS) stability of Amadori and Heyns compounds related to the endogenous opioid pentapeptides leucine- and methionine-enkephalin (Tyr-Gly-Gly-Phe-Leu/Met) were investigated.

Key words Endogenous opioid peptides, Leucine-enkephalin, Methionine-enkephalin, Maillard reaction, Amadori compounds, Heyns compounds, Glucose, Fructose

1 Introduction

The opioid alkaloids such as morphine isolated from the poppy plant extract have been used for a centuries for relief of severe pain. As well as having analgesic and euphoric properties, opiates have undesirable side-effects, such as tolerance, physical dependence, respiratory depression, hypotension, and diverse gastrointestinal effects. Discovery of the opioid receptors about 35 years ago put some light to the function of endogenous opioid-like compounds [1]. Approximately at the same time the first endogenous opioid peptides [Leu^5]- and [Met^5]-enkephalin were found [2]. Soon afterwards the whole new field of endogenous opioid peptides was discovered. These peptides are involved in variety of central and peripheral effects, such as stress, tolerance, dependence, nociception, gastrointestinal, renal and hepatic functions, cardiovascular and immunological responses, respiration and thermoregulation,

Predrag Cudic (ed.), *Peptide Modifications to Increase Metabolic Stability and Activity*, Methods in Molecular Biology, vol. 1081, DOI 10.1007/978-1-62703-652-8_9,

and neurological disorders [3]. It is well established that the functions of opioid peptides are mediated by a family of G-protein coupled receptors, designated μ, δ, and κ opioid receptors [4]. More extensive studies have suggested the possible existence of the other less well-characterized opioid receptors ε [5], λ [6], τ [7] and ζ [8].

Opioid peptides have been classified into two groups; "typical" and "atypical" opioid peptides [9]. The "typical" opioid peptides, including enkephalins, dynorphins, and β-endorphin, are derived from three different precursor proteins: pro-enkephalin (PENK), pro-dynorphin (PDYN), and pro-opiomelanocortin (POMC), and all have the tetrapeptide sequence Tyr-Gly-Gly-Phe at their *N*-terminus. None of these peptides binds exclusively to only one opioid receptor type. Endomorphins as a product of an as yet unidentified precursor can be classified as "atypical" opioids, which are characterized by a tetrapeptide sequence, Pro residue in the second position and amidated *C*-terminus. They bind to the μ-opioid receptor with high affinity and selectivity [10]. The structure of most naturally occurring peptide opioids can be divided into two components: the biologically important *N*-terminal tri- or tetrapeptide fragment named "message sequence," and the remaining *C*-terminal fragment named "address sequence" [11]. The "message sequence" provides information for signal transduction that leads to the biological response, whereas the "address domain" primarily influences binding affinities and accommodates the elements of selectivity. The *N*-terminal message sequence in typical opioid peptides is composed of two pharmacophoric amino acid residues, Tyr and Phe, in which the amino and phenolic groups of Tyr and the aromatic ring of Phe are required for opioid receptor recognition.

In spite of potent biological activity, reasons for their limited use are primarily based on physiological and biochemical nature of the endogenous opioid peptides themselves. The chemical adaptation and design considerations are necessary to produce viable acting peptide drugs. The use of conformational constraints, bond replacements, and other modifications must be weighed against stability, receptor affinity, membrane permeability, elimination, and various independent attributes of the human body (i.e., age, sex, pathology, etc). The goal is to prolong and/or enhance the biological activity of the peptide drug at the target site. This can be accomplished in several manners: increasing bioavailability, reducing elimination, altering biodegradation, increasing the time in the "receptor-compartment," and increasing selectivity and/or affinity of the peptide for the receptor [12].

The nonenzymatic reaction between reducing sugars, such as glucose, and reactive amino groups in proteins, lipids, and nucleic acids, known as the Maillard reaction, has received considerable attention in recent years, both in food and health science. The early stage is marked with the formation of a Schiff base which, through rearrangements, gives rise to an Amadori product (ketoamine) if the reducing sugar is an aldose or Heyns product if the reducing sugar is a ketose. The latter is then degraded to a variety of carbonyl compounds, being much more reactive then the original sugar. The main degradation pathway of Amadori products is through dehydration of the sugar moiety to form deoxyglucosones. In the late stage, these highly reactive carbonyl compounds lead to the formation of a group of substances known as advanced glycation end products or AGEs, often colored, fluorescent and prone to produce cross-links in proteins. These compounds include mutagens, carcinogens as well as beneficial products having antimicrobial and/or antioxidant activity.

The glycation, as a slow process, usually attacks structural proteins, but recent attention has been devoted to possible glycation and impaired function of relatively short-lived important peptide hormones. As a result of the increased understanding of the important roles that endogenous as well as exogenous peptides play in fundamental life-sustaining processes, and in an effort to better understand the reactivity of glycated products and the mechanism of their further reactions the preparation of model system is necessary.

The role of glucose as the most abundant sugar has received preferential attention in these reactions, but other sugars also deserve consideration. For example, the intake of the keto-sugar fructose, as "healthy sugar," has greatly increased during the past years. This led to increased interest in the Maillard reaction with fructose [13–17]. Heyns compounds as a products of reaction of ketoses and amines, have been detected in liver extracts and in human ocular lens proteins in measurable amounts, suggesting that they have more important role than realized so far [18].

2 Materials

1. All materials and reagents are commercially available and used as received.
2. Synthesis and analysis solvents, such as methanol (MeOH), acetic acid (AcOH), trifluoroacetic acid (TFA), phosphate buffer (PBS), DMSO-d_6, D_2O were synthesized as per analysis grade and can be obtained from Sigma-Aldrich Chemie GmbH (Taufkirchen, Germany), Fisher Scientific (Laughborough, UK) or other commercial sources.
3. Carbohydrates D-glucose and D-fructose for Amadori and Heyns compounds synthesis may be obtained from Sigma-Aldrich Chemie GmbH (Taufkirchen, Germany).
4. Peptides [Leu^5]- and [Met^5]-enkepahlin for the synthesis may be obtained from Bachem AG (Bubendorf, Switzerland).

5. Amadori and Heyns compounds synthesis reagents such as *N*-ethylmorpholine (NEM), KOH, glycerol, *o*-hydroxyphenyl acetic acid, NaN_3, $Na_2S_2O_5$ may be obtained from Sigma-Aldrich Chemie (GmbH, Taufkirchen, Germany).
6. Human serum for stability analysis can be obtained from Sigma-Aldrich Chemie GmbH (Taufkirchen, Germany) or other commercial sources.
7. Sephadex-G15 resin for purification of prepared compounds can be obtained from Sigma-Aldrich Chemie GmbH (Taufkirchen, Germany).
8. 0.1 % TFA solution: Add 0.67 mL TFA to a 1 L graduated cylinder containing ~500 mL water, and fill the cylinder up to 1 L with water.
9. 40 % MeOH/0.1 %TFA solution: Add 506 mL to a 1 L graduated cylinder then fill it up to 1 L with 0.1 % TFA solution.
10. 1 % Aqueous acetic acid: Add 1 mL of acetic acid to a 100 mL graduated cylinder and fill it up to 100 mL with water.
11. 48 % Aqueous TFA: Add 3.2 mL of TFA to a 10 mL graduated cylinder and fill it up to 10 mL with water.
12. 0.02 % NaN_3 in PBS: Add 20 mg of NaN_3 to a 100 mL graduated cylinder and fill it up to 100 mL with PBS.
13. 80 % Human serum: Add 8 mL of human serum to a teflon-lined screw-cap tube containing 2 mL of water.

3 Methods

3.1 Synthesis and Properties of Amadori Compounds Related to the Opioid Peptides

The main glycation products of the first step of Maillard reaction are Amadori (1-deoxy-D-fructos-1-yl)-Tyr-Gly-Gly-Phe-Leu/Met and imidazolidinone compounds in reaction with D-glucose in methanol and in reaction with D-fructose as main glycation products gave Heyns compounds, *N*-(2-deoxy-D-glucopyranos-2-yl) and *N*-(2-deoxy-D-mannopyranos-2-yl)-Tyr-Gly-Gly-Phe-Leu/Met and imidazolidinone compounds for both peptide studied (Fig. 1). While it is known that formation of Amadori compounds in the reaction with glucose in methanol does not require base or acid initiation, a base is necessary for the formation of imidazolidinone [19, 20] as well as Heyns compounds [20]. In order to generate imidazolidinone compounds, in the glucose glycation reaction *N*-ethylmorpholine (NEM) was used as a base, in the molar ratio sugar-peptide-base 15:1:5 (Subheading 3.2.2). Typically, two bases were used as catalysts with methanol as a solvent in the reaction of enekphalins and D-fructose: the inorganic base, KOH (Subheading 3.3.1), and the organic base, NEM (Subheading 3.3.2). The best results were obtained in the

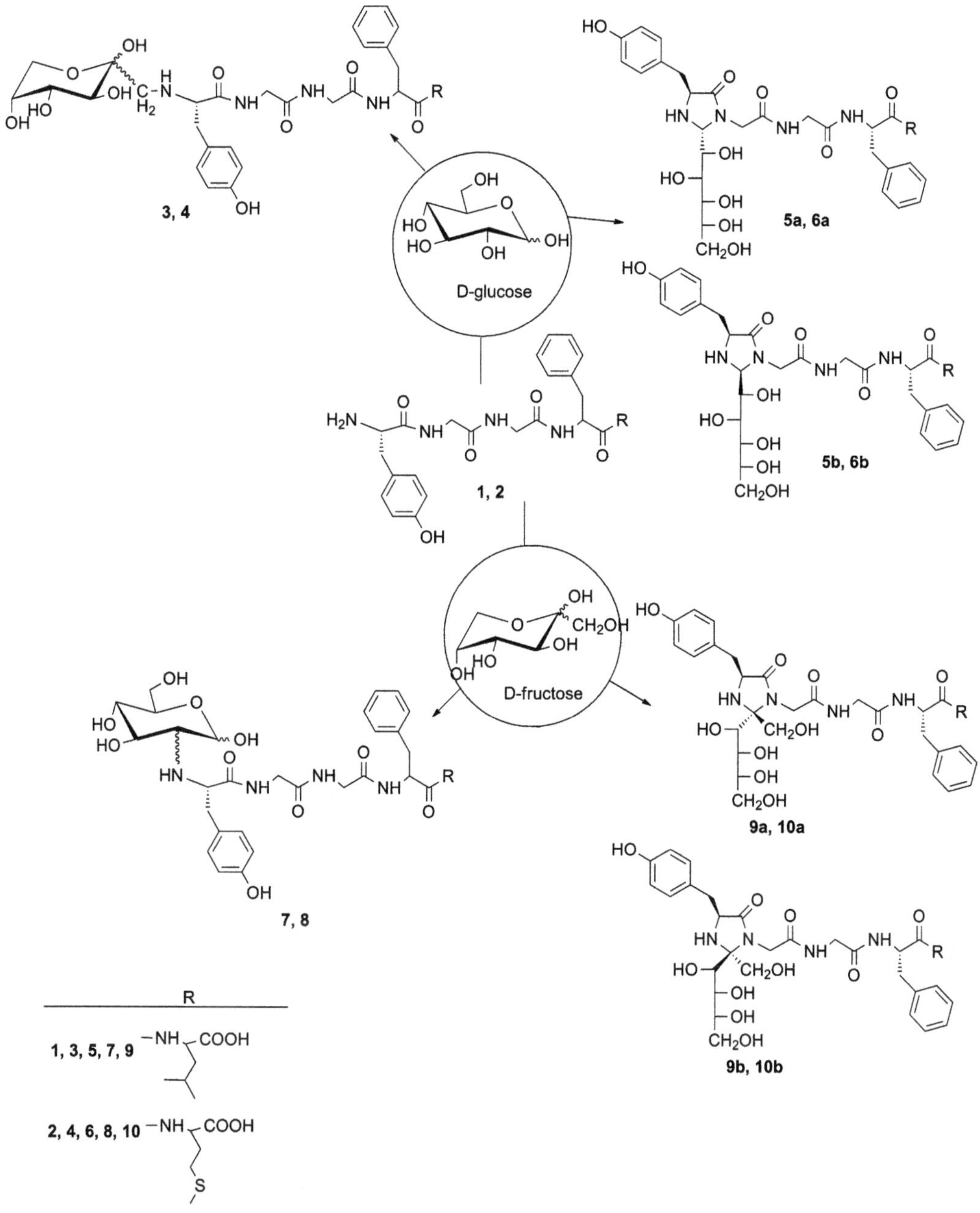

Fig. 1 Early Maillard products in the reaction of enkephalins with reducing sugars glucose or fructose

reactions with NEM when the molar ratio sugar-peptide-base was 15:1:15 (Subheading 3.3.2) (*see* **Note 1**). The amounts of Heyns or imidazolidinone compounds formed depend on the nature of the base used as a catalyst. Addition of NEM leads preferably to the formation of imidazolidinone compounds, while Heyns compounds were the major products in the presence of KOH (Table 1).

Table 1
Comparison of the glycation product formation from Leu- and Met-enkephalin in the presence of D-glucose and D-fructose

				Imidazolidinone product	
Peptide	**Sugar**	**Method**	**Amadori/Heyns products**	**2R*, 2S***	**2S*, 2S***
1	Glc	1A	50		
1	Glc	1B	39	10	
2	Glc	1A	69	1	
2	Glc	1B	61	5	
1	Fru	2A	35	11	9
1	Fru	2B	17	22	25
2	Fru	2A	35	12	10
2	Fru	2B	15	23	29

The Heyns compounds were accompanied by 5 % of Amadori compound in the reaction of fructose with Leu-enkephalin using KOH as a catalyst. This was detected by NMR analysis, since attempts to separate the Heyns and Amadori compound by RP HPLC failed (*see* **Note 2**).

The influence of *N*-glycation on the tautomeric distribution of Amadori compounds, explored by NMR analysis in D_2O, revealed the presence of β-pyranose, α-furanose, and β-furanose forms in solution, the β-pyranose tautomer being the most abundant at equilibrium (69 %) (Fig. 2, Table 2). The α-pyranose form and the open chain keto form were not detected. In contrast, in DMSO-d_6 solution, the equilibrium compositions of Amadori compounds were markedly shifted towards a high proportion of furanose forms, amounting to two-thirds of the mixture, which also contained a surprisingly high proportion of the acyclic hydrate form (~10 %) (Fig. 2). On the basis of NMR analysis the second product isolated from the reaction of fructose with enkephalins **1** and **2** was identified as imidazolidinone compounds **5** and **6** (Table 1). RP HPLC analysis revealed that main diastereoisomers were *trans*-isomers **5a**, as well as **6a** (~90 %) [19].

The Heyns compounds were isolated as inseparable mixtures of *gluco* and *manno* derivatives (*see* **Note 2**). The structure and the population of the isomers present in the mixtures of Heyns compounds were determined by NMR analysis (Table 2) [20]. The *gluco* derivatives were the major products (74–80 %) in the isolated Heyns compounds of Leu- and Met-enkephalin, as can be seen from data presented in Table 2. The estimated equilibrium composition of the tautomeric forms in aqueous solution reveals

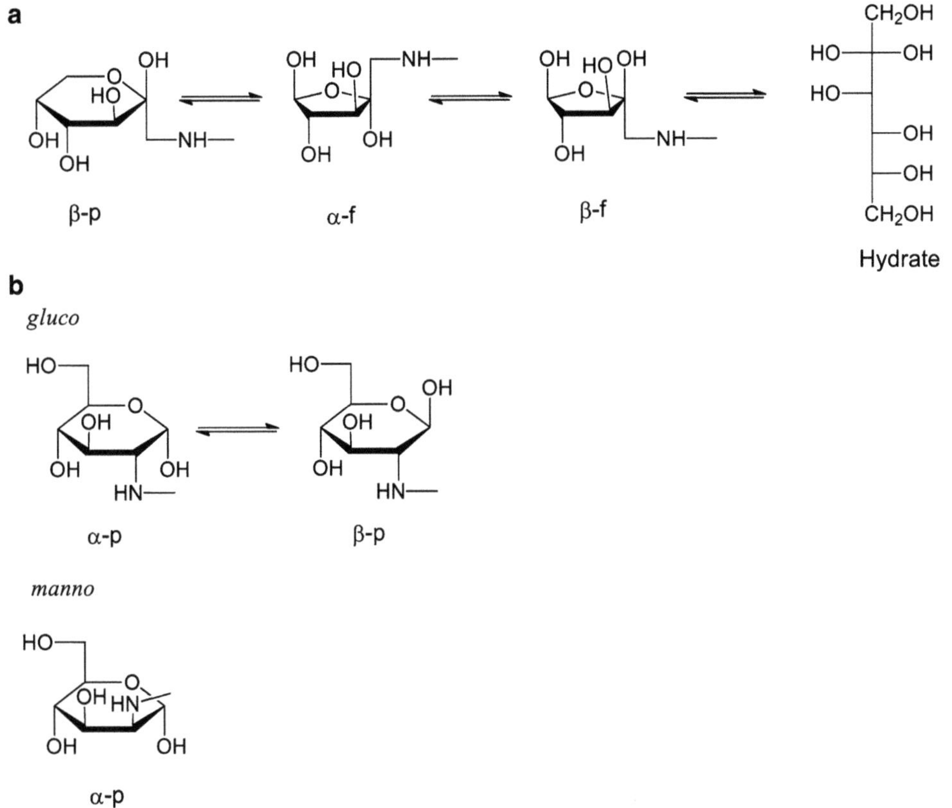

Fig. 2 Tautomers presented in Amadori compounds (**a**) and in Heyns compounds of enkephalins (**b**) in D_2O and DMSO-d_6

Table 2
Tautomeric composition of the Amadori and Heyns compounds in D_2O and DMSO-d_6 estimated from NMR data

		D_2O			DMSO-d_6		
Compound	**Sugar type**	$\delta_{C\text{-}1/C\text{-}2}$[a] (ppm)	$\delta_{H\text{-}1}$ (ppm)	%	$\delta_{C\text{-}1/C\text{-}2}$[a] (ppm)	$\delta_{H\text{-}1}$ (ppm)	%
3 β-p	*fructo*	96.1		69	97.6		32
3 α-f	*fructo*	102.3		18	103.6		33
3 β-f	*fructo*	99.8		13	101.6		24
3 hydr	*fructo*				98.6		11
7a α-p	*gluco*	89.39	5.42(2.8)	61	91.18	4.88	29
7a β-p	*gluco*	94.33	4.80(8.5)	19	97.48	4.28	38
7b α-p	*manno*	90.43	5.27(1.4)	20	91.83	4.49	33
8a α-p	*gluco*	88.97	5.24(3.6)	56	91.18	4.95	29
8a β-p	*gluco*	94.57	4.98(8.4)	18	97.45	4.32	32
8b α-p	*manno*	90.41	5.02(1.3)	26	91.83	4.58	39

[a]C-1 for *gluco* and *manno* derivatives, and C-2 for *fructo* derivatives

that the *gluco* compound in α-pyranose form is the major tautomer by far (60 %) in the mixture of Heyns compounds 7 and **8** [20]. NMR data obtained from DMSO solution indicated almost equal distribution between the detected forms of compounds 7 and **8**. Only the α-pyranose form was detected in D_2O and in DMSO-d_6 solution at equilibrium for *manno* derivatives **7b** and **8b** [20]. On the basis of NMR analysis the second main products isolated from the reaction of fructose with peptides **1** and **2** were identified as diastereoisomeric imidazolidinone compounds **9** and **10**. Diastereoisomers **9a** and **9b**, as well as **10a** and **10b** were separated by RP HPLC. Considering the relative amounts of imidazolidinone diastereoisomers obtained from peptides **1** and **2** (Table 1), it can be assumed that ring closure occurred with almost equal facility to either *Re* or *Si* face of the initially formed Schiff base [20].

3.2 Method 1

3.2.1 Method 1A

1. Dissolve D-glucose (270 mg, 1.5 mmol) and peptide **1** or **2** (0.1 mmol) in MeOH (40 mL) in round-bottom flask.
2. Keep the flask closed for 2 days at 50 °C.
3. Evaporate the solvent.
4. Apply the residue to a column (90 × 1.6 cm) of Sephadex G-15 and elute it with 1 % aqueous acetic acid.
5. Combine the fractions containing glycation products (UV detection λ 280 nm) and evaporate the solvent.
6. Purify resulting compounds **3–6** by RP HPLC using 40 % MeOH/0.1 % TFA.

3.2.2 Method 1B

1. Dissolve D-glucose (270 mg, 1.5 mmol) and peptide **1** or **2** (0.1 mmol) in dry MeOH (20 mL).
2. Add *N*-ethylmorpholine (NEM) (64 μL, 0.5 mmol) to the solution and keep the flask for 2 days at 50 °C (*see* **Note 1**).
3. Further treatment of the reaction mixture is the same as in Subheading 3.2.1.

3.3 Method 2

3.3.1 Method 2A

1. Dissolve D-fructose (810 mg, 4.5 mmol) in MeOH (10 mL) in round-bottom flask.
2. Add peptide **1** or **2** (0.06 mmol), KOH (3 mg, 0.06 mmol) $Na_2S_2O_5$ (0.4 mg, 0.0016 mmol) and glycerol (0.5 mL).
3. Stir reaction mixture for 8 h at 70 °C.
4. Evaporate the solvent and apply the residue to a column (90 × 1.6 cm) of Sephadex G-15 and elute with 1 % aqueous acetic acid.
5. Combine the fractions containing glycation products.
6. Further purify resulted compounds **7–10** by RP HPLC using 40 % MeOH/0.1 % TFA.

3.3.2 Method 2B

1. Dissolve D-fructose (270 mg, 1.5 mmol), in dry MeOH (20 mL).
2. Add peptide **1** or **2** (0.1 mmol) and *N*-ethylmorpholine (NEM) (192 μL, 1.5 mmol) (*see* **Note 1**).
3. Stir for 24 h at 70 °C.
4. Further treatment of the reaction mixture is the same as in Subheading 3.3.1.

3.4 PBS and Human Serum Stability Studies of Glycated Compounds

The study of glycated peptides in a variety of solvent media at different temperatures are likely to provide insights into the stability of the early glycation products and their metabolites and should also contribute to better understanding of the processes of peptide/protein modifications during heat treatment of biopharmaceuticals in the presence of carbohydrate excipients. The degradation of Leu-enkephalin Amadori and imidazolidinone compounds was investigated under oxidative conditions in phosphate buffer, pH 7.4, and human serum at 37 °C. *N*-glycated Leu-enkephalin **3** degraded slowly in phosphate buffer ($t_{1/2}$ ~9 days) producing the parent peptide compound as a major product. The Leu-enkephalin derived *trans*-isomer **5a** was degraded slower than the compound **3** with a half-life of 20.5 days in PBS at 37 °C also producing parent peptide as a main degrading product. As can be seen from the data presented in Table 3, Amadori compound **3** was slowly degraded in human serum with an estimated half-life of 14 h. The main metabolites were Tyr-Gly-Gly-related Amadori compound and Phe-Leu. The half-life determined for Amadori compound **3** in human serum is considerably shorter (by a factor of ~15) than

Table 3
The half-lives of hydrolysis of glycation products in PBS and human serum at 37 °C

				$t_{1/2}$	
Compound	**Original sugar**	**Peptide**	**Structure of the compound**	**PBS**	**Human serum**
1		YGGFL (**1**)	Peptide	Stable	14.8 min
3	Glc	YGGFL (**1**)	Amadori	8.7 days	14.0 h
5	Glc	YGGFM (**2**)	Imidazolidinone	20.5 days	6.5 days
7	Fru	YGGFL (**1**)	Heyns	Stable	13.5 h
8	Fru	YGGFM (**2**)	Heyns	Stable	4.7 h
9a	Fru	YGGFL (**1**)	Imidazolidinone	12.8 days	1.8 days
9b	Fru	YGGFL (**1**)	Imidazolidinone	Stable	5.9 days
10a	Fru	YGGFM (**2**)	Imidazolidinone	14.7 days	1.8 days
10b	Fru	YGGFM (**2**)	Imidazolidinone	Stable	3.0 days

the half-life in 0.05 M phosphate buffer at pH 7.4 and 37 °C. Enkephalins are rapidly degraded in organisms by various enzymes. At least seven enzymes hydrolyzing different peptide bonds in the enkephalin molecule have been found in human blood plasma [21]. These include aminopeptidases, dipeptidyl aminopeptidases, and dipeptidyl carboxypeptidases. The predominant (~80 %) route for the degradation of leucine-enkephalin is reported by hydrolysis of the *N*-terminal Tyr^1-Gly^2 bond by aminopeptidases [22]. In agreement with these results, we found rapid leucine-enkephalin disappearance ($t_{1/2}$ = 14.8 min) in 80 % human serum (Table 3), parallel by the appearance of Tyr and peptides whose amino acid composition is consistent with the *N*-terminal leucine-enkephalin hydrolysis by-products. The serum half-life of leucine-enkephalin determined here is higher than published earlier for human plasma. However, this difference can be related to the origin of the serum, since large variations in the plasma half-life of leucine-enkephalin, dependent on the batch and treatment of the plasma used, have been reported [23]. The results of the degradation of leucine-enkephalin-derived Amadori compound **3** in human serum revealed that glycation of leucine-enkephalin at the *N*-terminal position with a ketose moiety fully stabilizes the parent opioid peptide against cleavage by aminopeptidase. The consequences of this derivatization are that the *N*-terminal amino group is transformed into a secondary amino group and the Tyr^1-Gly^2 bond becomes *N*-alkylated, making the glycated pentapeptide a poor substrate for aminopeptidases. However, compound **3** turned out to be degraded by dipeptidyl carboxypeptidase, which cleaves the Gly^3-Phe^4 bond in leucine-enkephalin, playing a minor although significant role in degrading enkephalins in human plasma [22]. As well as sugar moiety the imidazolidinoyl moiety at the *N*-terminus of the peptide affected amino-terminal degradation by aminopeptidases and made imidazolidinone compound **5a** more available for carboxypeptidase activity. Thus the degradation process of **5a** in human serum followed two main pathways involving dipeptidyl carboxypeptidase cleavage of the Gly-Phe peptide bond as well as carboxypeptidase hydrolysis of Phe-Leu. The very slow increase in tyrosine formation over time reflects the bio-reversibility of **5a** to parent peptides by spontaneous hydrolysis followed by degradation at the Tyr-Gly bond by serum aminopeptidase.

The stability of Heyns compounds **7** and **8** and of imidazolidinones **9a**, **9b**, **10a,** and **10b** was investigated at 37 °C in 0.05 M PBS (pH 7.4) and in human serum (Table 3). The Heyns compounds derived from Leu- or Met-enkephalin were very stable in PBS, with 60–70 % of the starting compound being present in solution after 4 weeks of incubation. Even though the parent 2-amino-2-deoxyaldoses are very unstable in neutral aqueous solution, it appears that *N*-peptidyl glucosamine and mannosamine glycation products are greatly stabilized toward autoxidative rearrangements and degradation. Interestingly, under identical conditions, Amadori product **3** (Fig. 1), derived from glucose and Leu-enkephalin, undergoes faster degradation [24, 25] through 1,2- or 2,3-enolization reactions (Table 3). Surprisingly, the 2R*, 4S*-isomers of imidazolidinone compounds of Leu-enkephalin **9a** and Met-enkephalin **10a** are less stable than the corresponding 2S*, 4S*-isomers **9b** and **10b** having bulky benzyl and the tetritol moieties on the same side of the ring.

In human serum, all glycated products derived from either Leu- or Met-enkephalin were more stable than the native peptides. In general, the most susceptible to enzymatic degradation among the glycated products were Heyns compounds (especially **8**, Table 3) [20]. It was mentioned earlier that Amadori compound **3** and Leu-enkephalin derived imidazolidinone compound **5** (Fig. 1) are stable against degradation mediated by aminopeptidases [19, 20]. It was found that both compounds were degraded preferably by dipeptidylcarboxypeptidase, which cleaves the Gly^3-Phe^4 bond. Imidazolidinone **5** was also a substrate of carboxypeptidase, which targets the Phe^4-Leu^5 bond [19]. Similarly, the Heyns compound of Leu-enkephalin **7** and Met-enkephalin **8** were degraded in human serum by dipeptidylcarboxypeptidase to the same product—the Heyns compound of tripeptide [20]. The estimated half-life for Heyns compound **7** in human serum was almost the same as that found for Amadori compound **3**. The 2S* isomers of the imidazolidinone compounds **9b** and **10b** showed higher stability, in human serum, than the 2R* isomers (**9a**, **10a**), presumably as a result of steric hindrance in *cis*-isomers (Table 3) [20]. Leu-enkephalin derivative **9a** is approximately three times less stable than imidazolidinone compound **5**, indicating that the glucose-derived imidazolidinone is more stable than its fructose-derived analog. The main degradation product, as a result of the dipeptidylcarboxypeptidase activity on imidazolidinone compounds **9** and **10**, was detected as well as product of carboxypeptidase activity.

3.4.1 Stability Analysis in PBS

1. Prepare solutions of glycated compounds (8×10^{-4} M) containing the internal standard, *o*-hydroxyphenylacetic acid (40 μg/mL) and NaN_3 (0.02 %) in phosphate buffers/0.1 M NaCl (pH 7.4) (PBS) of concentration (0.05 M).
2. Prepare solutions in triplicate for every compound.
3. Sterilize the solutions by passage through a 0.45 μm nylon filter.
4. Incubate in the dark, at 37 °C, in teflon-lined screw-cap test tubes.
5. At appropriate times, take the samples of the reaction mixtures and chromatograph immediately by RP-HPLC using 40 % MeOH/0.1 % TFA.

3.4.2 Stability Analysis in Human Serum

1. Prepare solutions of glycated compounds (8×10^{-4} M) or peptide containing the internal standard, *o*-hydroxyphenylacetic acid (40 μg/mL) in human serum 80 % (v/v), diluted with H_2O.
2. Keep solution in a teflon-lined screw-cap test tube at 37 °C [24].
3. Take samples in triplicate (100 μL each) at appropriate intervals.
4. Deproteinize by the addition of 20 μL of 48 % aqueous TFA.
5. Briefly vortex the samples and froze them.
6. Thaw samples for the analysis and centrifuge them for 10 min ($15{,}000 \times g$).
7. Analyze supernatants by RP-HPLC (Eurospher 100 reversed-phase C-18 analytical column (250 × 4 ID, 5 μm) can be used), using 40 % MeOH/0.1 % TFA. Perform UV detection at 280 and 215 nm (*see* **Note 3**).
8. Determine the glycated compound or peptide concentration of samples in triplicate by electronic integration of peak areas and calculation of analyte/internal standard peak-areas.

4 Notes

1. Addition of NEM is necessary for the formation of imidazolidinone compounds. Replacement of NEM with *N*-methylmorpholine (NMM) does not give desired glycated product.
2. Attempts to separate the Heyns and Amadori compound by RP HPLC failed because Amadori compounds (containing fructose), and Heyns compounds (containing mannose/glucose) exhibit same hydrophile and they cannot be separated on RP column.
3. UV absorption at 215 nm results from peptide bond in compounds, while at 280 nm results from Tyr chromophore.

References

1. Pert CB, Snyder SH (1973) Opiate receptor: demonstration in nervous tissue. Science 179:1011–1014
2. Hughes J, Smith TW, Kosterlitz HW, Fothergill LA, Morgan BA, Morris HR (1975) Identification of two related pentapeptides from the brain with potent opiate agonist activity. Nature 258:577–579
3. Horvat Š (2001) Opioid peptides and their glycoconjugates: structure–activity relationship. Curr Med Chem CNS Agent 1: 133–154
4. Janecka A, Perlikowska R, Gach K, Wyrębska A, Fichna J (2010) Development of opioid peptide analogs for pain relief. Curr Pharmac Des 16:1126–1135
5. Narita M, Tseng LF (1998) Evidence for the existence of the β-endorphin-sensitive "ε-opioid receptor" in the brain: the mechanisms of ε-mediated antinociception. Jpn J Pharmacol 76:233–253
6. Grevel J, Yu V, Sadee W (1985) Characterization of a labile naloxone binding site (λ site) in rat brain. J Neurochem 44:1647–1656

7. Oka T (1980) Enkephalin receptor in the rabbit ileum. Br J Pharmacol 68:193–195
8. Zagon IS, Gibo DM, McLaughlin PJ (1991) Zeta (ξ), a growth-related opioid receptor in developing rat cerebellum: identification and characterization. Brain Res 551:28–35
9. Teschemacher H (1993) Opioids I. In: Hertz A (ed) Handbook of experimental pharmacology 104/I. Springer, Berlin, pp 499–528
10. Zadina JE, Hackler L, Ge LJ, Kastin AJ (1997) A potent and selective endogenous agonist for the μ-opiate receptor. Nature 386:499–502
11. Yamazaki T, Ro S, Goodman M, Chung NN, Schiller PW (1993) A topochemical approach to explain morphiceptin bioactivity. J Med Chem 36:708–719
12. Witt KA, Gillespie TJ, Huber JD, Egleton RD, Davis TP (2001) Peptide drug modifications to enhance bioavailability and blood–brain barrier permeability. Peptides 22:2329–2343
13. Gaby AR (2005) Adverse effects of dietary fructose. Altern Med Rev 10:294–306
14. Schalkwijk CG, Stehouwer CDA, van Hinsbergh VWM (2004) Fructose-mediated non-enzymatic glycation: sweet coupling or bad modification. Diabet Metab Res Rev 20:369–382
15. Hinton DJS, Ames JM (2006) Site specificity of glycation and carboxymethylation of bovine serum albumin by fructose. Amino Acids 30:425–433
16. Shipar MAH (2006) Formation of the Heyns rearrangement products in dihydoxyacetone and glycine Maillard reaction: a computational study. Food Chem 97:231–243
17. Suarez G, Rajaram R, Oronsky AL, Gawinowicz MA (1989) Nonenzymatic glycation of bovine serum albumin by fructose (fructation). Comparison with the Maillard reaction initiated by glucose. J Biol Chem 264:3674–3679
18. Ahmad N, Furth AJ (1992) Failure of common glycation assays to detect glycation by fructose. Clin Chem 38:1301–1303
19. Roščić M, Horvat Š (2006) Transformations of bioactive peptides in the presence of sugar-characterization and stability studies of the adducts generated via the Maillard reaction. Bioorg Med Chem 14:4933–4943
20. Jakas A, Vinković M, Smrečki V, Šporec M, Horvat Š (2008) Fructose-induced *N*-terminal glycation of enkephalin and related peptides. J Pept Sci 14:936–945
21. Marini M, Roscetti G, Bongiorno L, Urbani A, Roda LG (1990) Hydrolysis and protection from hydrolysis of enkephalins in human plasma. Neurochem Res 15:61–67
22. Bolacchi F, Marini M, Urbani A, Roda LG (1995) Enzymes and inhibitors in Leu-enkephalin in metabolism in human plasma. Neurochem Res 20:991–999
23. Ostreowska H (1997) Cathepsin A—like activity in thrombin—activated human platelets substrate specificity, pH dependence, and inhibitory profile. Thromb Res 86:393–404
24. Jakas A, Horvat Š (2004) The effect of glycation on the chemical and enzymatic stability of the endogenous opioid peptide, leucine-enkephalin, and related fragments. Bioorg Chem 32:516–526
25. Jakas A, Horvat Š (2003) The effect of glycation on the chemical and enzymatic stability of the endogenous opioid peptide, leucine-enkephalin, and related smaller fragents: stability, reactions, and spectroscopic peptides. Biopolymers 69:421–431

Chapter 10

Solid-Phase Guanidinylation of Peptidyl Amines Compatible with Standard Fmoc-Chemistry: Formation of Monosubstituted Guanidines

Nina Bionda and Predrag Cudic

Abstract

With the growing importance of peptides and peptidomimetics as potential therapeutic agents, a continuous synthetic interest has been shown for their modification to provide more stable and bioactive analogs. Among many approaches, peptide/peptidomimetic guanidinylation offers access to analogs possessing functionality with strong basic properties, capable of forming stable intermolecular H-bonds, charge pairing, and cation-π interactions. Therefore, guanidinium functional group is considered as an important pharmacophoric element. Although a number of methods for solid-phase guanidinylation reactions exist, only a few are fully compatible with standard Fmoc solid-phase peptide chemistry.

In this chapter we summarize the solid-phase guanidinylation methods fully compatible with standard Fmoc-synthetic methodology. This includes use of direct guanidinylating reagents such as 1-*H*-pyrazole-1-carboxamidine and triflylguanidine, and guanidinylation with di-protected thiourea derivatives in combination with promoters such as Mukaiyama's reagent, *N*-iodosuccinimide, and *N*,*N'*-diisopropylcarbodiimide.

Key words Guanidinylation, Fmoc solid-phase peptide synthesis, Depsipeptides, Triflylguanidine, 1-*H*-pyrazole-1-carboxamidine, Thiourea, Mukaiyama's reagent, *N*-Iodosuccinimide, *N*,*N'*-Diisopropylcarbodiimide

1 Introduction

Guanidinium group is considered as an important pharmacophoric element due to its strong basic property and capabilities of forming intermolecular H-bonds, charge pairing, and cation-π interactions [1, 2]. Many synthetic and natural products possessing guanidinium functionality exhibit a wide range of biological activities [3] including antimicrobial activity [4–7], Na-H exchange (NHE) inhibition [8, 9], thrombin inhibition [10], gene delivery [11, 12], and cell-penetrating properties [13, 14]. Basic amino acids, such as arginine and lysine, are essential for biological activities of many cell-penetrating peptides (CPPs) and cationic antimicrobial peptides (CAMPs). Among CPPs, arginine-rich peptides have been

Predrag Cudic (ed.), *Peptide Modifications to Increase Metabolic Stability and Activity*, Methods in Molecular Biology, vol. 1081, DOI 10.1007/978-1-62703-652-8_10, © Springer Science+Business Media New York 2013

the most extensively studied [15, 16]. Examples include TAT peptide originating from the HIV-1 TAT protein, penetratin derived from the transcription factor of Antennapedia of *Drosophila*, and designed oligoarginine peptides. For these CPPs internalization efficacy was shown to depend on the length of the peptide backbone, with peptides possessing 6–15 arginine residues exhibiting the highest internalization potentials [17, 18]. The production of CAMPs has been reported for a wide variety of organisms, including bacteria, fungi, plants, and animals. CAMPs share cationic (net charge of +2 to +9) and amphiphilic properties, but differ in size and secondary structure [19–21]. They have attracted a great deal of interest as promising candidates for a novel class of antibiotics active against multidrug-resistant (MDR) bacteria. Among CAMPs, fusaricidin or LI-F family of natural products represents a particularly attractive source of candidates for the development of new antibacterial agents. Fusaricidins/LI-Fs are positively charged cyclic lipodepsipeptide antifungal antibiotics isolated from *Paenibacillus* sp., Fig. 1 [4–6]. Interestingly, in contrast to typical CAMPs, whose sequences contain basic amino acids such as arginine, lysine, and histidine [19–21], fusaricidins/LI-Fs have a net positive charge of +1 due to the presence of the guanidinium group located at the termini of their lipidic tail [4–6]. Among isolated fusaricidin/LI-F antibiotics, fusaricidin A or LI-F04a, Fig. 1, showed the most promising antimicrobial activity against a variety of fungi, and Gram-positive bacteria such as *Staphylococcus aureus* (MICs ranging from 0.78 to 3.12 μg/mL). However, full exploitation of this class of natural product potentials strongly depends on unlimited synthetic access to their analogs.

We have previously reported Fmoc solid-phase synthesis (Fmoc SPPS) of fusaricidin analogs containing 12-guanidinododecanoic acid, Fig. 2, and the structure–activity relationship studies revealing key structural requirements for antibacterial activity and decreasing nonselective cytotoxicity [22, 23]. The positively charged

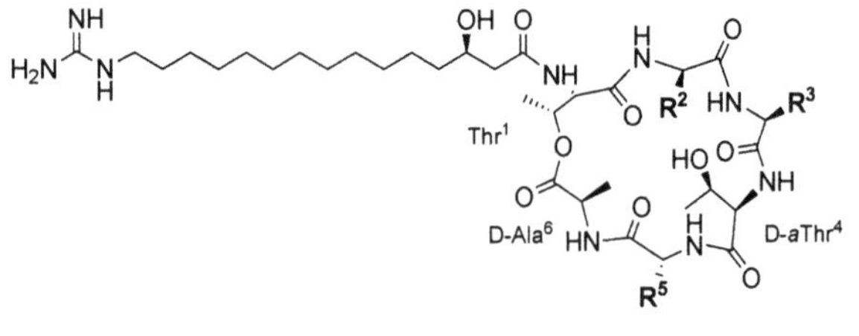

Natural product	R^2	R^3	R^5
LI-F03a/Fusaricidin C	D-Val	Tyr	D-Asn
LI-F03b/Fusaricidin D	D-Val	Tyr	D-Gln
LI-F04a/Fusaricidin A	D-Val	Val	D-Asn
LI-F04b/Fusaricidin B	D-Val	Val	D-Gln
LI-F05a	D-Val	Ile	D-Asn
LI-F05b	D-Val	Ile	D-Gln
LI-F06a	D-alle	Val	D-Asn
LI-F06b	D-alle	Val	D-Gln
LI-F07a	D-Val	Phe	D-Asn
LI-F07b	D-Val	Phe	D-Gln
LI-F08a	D-Ile	D-Ile	D-Asn
LI-F08b	D-Ile	D-Ile	D-Gln

Fig. 1 Fusaricidin/LI-F family of natural products

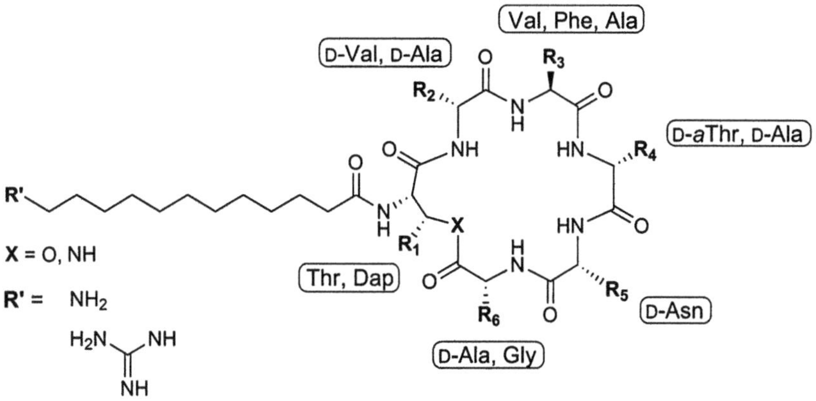

Fig. 2 Schematic representation of synthetic fusaricidin A/LI-F04a analogs

guanidinium group at the end of the 12-carbon atom lipid tail and the presence of hydrophobic amino acids in the depsipeptide/peptide sequence were shown to be crucial for antibacterial activity. Similar to the parent natural product, guanidinylated cyclic lipodepsipeptide analogs showed in Fig. 2 exhibited potent activity against MDR Gram-positive bacteria (MIC ~ 8 μg/mL), whereas removal of the guanidinium group led to a complete loss of antibacterial activity [23]. Therefore, introduction of the guanidinium functionality into fusaricidin synthetic analogs is of utmost importance.

Although a number of methods for solid-phase guanidinylation have been reported [1, 2, 24], only a few described guanidinylation reactions are fully compatible with standard Fmoc SPPS [24–30]. The main interest behind peptide guanidinylation was to prepare arginine- and homoarginine-containing peptides from a single ornithine- and lysine-containing precursor peptide, and to avoid potential problems associated with the use of conventionally protected arginine or homoarginine building blocks [31]. The di-Boc-protected thiourea in combination with Mukaiyama's reagent and 1-*H*-pyrazole-1-carboxamidine derivatives are the most commonly used reagents for this purpose [22, 23, 25–30, 32]. However, these reagents are not without shortcomings. 1-*H*-pyrazole-1-carboxamidine failed to completely guanidinylate resin-bound amines even after prolonged reaction time [22, 26, 27] and it needs to be used in a large excess, whereas the efficacy of di-Boc-protected thiourea under Mukaiyama's reaction conditions depends on the steric hindrance of the amino group and the solvent [26, 32].

In this chapter we summarize the methods for guanidinylation of resin-bound amines, including peptides, fully compatible with standard Fmoc-synthetic strategy, Fig. 3. The use of direct guanidinylating reagents such as 1-*H*-pyrazole-1-carboxamidine and triflylguanidines, as well as guanidinylation with di-protected

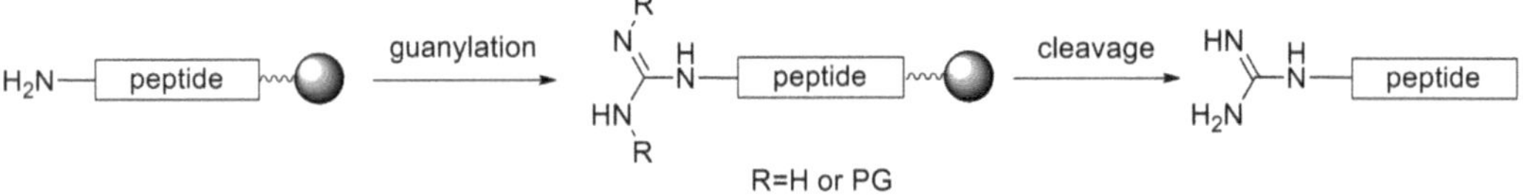

Fig. 3 Guanidinylation of peptides on solid support

thiourea derivatives in combination with promoters, Mukaiyama's reagent, *N*-iodosuccinimide (NIS), and diisopropylcarbodiimide (DIC), is described.

2 Materials

1. All materials and reagents are commercially available and used as received.
2. Solvents for synthesis, including acetone, acetonitrile (ACN), chloroform ($CHCl_3$), dichloroethane (DCE), dichloromethane (DCM), *N*,*N*-dimethylformamide (DMF), dimethylsulfoxide (DMSO), 1,4-dioxane, methanol (MeOH), tetrahydrofuran (THF), trifluoroacetic acid (TFA), and water were of high-performance liquid chromatography (HPLC) or peptide synthesis grade and can be obtained from Sigma-Aldrich, Fisher Scientific, VWR, or other commercial sources.
3. HPLC solvents (CHROMASOLV gradient grade), including ACN and water were purchased from Sigma-Aldrich.
4. Guanidinylation reagents including 1*H*-pyrazole-1-carboxamidine hydrochloride, *N*,*N*′-di-Boc-1*H*-pyrazole-1-carboxamidine, 4-nitro-*N*, *N*′-di-Boc-1*H*-pyrazole-1-carboxamidine, *N*,*N*′-di-Boc-N″-triflylguanidine, NIS, mercury(II) chloride, copper(II) chloride, ethyl-(*N*′,*N*′-dimethylamino)propylcarbodiimide hydrochloride (EDC), 1-methyl-2-chloropyridinium iodide (Mukaiyama's reagent), DIC, and *N*,*N*′-di-Boc-thiourea can be purchased from Sigma-Aldrich, Fisher Scientific, VWR, or other commercial sources.
5. Pyrazole, cyanamide, hydrochloric acid (HCl), *N*,*N*-diisopropylethylamine (DIEA), di-*tert*-butyl dicarbonate (Boc_2O), lithium hydride, benzyl chloroformate (Cbz-Cl), guanidine hydrochloride, triflic anhydride, triethylamine (TEA), thioanisole (TA), and other chemicals can be purchased from Sigma-Aldrich, Fisher Scientific, VWR, or other commercial sources.
6. Kaiser ninhydrin test kit is commercially available from AnaSpec.

3 Methods

3.1 Direct Guanidinylation

3.1.1 Guanidinylation Using 1-H-Pyrazole-1-Carboxamidine Hydrochloride and Derivatives

1*H*-pyrazole-1-carboxamidine hydrochloride **1** was first explored for the use as a guanidinylating agent by Bernatowicz et al. [28]. This guanidinylating reagent can easily be prepared in high yield from pyrazole and 1 eq of cyanamide in refluxing anhydrous HCl/1,4-dioxane. The product crystalizes from the reaction mixture in a form of a hydrochloric salt without the need for further purification [28]. The guanidinylating reagent **1** is also commercially available. When neutralized with 1 eq of DIEA, **1** is soluble in DMF and can be used for guanidinylation of resin-bound amines, Fig. 4. Neutralized **1** has also very good solubility in water and water-miscible solvents such as ACN, acetone, MeOH, and THF. Interestingly, it was found that the reactivity of **1** is enhanced in water, unclear whether the reason lies in the difference in stability of intermediates, mechanism, or better solvation and accessibility of resin-bound amino groups [28].

Self-condensation of **1** in DMF in the presence of 1 eq of DIEA to form the diguanidine derivative **2** was observed (Fig. 5), giving rise to a concern about formation of undesired *N*-alkyldiguanidines [28]. However, taking into consideration that the formation of diguanidine **2** from **1** is very slow at room temperature, and that **1** reacts much faster with primary amines than **2**, possibility that this undesired side reaction will occur under standard Fmoc SPPS conditions is low.

The need for synthesis of protected guanidines led to modification of **1** and preparation of guanidinylating reagents such as *N*,*N'*-di-Boc-1*H*-pyrazole-1-carboxamidine **3** [33] and 4-nitro-N,*N'*-di-Boc-1*H*-pyrazole-1-carboxamidine **4** (Fig. 6) [27]. Despite their greater steric bulk, the reactivity of *N*,*N'*-di-Boc-protected carboxamidines **3** and **4** remains unchanged compared to the

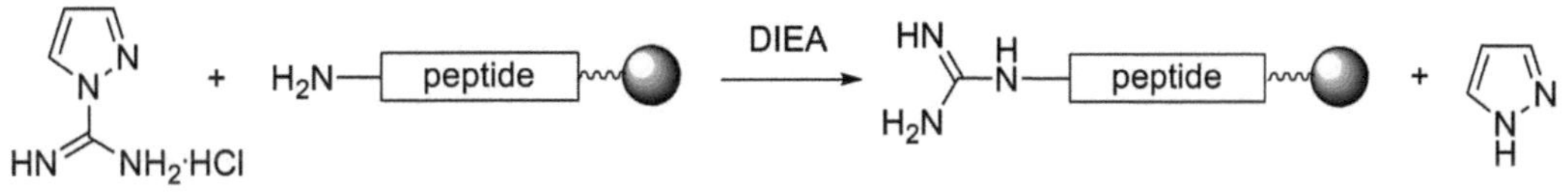

Fig. 4 Guanidinylation using 1-*H*-pyrazole-1-carboxamidine **1**

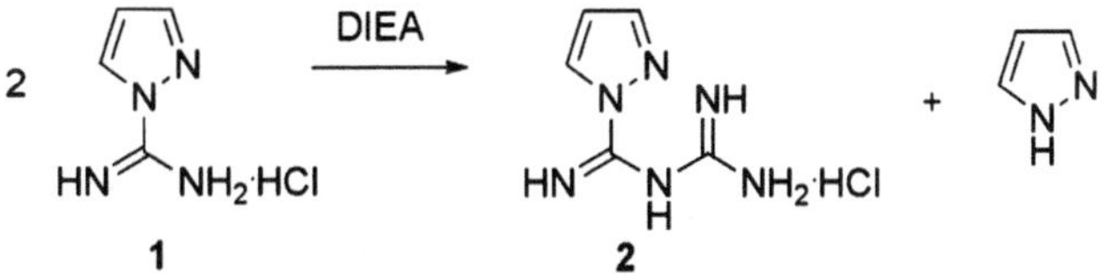

Fig. 5 Self-condensation of 1-*H*-pyrazole-1-carboxamidine **1**

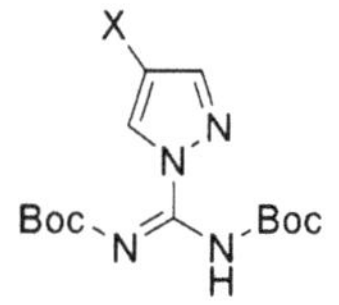

Fig. 6 Di-Boc-protected derivatives of **1**

parent compound **1**. This can be attributed to an electrophilicity increase of the amide carbon relative to the unprotected **1** due to the presence of two electron-withdrawing Boc groups [33]. The guanidinylating agent **3** is easily accessible by reacting **1** with di-*tert*-butyl dicarbonate (Boc_2O, 3 eq) and lithium hydride (3 eq) [33] and is also commercially available. In some cases, failure to drive the guanidinylation to completion even with the use of large excess of **3** and prolonged reaction time prompted modification of this guanidinylating agent. Most successful modification **4** includes incorporation of a nitro group in position 4 of the pyrazole ring, rendering it more electrophilic. Use of 4 eq of **4** resulted in complete guanidinylation of amines on the Dab- and Lys-side chains of resin-bound peptides within 20 h at room temperature [27]. It is important to note that both **4** and 2-nitropyrazole side products formed during the reaction are fully soluble in a variety of organic solvents and thus can easily be removed from the resin after the reaction [27].

Guanidinylation of Resin-Bound Amines Using Pyrazole 1

This protocol was adapted from ref. 28.

1. Selectively remove the protecting group from the amino group of the resin-bound peptide (*see* **Note 1**).
2. Swell the resin in DMF for 20 min.
3. Dissolve **1** (3 eq based on the synthesis scale) in DMF and add DIEA (3 eq). Add this mixture to the resin.
4. Agitate the resin at 45–50 °C for 2.5 h.
5. The reaction progress can be monitored by Kaiser ninhydrin test (*see* **Note 2**).
6. Wash the resin 3× with DMF, and then 3× with DCM.

Guanidinylation of Resin-Bound Amines Using Pyrazole Derivatives 3 and 4

This protocol was adapted from refs. 33, 27.

1. Selectively remove the protecting group from the amino group of the resin-bound peptide (*see* **Note 1**).
2. Swell the resin in DMF for 20 min.
3. Dissolve **3** or **4** (4 eq based on the synthesis scale) in anhydrous DMF and add this solution to the resin.
4. Agitate the resin at room temperature for up to 20 h.

5. The reaction progress can be monitored by Kaiser ninhydrin test (*see* **Note 2**).

6. Wash the resin 3× with DMF, and then 3× with DCM.

3.1.2 Guanidinylation Using Triflyldiurethane-Protected Guanidines

Goodman et al. [34, 35] described preparation and use of di-urethane-protected triflylguanidines (Fig. 7) in direct guanidinylation of primary and secondary amines under mild conditions in solution, as well as on solid support. Both *N,N'*-di-Boc-*N''*-triflylguanidine 5 and *N,N'*-di-Cbz-*N''*-triflylguanidine 6 are readily available in two steps starting from guanidine hydrochloride 7, Fig. 8 [34]. Protection of the staring material 7 with Boc_2O or benzyl chloroformate (Cbz-Cl) was carried out under strong alkaline conditions, yielding intermediates **8** and **9**, which are in the subsequent step converted into desired guanidinylating agents **5** and **6** by treatment with triflic anhydride (Fig. 8) [34, 35].

Although guanidinylation using di-protected triflylguanidines **5** and **6** proceeds much faster in nonpolar solvents such as DCM or $CHCl_3$, it can be successfully carried out in polar DMF or MeOH [34]. The difference in reactivity, depending on the solvent polarity, comes from the properties of the reagents themselves. Namely, **5** and **6** do behave not only as electrophiles but also as weak acids. Polar solvents may facilitate their deprotonation by the free amine, thereby slowing the guanidinylation reaction.

The guanidinylating reagent **5** was employed for conversion of side chain amino groups on resin-bound peptides using both Boc- and Fmoc-chemistry [35]. Also, **5** was found to successfully guanidinylate free amines formed in situ from azides [36].

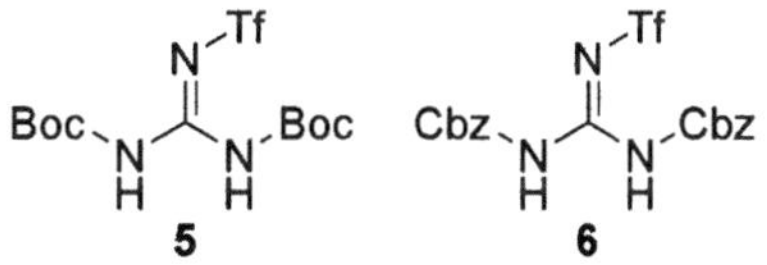

Fig. 7 Structures of di-urethane-protected triflylguanidines **5** and **6**

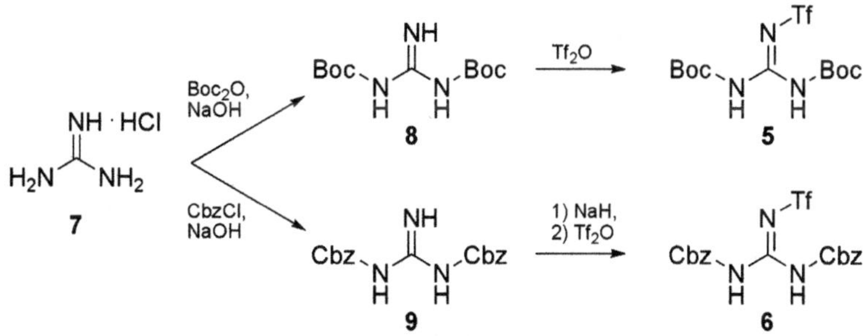

Fig. 8 Synthesis of di-protected triflylguanidines **5** and **6**

Guanidinylation of Resin-Bound Amines Using Di-Urethane-Protected Triflylguanidines 5 and 6

This protocol was adapted from ref. 35.

1. Selectively remove the protecting group from the amino group of the resin-bound peptide (*see* **Note 1**).
2. Swell the resin in DCM for 20 min.
3. Add solution of **5** or **6** (5 eq) and TEA (5 eq) in DCM to the resin (*see* **Note 3**).
4. Agitate the resin at room temperature for up to 8 h (*see* **Note 4**).
5. The reaction progress can be monitored by Kaiser ninhydrin test (*see* **Note 2**).
6. Wash the resin 3× with DCM.

3.2 Guanidinylation Using Thiourea and Activators

Although there are reports of successful guanidinylation of primary amines in solution using only di-protected thiourea and base [37], in general, the conversion of amines into guanidines with thioureas requires initial activation [1]. For this purpose many different activators can be used, including mercuric and copper salts [1, 38], 1-methyl-2-chloropyridinium iodide (Mukaiyama's reagent) [26], NIS [39], carbodiimides such as diisopropylcarbodiimide (DIC) [24] and EDC [32]. In general, the formation of the activated intermediates proceeds quickly and the intermediate is not isolable. The presence of a tertiary amine in the reaction mixture (e.g., triethylamine) and electron-withdrawing substituents in the thiourea moiety accelerates the desulfurization [1]. Here, Fmoc-chemistry-compatible solid-phase guanidinylation reaction using thiourea is described.

3.2.1 Mukaiyama's Reagent

In some cases, low efficiency of pyrazole derivatives **3** and **4** and requirement for large excess of these reagents to complete the guanidinylation reaction prompted an investigation of new alternatives. It has been reported that treatment of amines with mercuric chloride ($HgCl_2$) and di-Boc-thiourea **10** (Fig. 9) is a very fast and efficient method for guanidinylation [37, 40]. The efficiency of this reaction is most likely a result of formation of a highly electrophilic intermediate, *N,N'*-di-Boc-carbodiimide **11**. However, due to formation of insoluble mercuric sulfide, the activation of di-Boc-thiourea with mercuric chloride is not compatible with solid-phase guanidinylations [40].

On the other hand, replacement of mercuric chloride with Mukaiyama's reagent (1-methyl-2-chloropyridinium iodide, **12**)

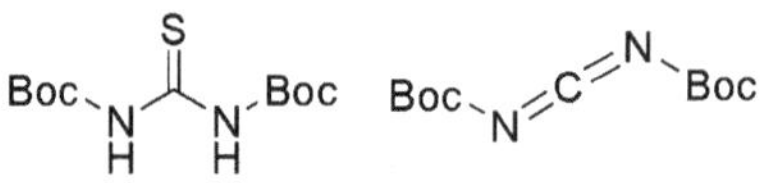

Fig. 9 Structures of di-Boc-protected thiourea **10** and the corresponding carbodiimide intermediate **11**

resulted in complete homogeneous solutions, allowing its utilization in solid-phase guanidinylation reactions [26, 41]. Lipton et al. [26] showed that primary and unhindered secondary amines react readily and give high yield using an excess of the guanidinylating reagents in anhydrous DMF. Hindered and unreactive amines usually gave much higher yields if the reaction was carried out in DCM. The effect of solvent on reaction yield was explained by the instability of the carbodiimide intermediate **11** (Fig. 9).

In the case of slow nucleophilic attack by an amine, the competitive carbodiimide intermediate decomposition may occur resulting in a loss of available reagents. In DCM, formation of the carbodiimide intermediate is slower due to lower solubility of the di-protected thiourea **10**, allowing its more efficient consumption by hindered and relatively unreactive amines. Utilization of Mukaiyama's reagent in solid-phase guanidinylation reaction is not without shortcomings. It has been reported that resin-bound amines in some cases fail to guanidinylate under Mukaiyama's reaction conditions and instead react with Mukaiyama's reagent giving an adduct **13** (Fig. 10) [27, 39, 42].

Guanidinylation of Resin-Bound Amines Using Di-Boc-Thiourea 10 and Mukaiyama's Reagent 12

This protocol was adapted from ref. 26.

1. Selectively remove the protecting group from the amino group of the resin-bound peptide (*see* **Note 1**).
2. Swell the resin in DCM for 20 min.
3. Add solution of TEA (4 eq) and di-Boc-thiourea **10** (3 eq) to the resin.
4. Agitate the resin at room temperature for 15 min.
5. Dissolve Mukaiyama's reagent **12** in DCM (3 eq) and add this solution to the resin (*see* **Note 5**).
6. Agitate the resin at room temperature for 3 h.
7. The reaction progress can be monitored by Kaiser ninhydrin test (*see* **Note 2**).
8. If the reaction is not completed after 3 h repeat **steps 2–5**.
9. Wash the resin 3× with DMF and then 3× with DCM (*see* **Note 6**).

3.2.2 NIS-Promoted Guanidinylation

In order to expend the spectrum of available guanidinylation reagents and to overcome shortcomings of the above-mentioned

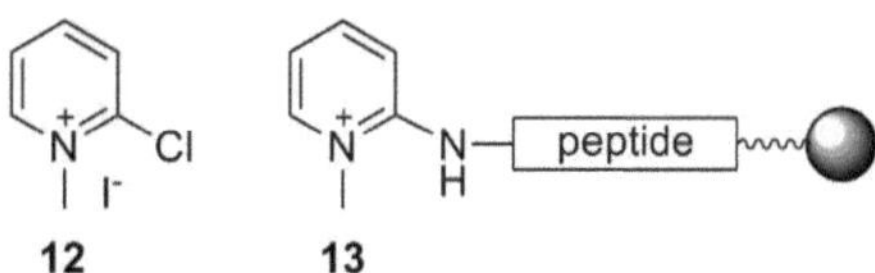

Fig. 10 Mukaiyama's reagent **12** and the side product **13** observed during guanidinylation reaction

methods [27, 39, 42], Smietana et al. [39] introduced NIS **14** as a promoter for the guanidinylation reaction. NIS behaves as a soft Lewis acid, and thus is capable of coordinating the protected thio- and *S*-methylisothioureas. Addition of TEA to the reaction mixture leads to formation of a carbodiimide intermediate and consequently to the desired protected guanidine. This approach was shown to be successful for guanidinylation in-solution of primary and secondary amines, with primary amines giving the highest yields. Sterically hindered and less reactive amines require longer reaction time and higher temperature for guanidinylation. It has been suggested that NIS works solely for activation of the sulfur-leaving group as no iodinated by-products had been detected during the reaction [39]. On the other hand, significant amount of a side product corresponding to *S*-aminoisothiourea was isolated after guanidinylation of hindered secondary amines, such as *N,N-dibutylamine*, using di-Boc-*S*-methylisothiourea **15** in the presence of NIS. This side product is most likely the result of competitive nucleophilic attack of the amine on the *S*-iodoisothiourea intermediate [43, 44]. In the case of *N,N*-dibutylamine, the side product **16** is formed in 25 % yield, Fig. 11 [39]. Formation of *S*-aminoisothiourea side products can be avoided by using slight excess of the di-Boc-*S*-methylisothiourea **15** in the presence of 2 eq of TEA.

Guanidinylation of Resin-Bound Amines Using Di-Boc-Thiourea 10 and NIS 14

This protocol was adapted from ref. 39.

1. Selectively remove the protecting group from the amino group of the resin-bound peptide (*see* **Note 1**).
2. Swell the resin in DCM.
3. Dissolve NIS **14** (1 eq), di-Boc-thiourea **10** (1.5 eq), and TEA (1.5 eq) in DCM.
4. Add this solution to the resin.
5. Agitate the resin at room temperature for 3 h.
6. The reaction progress can be monitored by Kaiser ninhydrin test (*see* **Note 2**).
7. If the reaction is not completed after 3 h repeat **steps 2–5** (*see* **Note 7**).
8. Wash the resin 3× with DCM.

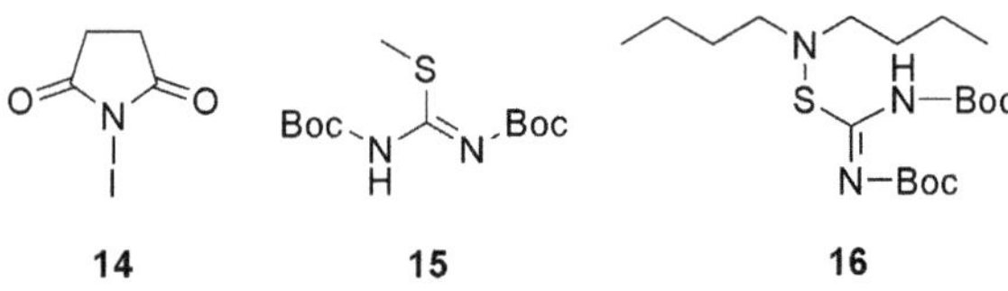

Fig. 11 Structures of *N*-iodosuccinimide **14**, di-Boc-*S*-methylisothiourea **15**, and the side product **16** of *N,N*-dibutylamine guanidinylation

3.2.3 DIC-Promoted Guanidinylation

Robinson and Roskamp reported the use of di-Boc-thiourea **10** in combination with DIC **17**, Fig. 12, for guanidinylation of primary and secondary amines in-solution and on solid phase [24]. In general, DIC-promoted guanidinylation of primary and secondary amines resulted in satisfactory yields under both solution and solid-phase conditions. However, the reaction was slow, requiring in some cases up to 9 days for completion. Besides DIC, other carbodiimides can be used as guanidinylation promoters as well [32].

Guanidinylation of Resin-Bound Amines Using Di-Boc-Thiourea 10 and DIC 17

This protocol was adapted from ref. 24.

1. Selectively remove the protecting group from the amino group of the resin-bound peptide (*see* **Note 1**).
2. Swell the resin in DCE.
3. Add to the resin solution of di-Boc-thiourea **10** (2 eq) and DIC (2 eq) in DCE.
4. Agitate the resin at room temperature for 4–9 days.
5. The reaction progress can be monitored by Kaiser ninhydrin test (*see* **Note 2**).
6. Wash the resin with the following solvent sequence after guanidinylation reaction: DCM, MeOH, DMF, MeOH, DCM, MeOH, and DCM.

3.3 Solid-Phase Guanidinylation of Fusaricidin Class of Cyclic Lipodepsipeptides

Taking into consideration that the Fmoc SPPS is the method of choice for peptide synthesis, as well as the lack of commercially available appropriate guanidino fatty acids, we have developed and optimized a synthetic strategy for fusaricidin class of cyclic lipodepsipeptides and their amide analogs, Fig. 2, that includes solid-phase assembly of cyclic lipopeptide and in the final step conversion of fatty acid's primary amine into a guanidine [22, 23]. To find an optimal method for fusaricidin class of cyclic lipopeptide solid-phase guanidinylation, we have applied most frequently used methods described in this chapter and results were compared. Since the Fmoc-SPPS of this class of cyclic lipopeptides was optimized for TentaGel S RAM resins [22, 23], this resin was used in guanidinylation reactions as well. Guanidinylation protocols were used as described in the chapter. At 1-h time intervals, samples of the peptidyl-resin (cca 5 mg) were taken from the reaction vessel and peptide cleaved using standard cleavage protocol (*see* **Note 8**), and progress of the reactions monitored by RP-HPLC and MALDI-TOF MS analysis. As shown in Fig. 13, the best results were

N=C=N

17

Fig. 12 Structure of *N*,*N'*-diisopropylcarbodiimide **17**

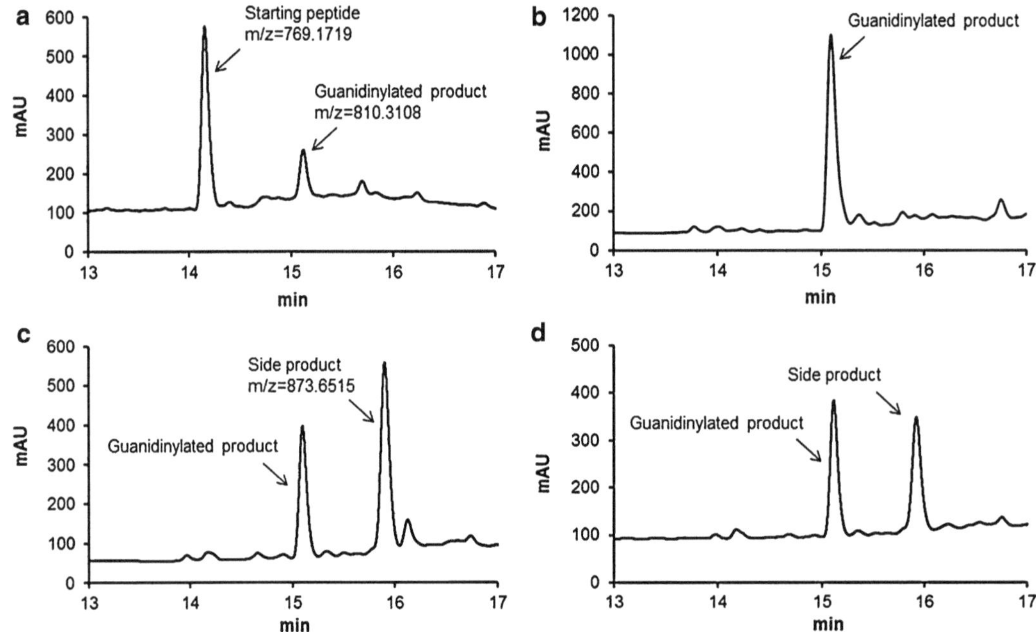

Fig. 13 RP HPLC chromatograms of cleaved crude peptide after guanidinylation using (**a**) 1-*H*-pyrazole-1-carboxamidine for 18 h; (**b**) *N,N'*-di-Boc-*N''*-triflylguanidine for 5 h; (**c**) di-Boc-thiourea and Mukaiyama's reagent for 18 h; (**d**) di-Boc-thiourea and *N*-iodosuccinimide for 5 h. All reactions were carried out on TentaGel S RAM resin

obtained using *N,N'*-di-Boc-*N''*-triflylguanidine **5**. The solid-phase guanidinylation with **5** was completed within 5 h, resulting in desired product in high yields with no detectable side products, Fig. 13b. On the other hand, use of other guanidinylation methods described herein gave unsatisfactory results. Guanidinylation with 1*H*-pyrazole-1-carboxamidine **1** did not go to completion even after 18 h, and ~80 % of the starting material was recovered, Fig. 13a. In guanidinylation reaction using di-Boc-thiourea **10** two different activators were tested, Mukaiyama's reagent **12** and NIS **14**. In both cases consumption of the starting material was completed within 5–8 h, but besides desired guanidinylated cyclic lipopeptide, a large amount of unidentified side product was obtained as well, Fig. 13c, d. Our experimental results indicate that selection of appropriate method for guanidinylation of resin-bound amines is critical in order to minimize formation of undesired side products and increase guanidinylation reagent efficacy.

4 Notes

1. In Fmoc SPPS the protecting groups are, for example, Fmoc, Alloc, ivDde, and Mtt. Standard protocols for peptide-protecting group removal should be applied. The resin should be rinsed with the following sequence after deprotection: 3× DMF, 3× MeOH, and 3× DCM.

2. Kaiser ninhydrin test should be negative upon reaction completion.
3. Due to the high reactivity of the guanidinylating reagents **5** and **6**, the amount can probably be reduced to 2–3 eq.
4. Check the reaction progress by Kaiser ninhydrin test every hour. In some cases the reaction goes to completion within 2 h.
5. In the case of primary or unhindered secondary amines add dropwise suspension of Mukaiyama's reagent in anhydrous DMF into the reaction mixture.
6. Washing with DMF removes any undissolved reagent.
7. The reaction proceeds better if lower amounts of NIS are used and the reaction repeated with fresh reagents.
8. Peptide cleavage conditions: 90:5:5 = TFA:TA:H_2O (v/v), room temperature, 2 h. Peptides were precipitated with cold methyl-*tert*-butyl ether, supernatant removed, precipitate dissolved in DMSO, and injected on analytical RP HPLC (analytical RP HPLC; method: 2 % solvent B for 0.5 min followed by linear gradient 2 → 98 % solvent B over 30 min, for which solvent A is 0.1 % TFA in H2O, and B is 0.08 % TFA in ACN).

References

1. Katritzky AR, Rogovoy BV (2005) Recent developments in guanidinylating agents. Arkivoc 4:49–87
2. Manimala JC, Eric V, Anslyn EV (2002) Solid-phase synthesis of guanidinium derivatives from thiourea and isothiourea functionalities. Eur J Org Chem 2002:3909–3922
3. Heys L, Moore CG, Murphy PJ (2000) The guanidine metabolites of and related compounds; Isolation and synthesis. Chem Soc Rev 29:57–67
4. Kajimura Y, Kaneda M (1996) Fusaricidin A, a new depsipeptide antibiotic produced by *Bacillus polymyxa* KT-8. Taxonomy, fermentation, isolation, structure elucidation and biological activity. J Antibiot 49:129–135
5. Kajimura Y, Kaneda M (1997) Fusaricidins B, C and D, new depsipeptide antibiotics produced by *Bacillus polymyxa* KT-8: Isolation, structure elucidation and biological activity. J Antibiot 50:220–228
6. Kurusu K, Ohba K (1987) New peptide antibiotics LI-F03, F04, F05, F07, and F08, produced by *Bacillus polymyxa*. I. Isolation and characterization. J Antibiot 40:1506–1514
7. Jenssen H, Hamill P, Hancock REW (2006) Peptide antimicrobial agents. Clin Microbiol Rev 19:491–511
8. Yamamoto T, Hori M, Watanabe I, Tsutsi H, Harada K, Ikeda S, Ohtaka H (1997) Structural requirements for potential Na/H exchange inhibitors obtained from quantitative structure-activity relationships of monocyclic and bicyclic aroylguanidines. Chem Pharm Bull 45:1282–1286
9. Yamamoto T, Hori M, Watanabe I, Tsutsi H, Harada K, Ikeda S, Maruo T, Ohtaka H (1998) Synthesis and quantitative structure-activity relationships of N-(3-oxo-3,4-dihydro-2H-benzo[1,4]oxazine-6-carbonyl)guanidines as Na/H exchange inhibitors. Chem Pharm Bull 46:1716–1723
10. Adang AE, Lucas H, de Man AP, Engh RA, Grootenhuis PD (1998) Novel acylguanidine containing thrombin inhibitors with reduced basicity at the P1 moiety. Bioorg Med Chem Lett 8:3603–3608
11. Sainlos M, Belmont P, Vigneron JP, Lehn P, Lehn JM (2003) Aminoglycoside-derived cationic lipids for gene transfection: synthesis of Kanamycin A derivatives. Eur J Org Chem 2003:2764–2774
12. Sainlos M, Hauchecorne M, Oudrhir IN, Zertal-Zidani S, Aissaoui A, Vigneron JP, Lehn JM, Lehn P (2005) Kanamycin A-derived cationic lipids as vectors for gene transfection. Chembiochem 6:1023–1033

13. Schmidt N, Mishra A, Lai GH, Wong GC (2010) Arginine-rich cell-penetrating peptides. FEBS Lett 584:1806–1813
14. Milletti F (2012) Cell-penetrating peptides: classes, origin, and current landscape. Drug Discov Today 17(15–16):850–860
15. El-Sayed A, Futaki S, Harashima H (2009) Delivery of macromolecules using arginine-rich cell-penetrating peptides: ways to overcome endosomal entrapment. AAPS J 11:13–22
16. Wender PA, Galliher WC, Goun EA, Jones LR, Pillow TH (2008) The design of guanidinium-rich transporters and their internalization mechanisms. Adv Drug Deliv Rev 60: 452–472
17. Mitchell DJ, Kim DT, Steinman L, Fathman CG, Rothbard JB (2000) Polyarginine enters cells more efficiently than other polycationic homopolymers. J Pept Res 56:318–325
18. Futaki S, Suzuki T, Ohashi W, Yagami T, Tanaka S, Ueda K, Sugiura Y (2001) Arginine-rich peptides. An abundant source of membrane-permeable peptides having potential as carriers for intracellular protein delivery. J Biol Chem 276:5836–5840
19. Wu Z, de Leeuw E, Ericksen B, Lu W (2005) Why is the Arg5-Glu13 salt bridge conserved in mammalian alpha-defensins? J Biol Chem 280:43039–43047
20. Hancock REW, Lehrer R (1998) Cationic peptides: a new source of antibiotics. Trends Biotechnol 16:82–88
21. Peschel A, Sahl HG (2006) The co-evolution of host cationic antimicrobial peptides and microbial resistance. Nat Rev Microbiol 4: 529–536
22. Stawikowski M, Cudic P (2006) A novel strategy for the solid-phase synthesis of cyclic lipodepsipeptides. Tetrahedron Lett 47: 8587–8590
23. Bionda N, Stawikowski M, Stawikowska R, Cudic M, López-Vallejo F, Treitl D, Medina-Franco J, Cudic P (2012) Effects of cyclic lipodepsipeptide structural modulation on stability, antibacterial activity, and human cell toxicity. ChemMedChem 7:871–882
24. Robinson S, Roskamp EJ (1997) Solid phase synthesis of guanidines. Tetrahedron 53:6697–6705
25. Kowalski J, Lipton MA (1996) Solid phase synthesis of a diketopiperazine catalyst containing the unnatural amino acid (S)-norarginine. Tetrahedron Lett 37:5839–5840
26. Yong YF, Kowalski JA, Lipton MA (1997) Facile and efficient guanidinylation of amines using thioureas and mukaiyama's reagent. J Org Chem 62:1540–1542
27. Yong YF, Kowalski JA, Thoen JC, Lipton MA (1999) A new reagent for solid and solution phase synthesis of protected guanidines from amines. Tetrahedron Lett 40:53–56
28. Bernatowicz MS, Wu Y, Matsueda GR (1992) 1H-Pyrazole-1-carboxamidine hydrochloride an attractive reagent for guanidinylation of amines and its application to peptide synthesis. J Org Chem 57:2497–2502
29. Zakhariev S, Szekely Z, Guarnaccia C, Antcheva N, Pongor S (2000) A highly effective method for synthesis of *N*ω-substituted arginines. Peptides for new millenium. In: Fields GB, Tam JP, Barany G (ed) Proceedings of the 16th American peptide symposium. Kluver Academic, Dordrecht, pp 74–75
30. Thamm P, Kolobeck W, Musiol HJ, Moroder L (2004) Other side-chain protections, guanidino group. In: Goodman M et al (eds) Houben-Weyl: Synthesis of peptides and peptidomimetics, vol E22a. Georg Thieme Verlag, New york, pp 315–333
31. Fields GB, Noble RL (1990) Solid phase peptide synthesis utilizing 9-fluorenylmethoxycarbonyl amino acids. Int J Peptide Protein Res 35: 161–214
32. Schneider SE, Bishop PA, Salazar MA, Bishop OA, Anslyn EV (1998) Solid phase synthesis of oligomeric guanidiniums. Tetrahedron 54:15063–15086
33. Drake B, Patek M, Lebl M (1994) A convenient preparation of monosubstituted *N*,*N*′-di(Boc)-protected guanidines. Synthesis 6:579–582
34. Feichtinger K, Zapf C, Sings HL, Goodman M (1998) Diprotected triflylguanidines: a new class of guanidinylation reagents. J Org Chem 63:3804–3805
35. Feichtinger K, Sings HL, Baker TJ, Matthews K, Goodman M (1998) Triurethane-protected guanidines and triflyldiurethane-protected guanidines: new reagents for guanidinylation reactions. J Org Chem 63:8432–8439
36. Santana AG, Francisco CG, Suarez E, Gonzalez CC (2010) Synthesis of guanidines from azides: a general and straightforward methodology in carbohydrate chemistry. J Org Chem 75:5371–5374
37. Poss MA, Iwanowicz E, Reid JA, Lin J, Gu Z (1992) A mild and efficient method for the preparation of guanidines. Tetrahedron Lett 33:5933–5936
38. Levallet C, Lerpiniere J, Ko SY (1997) The $HgCl_2$-promoted guanidinylation reaction: the scope and limitations. Tetrahedron 53:5291–5304
39. Ohara K, Vasseur JJ, Smietana M (2009) NIS-promoted guanidinylation of amines. Tetrahedron Lett 50:1463–1465

40. Kim KS, Qian L (1993) Improved method for the preparation of guanidines. Tetrahedron Lett 34:7677–7680
41. Shibanuma T, Shiono M, Mukaiyama T (1977) A convenient method for the preparation of carbodiimides using 2-chloropyridinium salt. Chem Lett 5:575–576
42. Convers E, Tye H, Whittaker M (2004) Preparation and evaluation of a polymer-supported Mukaiyama reagent. Tetrahedron Lett 45:3401–3404
43. Ley K, Eholzer U (1966) S-amination of thioureas and thiourethanes. Angew Chem Int Ed 5:674–674
44. Ottmann G, Hooks H (1967) Preparation of S-aminoisothioureas by nucleophilic substitution of S-chloroisothiocarbamoyl chlorides. Angew Chem Int Ed 6:1072–1073

Chapter 11

Stabilization of Collagen-Model, Triple-Helical Peptides for In Vitro and In Vivo Applications

Manishabrata Bhowmick and Gregg B. Fields

Abstract

The triple-helical structure of collagen has been accurately reproduced in numerous chemical and recombinant model systems. Triple-helical peptides and proteins have found application for dissecting collagen-stabilizing forces, isolating receptor- and protein-binding sites in collagen, mechanistic examination of collagenolytic proteases, and development of novel biomaterials. Introduction of native-like sequences into triple-helical constructs can reduce the thermal stability of the triple-helix to below that of the physiological environment. In turn, incorporation of nonnative amino acids and/or templates can enhance triple-helix stability. We presently describe approaches by which triple-helical structure can be modulated for use under physiological or near-physiological conditions.

Key words Collagen, Triple-helix, Template, Peptide-amphiphile, Proline analogs, Disulfide knot

1 Introduction

1.1 Collagen

Collagen is the most abundant protein in animals, and is the major structural protein found in the connective tissues such as basement membranes, tendons, ligaments, cartilage, bone, and skin. The collagen family consists of at least 28 members [1–5]. The most defining feature of collagen is the supersecondary structure, composed of three parallel extended left-handed polyproline type II alpha chains of primarily repeating Gly-Xxx-Yyy triplets. Three left-handed strands intertwine in a right-handed fashion around a common axis to form a triple-helix (Fig. 1). The formation of the triple-helix creates a shallow superhelical pocket inside it. The collagen triple-helix is essential for the integrity of multiple connective tissues.

Depending on the primary structure of the alpha chains [4], collagens have been classified into two main categories, homotrimeric and heterotrimeric. Homotrimeric collagens have three identical alpha chain sequences (designated α1). Examples of such type of collagens are types II and III. Alternatively, heterotrimeric

Predrag Cudic (ed.), *Peptide Modifications to Increase Metabolic Stability and Activity*, Methods in Molecular Biology, vol. 1081, DOI 10.1007/978-1-62703-652-8_11,

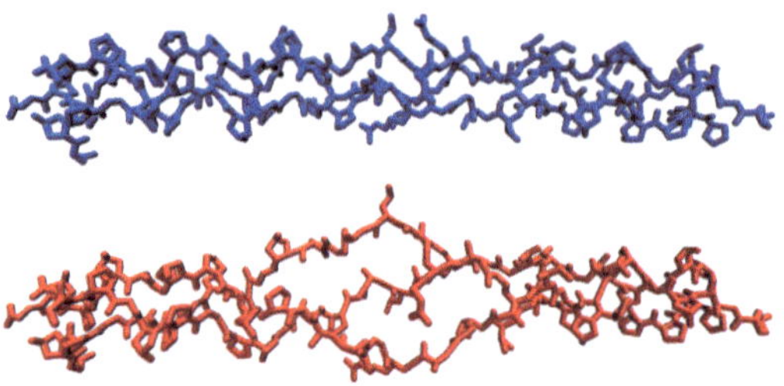

Fig. 1 "Ball and stick" computer-generated models of (*top*) a continuous collagen triple-helix (peptide T3-785, 3[(Pro-Hyp-Gly)$_3$-Ile-Thr-Gly-Ala-Arg-Gly-Leu-Ala-Gly-(Pro-Hyp-Gly)$_4$]) and (*bottom*) an unwound (heat denatured) version of the same sequence

collagens have either three alpha chains of different sequence, designated as α1, α2, and α3 (i.e., type V) or two alpha chains of identical sequence (α1) and third alpha chain of different sequence (α2), such as types I and VI [6]. Furthermore, based on their quaternary structure, collagens are grouped into subfamilies such as fibrillar, collagen associated with banded fibrils (i.e., fibril-associated collagens with interrupted triple-helices—FACITs), network-forming (i.e., basement membrane and short chain), transmembranous, and membrane associated with interrupted triple helices (MACITs) [6].

1.2 Stability of Collagen-Model Triple-Helical Peptides

For several decades triple-helical peptides (THPs) or "mini-collagens" consisting of collagen-model sequences and their three-dimensional folds have been constructed and studied to fully investigate the structural and biological roles of collagenous proteins [5, 7–13]. A model for the collagen triple-helix was first proposed in the 1950s [14, 15] and subsequently refined over time [16–20]. To understand the stability of the collagen triple-helix, it is important to understand its compositional elements. In the collagen Gly-Xxx-Yyy triplet, the residue in the Xxx position is often L-proline (Pro) and the residue in the Yyy position is often 4(R)-hydroxy-L-proline (Hyp), accounting for 20 % of the total amino acid composition in collagen [21]. The other commonly found amino acids are Ala, Lys, Arg, Leu, Val, Ser, and Thr [21]. The packing of the triple-helical coiled-coil structure requires Gly in every third position. Because of its compact structure, assembly of the triple-helix puts this residue at the interior of the helix and the side chain of the Gly, an H atom, is small enough to fit into the center of the helix.

Several factors influence the stability of triple-helical peptides. The chains are held together by hydrogen bonds that form between the peptide amine of Gly residues and peptide carbonyl groups in an adjacent polypeptide chain (Fig. 2) [9, 18]. Interstrand hydrogen bonds formed between the GlyN-H and XxxC=O are critical

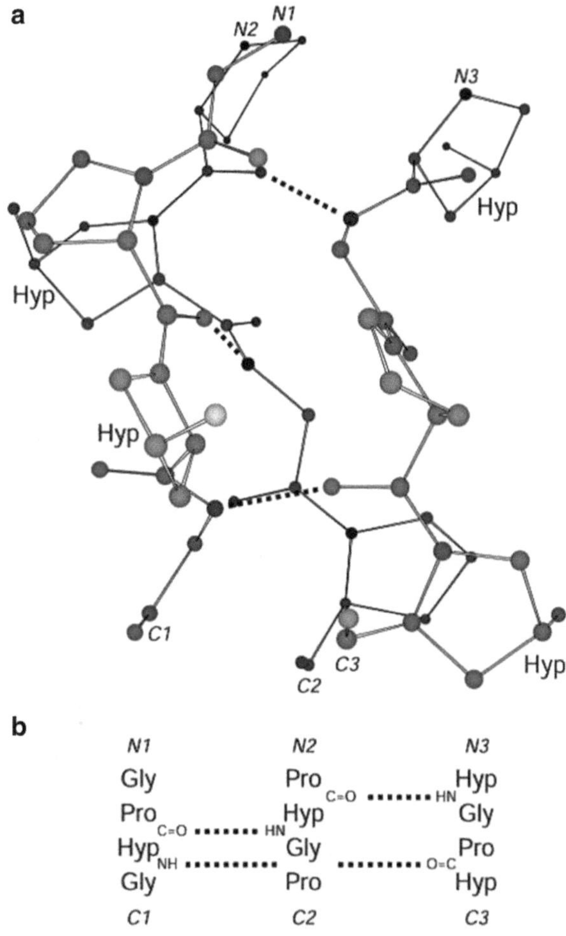

Fig. 2 Segment of a (Pro-Hyp-Gly)$_n$ triple-helix. (**a**) Ball-and-stick representation indicating 4-hydroxy-L-proline residues and XaaC=O–H–NGly hydrogen bonds. (**b**) Register of the residues in the three strands of panel (**a**). Atomic coordinates are from *Science* 1994, 266:75 (PDB entry 1CAG)

for triple-helix stability, as replacement of a central amide bond of (Pro-Pro-Gly)$_{10}$ with either an ester or an (*E*)-alkene, or replacement of the central Pro-Pro with a Pro-*trans*-Pro isostere, substantially destabilizes the THP [22–24].

The role of hydrogen bonding versus inductive effects in triple-helix stabilization is an intensive area of THP research [5]. Incorporation of several nonnative amine acids in the proper position within the peptide sequence often remarkably enhance triple-helix stability. One such example of a stabilizing nonnative amino acid is 4*R*-fluoroproline (4*R*-Flp) (Fig. 3). Incorporation of 4*R*-fluoroproline has been shown to induce hyperstability in the triple-helix of (Pro-4*R*Flp-Gly)$_{10}$ compared to (Pro-4*R*Hyp-Gly)$_{10}$ [25, 26]. Alternatively, 4*S*-Flp destabilizes the triple-helix of

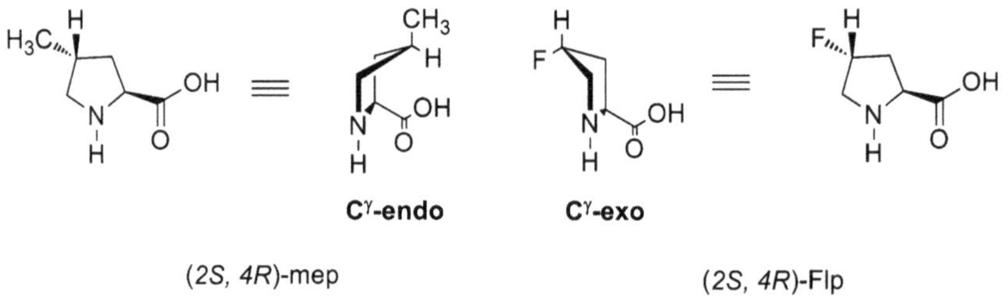

Fig. 3 Ring conformations of 4-substituted L-prolines

$(\text{Pro-4}S\text{Flp-Gly})_7$ compared to $(\text{Pro-4}R\text{Hyp-Gly})_7$ [27]. More specifically, substitution of all of the 4*R*-Hyp residues in $(\text{Pro-4}R\text{Hyp-Gly})_{10}$ or $(\text{Pro-4}R\text{Hyp-Gly})_7$ by 4*R*-Flp increased T_m by 22 and 9 °C, respectively [25–27]. Conversely, an analogous substitution in $(\text{Pro-4}R\text{Hyp-Gly})_7$ by 4*S*-Flp dramatically decreased T_m by greater than 26 °C [27]. The relative effects of 4*R*-Hyp, 4*R*-Flp, and 4*S*-Flp coincide with their propensity for forming *trans*-peptide bonds compared to *cis*-peptide bonds [27, 28]. Studies of collagen mimics have indicated that this increased stability arises from the inductive effects of an electronegative substituent in the 4*R* position, as in 4(*R*)-fluoroproline [9, 25, 26].

It has long been noted that the thermal stability of the collagen triple-helix is enhanced by Hyp residues. It was hypothesized that this stability is due to Hyp residues involved in the hydrogen-bonded network mediated by water molecules, which connect the hydroxyl group of Hyp in one strand to the main chain amide carbonyl of another chain [29]. Substituting Pro for Hyp in the chain significantly decreased the T_m in collagen-model peptides [29]. However, collagen peptides containing the nonnative amino acid 4*R*-Flp exhibited higher stability than peptide with Hyp residues [25, 26] and thus the stabilizing effect arising from Hyp might not be due to hydrogen bonding but rather from the electronegative oxygen preorganizing the main chain in the proper conformation for triple-helix formation [30]. X-ray crystallographic analysis indicated that the 4-OH group in Hyp has no effect on the hydration pattern and the resulting molecular structures [31]. The *O*-methylation of hydroxyproline in Yyy stabilizes the triple-helix more than Hyp itself, most likely because the pyrrolidine ring of 4-methoxyproline adopts a Cγ-exo ring pucker. The conformational stability of the triple-helix arising from *O*-methylation provides strong evidence that the hydroxyl group of Hyp acts primarily through stereoelectronic effects [30].

Both the position of the residue containing the electronegative substituent and the stereochemistry of that electronegative substituent play a role in triple-helix stability. The positional effects of γ-substitution are illustrated by the stabilities, relative to $(\text{Pro-Pro-Gly})_{10}$, of $(\text{Pro-4}R\text{Hyp-Gly})_{10}$ (increased triple-helix stability) and

(4*R*Hyp-Pro-Gly)$_{10}$ (decreased stability) [32]. For many years, it was believed that peptides containing 4*R*-Hyp in the Xxx position of Gly-Xxx-Yyy repeats do not form collagen-like triple-helices. However, incorporation of 4*R*-Hyp in the Xxx position and Thr or Val in the Yaa position triggers triple-helix formation [33, 34]. For example, acetyl-(Gly-4*R*Hyp-Thr)$_{10}$-NH_2 forms a stable triple-helix in water [33]. Both the hydroxyl and methyl groups of Thr with their stereochemical configuration appear to induce the triple-helix stability. Molecular modeling showed that the Thr methyl group shields the interchain hydrogen bond between the carbonyl group of Hyp of the adjacent chain and the amino group of the next Gly residue in the same chain.

Other significant factors that influence the stability of collagen triple helix are the *trans/cis* ratio of the Xaa_{i-1}-Pro_i peptide bond and the ring pucker of Pro_i [35, 36]. The *trans/cis* ratio is crucial as all of the peptide bonds in the triple-helix prefer the *trans* conformation. Pro in solution can adopt either exo or endo ring puckers (Fig. 3). The first Pro in the Pro-Pro-Gly triplet prefers an endo ring pucker conformation while the second Pro prefers the exo ring pucker conformation [35]. This observation suggested that Pro derivatives preferring the Cγ-endo ring pucker could pre-organize triple-helix formation when in the Xaa position, whereas those that prefer the Cγ-exo ring pucker could preorganize triple-helix formation when in the Yaa position [37].

As discussed earlier, incorporation of 4*R*-Flp (Fig. 3) in the Yyy position of Xxx-Yyy-Gly triplets induces hyperstability in the triple-helix. This is because the 4*R*-fluoro group in 4*R*-fluoroproline stabilizes the Cγ-exo pucker conformation. These stereoelectronic effects markedly enhance the conformational stability of a collagen triple-helix [38].

Similar to 4*R*-Flp, incorporation of 4*R*-methylproline (4*R*-mep) (Fig. 3) within the peptide sequence significantly increases triple-helix stability. Raines and colleagues studied the effect of 4-methylproline in conformational stability of (Xxx-Yyy-Gly)$_7$. 4*S*-Methylproline (4*S*-Mep) in the Yyy position enhanced triple-helix stability more so than 4*R*-mep in the Xxx position [38]. A (4*R*-mep-4*S*-Mep-Gly)$_7$ triple-helix is more stable than (4*R*-mep-Pro-Gly)$_7$ or (Pro-4*S*-Mep-Gly)$_7$, simply because the steric effects are additive. In contrast, (4*S*-flp-4*R*-Flp-Gly)$_7$ is less stable than (4*S*-flp-Pro-Gly)$_7$ or (Pro-4*R*-Flp-Gly)$_7$ where the stereoelectronic effects induced by the hetero atom are not additive [38].

The combination of stereoelectronic and steric effects can also induce the preorganization of a polypeptide chain. Integration of both effect results in the most stable triple-helix known so far. Among four triple-helices formed from the four different polypeptide chains, (Pro-Pro-Gly)$_7$, (Pro-4*R*-Hyp-Gly)$_7$, (4*S*-flp-4*S*-Mep-Gly)$_7$, and (4*R*-mep-4*R*-Flp-Gly)$_7$, only the latter two showed hyperstability with T_m values of 51 and 58 °C, respectively [37].

The stability of THPs may also be regulated by pH. To create a triple-helix that was pH dependent, Hyp was modified by *O*-alkylation to a carboxylate group [39]. The incorporation of 1 or 3 Hyp(CO_2) residues within a Pro-Hyp-Gly template did not result in a pH-sensitive triple-helix. However, acetyl-[Pro-Hyp(CO_2)-Gly]$_7$-OH formed a triple-helix with a $T_m = 17$ °C at pH 2.7 but had no triple-helical structure at pH 7.2 [39]. pH-dependent triple-helical stability was also observed for (Pro-4*R*-Amp-Gly)$_6$ sequences, where 4*R*-Amp is (2*S*,4*R*)-4-aminoproline [40, 41].

Self-association of triple-helical structures has been used to "sandwich" a collagen-model sequence between repeats of Gly-Pro-Hyp to obtain THPs of reasonable stability (Fig. 4). Self-associated triple-helical peptides were used to study the structural aspects of collagen via the "host–guest" approach, using sequences such as (Pro-Hyp-Gly)$_n$-Xxx-Yyy-Gly-(Pro-Hyp-Gly)$_n$ [42–44]. Compared with Gly-Pro-Hyp, none of the 20 natural amino acids provides enhanced thermal stability. The most stable residues in the Yyy position were Hyp > Arg > Met, while for the Xxx position Pro > charged residues > Ala > Gln [43–45]. Within a host–guest THP 4*R*-Flp in the Yyy position is slightly destabilizing compared with Hyp, in contrast to (Pro-4*R*-Flp-Gly)$_n$ and (Pro-Hyp-Gly)$_n$ THPs [46]. It has been suggested that different mechanisms are in place when Pro-4*R*-Flp-Gly is inserted within (Pro-Hyp-Gly)$_n$ compared with (Pro-4*R*-Flp-Gly)$_n$ [46].

Electrostatic effects can also contribute favorably to the stability of THPs, as found in host–guest studies [45, 47, 48]. Favorable electrostatic interactions were observed for Gly-Lys-Asp and

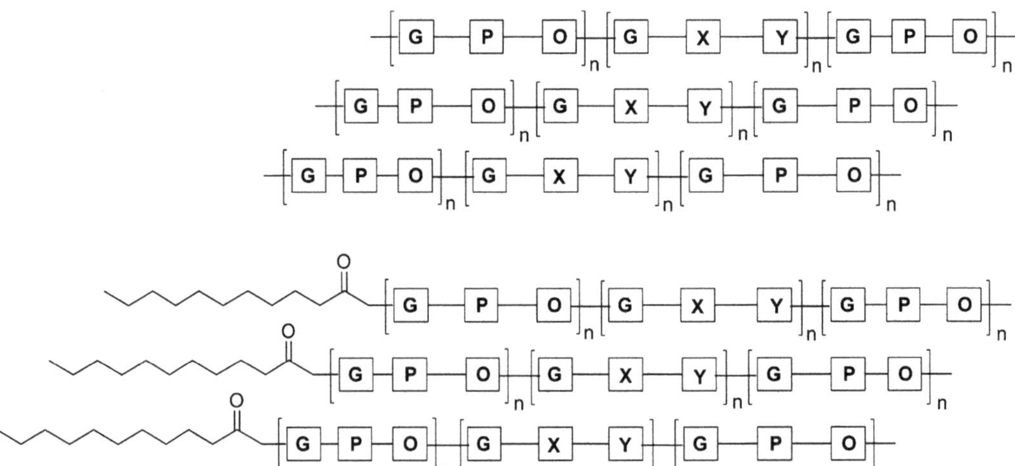

Fig. 4 Modular structures of (*top*) sandwiched associated triple-helical peptide and (*bottom*) sandwiched associated triple-helical peptide-amphiphile. The associated THP features repeats of Gly-Pro-Hyp [(GPO)$_n$] on both the *N*- and *C*-termini to induce or stabilize triple-helical structure, and a diverse collagen-like sequence [(GXY)$_n$] in the middle for structural and/or biological studies. The peptide-amphiphile additionally possesses a pseudo-lipid attached to the *N*-terminus to further enhance triple-helical stability via hydrophobic interactions

Gly-Arg-Asp within the acetyl-(Gly-Pro-Hyp)$_3$-Gly-Xxx-Yyy-(Gly-Pro-Hyp)$_4$-Gly-Gly-NH_2 host [49]. Conversely, guests Gly-Arg-Lys, Gly-Lys-Arg, and Gly-Glu-Asp exhibit charge repulsion [49]. The sequence Gly-Pro-Lys-Gly-[Asp/Glu]-Hyp is found to be as stabilizing as (Gly-Pro-Hyp)$_2$ due to interchain axial ion pair interactions [50, 51]. Arg in the Yyy position can offer high stability, possibly based on its ability to form a hydrogen bond with C=O in a neighboring strand [45, 52]. Heterotrimeric host–guest peptide approaches show a decreased thermal stability for an acetyl-(Gly-Pro-Hyp)$_3$-Gly-Pro-Yyy-(Gly-Pro-Hyp)$_4$-Gly-Gly-NH_2 sequence with an increased number of Arg residues [53]. For example, Hyp residues in all Yyy positions results in a THP with a $T_m = 47.2$ °C compared to 44.5, 40.8, and 37.5 °C, respectively, when all Hyp is replaced by Arg residues in one chain, two chains, or three chains [53].

1.3 Synthesis of Fmoc-Pro Derivatives

To efficiently incorporate (*2S*, *4R*)-4-fluoroproline (Flp) and (*2S*, *4R*)-4-methylproline (mep) into peptide sequences, convenient synthetic routes for the preparation of derivatized Flp and Mep residues such as Fmoc-*trans*-Flp **6** and Fmoc-*trans*-Mep **14** are required (Figs. 5 and 6). Several methods have been reported in the literature for the synthesis of *N*-protected-4(*R*)-Flp from *N*-protected-4(*R*)-Hyp derivatives [54–63]. Fields et al. previously reported the synthesis of Fmoc-*trans*-Flp starting from expensive *cis*-4(*S*)-hydroxy-L-proline [64]. The removal of benzyl ester was problematic under standard Pd/C hydrogenation, as the Fmoc group started decomposing upon prolonged exposure to Pd/C-hydrogen [65, 66]. Recently, we developed a more convenient synthetic route for the preparation of Fmoc-*trans*-Flp from readily available (and inexpensive) *trans*-4(*R*)-hydroxy-L-proline (Fig. 5). A one-pot, three-step procedure is deployed involving a bis-deprotection of the *N*- and *C*-termini under catalytic hydrogenation conditions followed by selective capping of the *N*-terminus with an Fmoc group to yield Fmoc-*trans*-Flp.

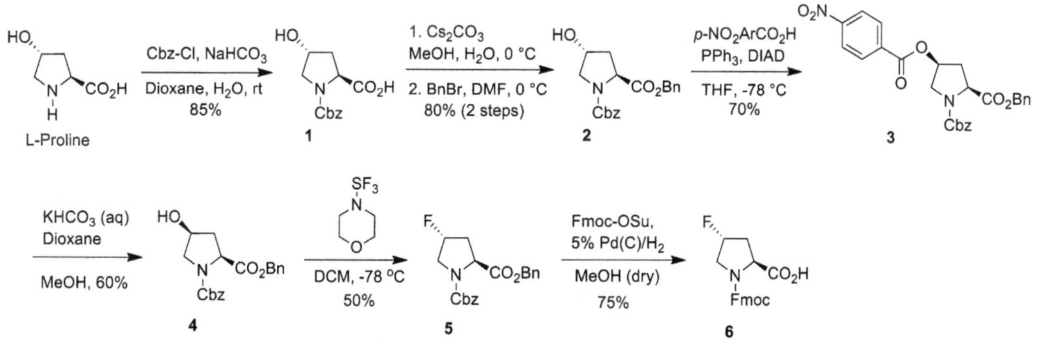

Fig. 5 Synthesis of Fmoc-(2*S*,4*R*)-4-fluoroproline (Flp)

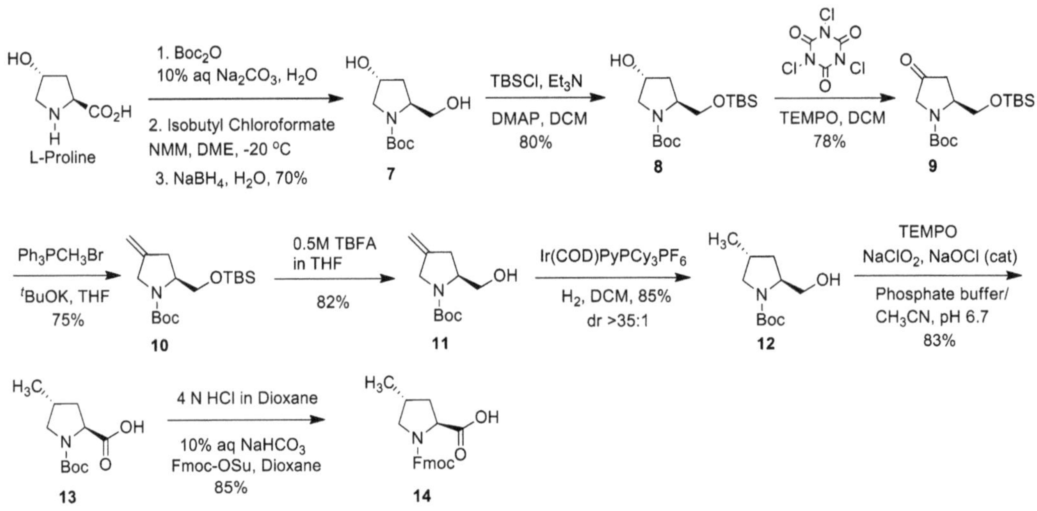

Fig. 6 Synthesis of Fmoc-(2*S*,4*R*)-4-methylproline (mep)

The strategy for the synthesis of the Fmoc-*trans*-Flp entails initial Cbz-protection of the amine nitrogen followed by benzylation of the carboxylic acid group. Cbz protection of *trans*-4(*R*)-hydroxy-L-proline is performed by using standard protocols to obtain **1** in 85 % of yield (Fig. 5) [67]. The conversion of *N*-Cbz-4(*R*)-Hyp **1** to *N*-Cbz-4(*R*)-Hyp-OBzl **2** is achieved in 80 % yield by reaction with $CsCO_3$ and BnBr [68]. Mitsunobu reaction of *N*-Cbz-4(*R*)-Hyp-OBzl **2** followed by hydrolysis results in the *N*-Cbz-4-Hyp derivative **4** in 42 % (two-step overall yield) yield with inversion of configuration at C-4 position [69]. The next step is the synthesis of diastereomeric fluoroproline by a stereospecific displacement of the hydroxyl group of Hyp by fluorine. Fluorination of *N*-Cbz-4(*S*)-Hyp-OBzl **4** is performed using diethylaminosulfur trifluoride (DAST) as fluorinating agent to yield *N*-Cbz-4(*R*)-Flp-OBzl **5** in 50 % yield [69]. Finally, catalytic hydrogenation of **5** with 5 % Pd/C under hydrogen (atm. pressure) in the presence of Fmoc-OSu affords desired *N*-Fmoc-4-Flp **6** in good yields (Fig. 5) [70]. This one-pot reaction involved the simultaneous removal of the Cbz- and benzyl-protecting groups and re-protection of amine nitrogen with Fmoc group. Optimal yield (~75 %) is observed with a reaction time of 2 h, whereas prolonged exposure of the reaction mixture to hydrogenation resulted in partial Fmoc deprotection.

To synthesize Fmoc-*trans*-mep **14** (Fig. 6), initially *N-Boc-trans*-mep **13** is synthesized by following a straightforward method reported by Goodman et al. [71]. The reaction starts with initial Boc protection of commercially available *trans*-4(*R*)-hydroxy-L-proline followed by carboxylic acid group reduction to provide Boc-protected alcohol **7**. Selective protection of the primary alcohol with TBDMSCl gave compound **8**, which is then oxidized with trichlorocyanuric acid and catalytic TEMPO to afford pyrrolidinone **9** in 78 % yields. Reaction of pyrrolidinone **9** with methyltriphenylphosphonium bromide gave olefin **10** in 75 % yields. Silyl-ether deprotection of **10** was first carried out to unmask the hydroxyl directing groups. Treatment of the resulting olefin **11** with 3 mol% of the Crabtree catalyst, under H_2 atm, gave excellent selectivity of *trans*-4-methyl pyrrolidine **12** in 85 % yields. Oxidation of the indolylprolinol derivative **12** in the presence of TEMPO, bleach, and sodium chlorite provided *N*-Boc-*trans*-mep **13**. Finally, deprotection of Boc group with 4 N HCl followed by Fmoc installation using Fmoc-OSu and aqueous $NaHCO_3$ in dioxane provided Fmoc-*trans*-mep **14** in 85 % yield [38].

1.4 Covalent Modification of Triple-Helical Peptides

Several covalent modification strategies have been employed to induce triple-helix structure formation and/or enhance their stability. Substantial stabilization can be achieved by use of a self-assembly approach where alignment of amphiphilic compounds at the lipid–solvent interface facilitates peptide alignment and structure initiation and propagation. Addition of lipophilic molecules at the *N*-terminus of the peptide, known as the "peptide-amphiphile" (PA) approach, often stabilizes self-associated peptides. The term peptide-amphiphile was first used in 1984, when an alanine residue was interposed between a charged head group and a double-chain pseudo-lipid tail [72]. In this approach pseudo-lipids such as mono- and di-alkyl chains are covalently attached to peptides to create peptide-amphiphiles that then associate via hydrophobic interactions (Fig. 4) [73–77]. Peptide-amphiphiles are advantageous in that they present a multivalent ligand [78] that is chemically well defined, avoiding loss of activity that can occur during nonspecific coupling of peptides to lipids [79]. The amphiphilic character of peptide-amphiphiles allows for the control of assembled structures by manipulating their molecular composition [80]. The thermal stability of triple-helical peptide-amphiphile head groups can be modulated by the length of the lipophilic moiety [73, 74, 76, 81, 82]. The thermal stability of the triple-helical structure in the peptide-amphiphile was found to increase as the monoalkyl tail chain length was increased over a range of C_6–C_{16} [74]. Conversely, alterations in the pseudo-lipid tail composition affect peptide-amphiphile aggregate structures [76, 80]. Desirable peptide head group melting temperature values can be achieved for in vivo use, as triple-helical PAs have been constructed with T_m values ranging from 30 to 70 °C [64, 73, 74, 76, 81–85].

The stability of associated THPs may also be regulated by photolysis. Modification of the sequence (Gly-Pro-Hyp)$_3$-Gly-Cys-Hyp-Gly-Pro-Hyp-Gly-Pro-Cys-(Gly-Pro-Hyp)$_5$-Gly-Gly-NH_2 with an azobenzene bridge between the Cys residues resulted in a THP whose stability decreased up irradiation at $\lambda = 330$ nm at

27 °C [86]. Unfortunately, the poor solubility of this peptide prohibited quantitative analysis of its stability [86].

The use of template-assisted approaches is another important strategy to create stable triple-helical conformation. In recent years much progress has been made in template-assisted synthetic methodologies. These include orthogonal protection schemes, chemoselective ligation, and the design of novel and increasingly flexible templates. Many templates have been used to create stable THPs. The main categories of template strategies involve a scaffold to which the peptides are covalently attached or the use of metal ions for non-covalent association. The covalent templates include (1) multi-Lys branching [87–91], (2) a double-disulfide "knot" (cystine knots) [53, 92–96], (3) *cis,cis*-1,3,5-trimethylcyclohexane-1,3,5-tricarboxylic acid (Kemp triacid; KTA) [97–101], (4) tris(2-aminoethyl)amine (TREN) [102], (5) cyclotriveratrylene (CVT) [103], (6) macrocyclic scaffold [104], or (7) carbohydrates (Fig. 7) [105, 106]. Alternatively, the metal ions Ca^{2+}, Ru^{2+}, Ni^{2+}, Fe^{2+}, Cd^{2+}, and Hg^{2+} have been used as templates in conjugation with amino acids or *N*-terminal pyridyl or bipyridyl functionalities to create multistrand linkages [107–109].

A significant strategy to stabilize THPs is the covalent attachment of three strands via a *C*-terminal branch. The *C*-terminal branch is expected to align and entropically stabilize the *C*-terminus of the THP and thus enhance triple-helical thermal stability [90] and to provide a model of the disulfide-linked *C*-terminus of type III collagen [110, 111]. Branching can be achieved by selective deprotection of Lys N^{α}- and N^{ε}-amino groups.

Solid-phase assembly of triple-helical collagen-model peptides using a Lys-Lys C-terminal branch requires three different protecting group strategies (Fig. 8): N^{α}-amino protection (**A**); Lys N^{ε}-amino side-chain protection (**B**), which must be stable to the N^{α}-amino group removal conditions; and C^{α}-carboxyl protection (*linker*), which must be stable to both the N^{α}- and N^{ε}-amino-protecting group removal conditions. Based on the three different protecting group strategies, four different branching combinations have been developed [90, 112]. Branching is achieved by synthesizing **A**-[Lys(**B**)]$_2$-Tyr(**C**)-Gly-*linker* resin and deprotecting the N^{α}- and N^{ε}-amino groups. Tyr was incorporated prior to branching to provide a convenient chromophore for eventual concentration determination. Incorporation of Ahx (6-aminohexanoic acid) onto the *N*-termini of all three strands provided a flexible spacer. All solid-phase methods to prepare THPs were based on Fmoc N^{α}-amino group protection. For example, the type (IV) collagen-derived α(IV)1263-1277 branch was assembled by a three-dimensional orthogonal strategy, where **B** was *tert*-butyloxycarbonyl (Boc) and *linker* was 4-trityloxy-*Z*-but-2-enyloxyacetic acid [cleaved by $(Ph_3P)_4Pd$-catalyzed nucleophilic transfer] [91]. A (Gly-Pro-Hyp)$_8$ branch has also been assembled by using the

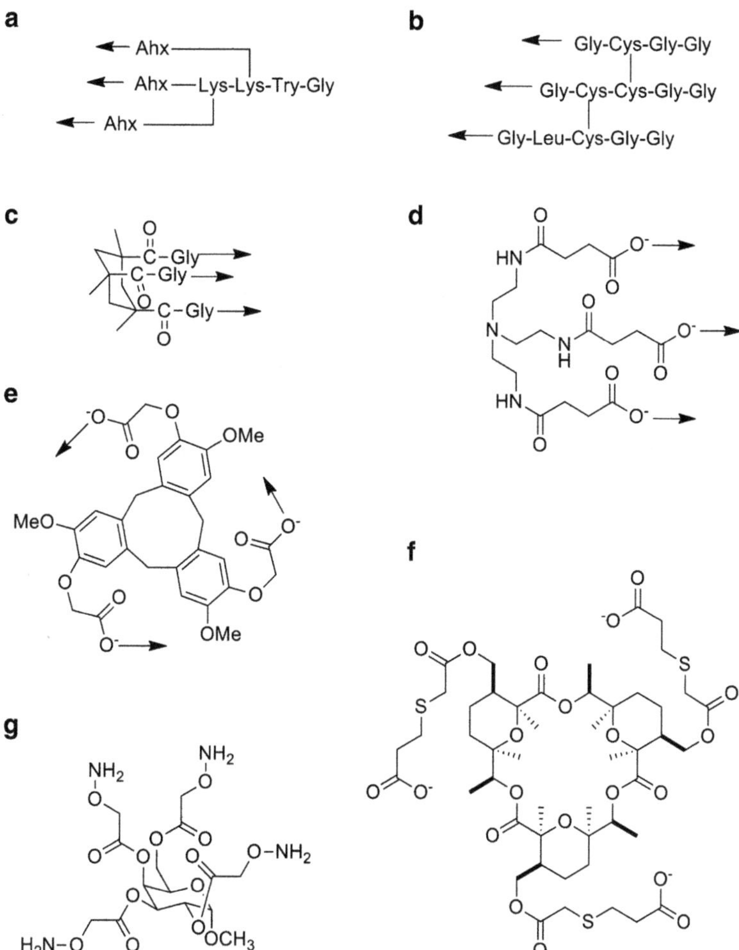

Fig. 7 Templates used for the chemical synthesis of triple-helical peptides: (**a**) di-Lys branch after coupling to 6-aminohexanoic acid (Ahx); (**b**) disulfide bridge (cystine knot); (**c**) *cis,cis*-1,3,5-trimethylcyclohexane-1,3,5-tricarboxylic acid (KTA) after coupling to Gly; (**d**) tris(2-aminoethyl)amine (TREN) after coupling to succinic acid; (**e**) cyclotriveratrylene (CTV) after coupling to a bromoethanoic acid; (**f**) macrocyclic; and (**g**) methyl 2,3,4,6-tetra-O-Aoa-α-D-Galp. The *arrows* indicate the direction of collagen-like sequence incorporation

same orthogonal strategy, where **B** was allyloxycarbonyl (Aloc) [cleaved by $(Ph_3P)_4Pd$-catalyzed nucleophilic transfer] and linker was hydroxymethylphenoxy (HMP; cleaved by TFA). The third three-dimensional orthogonal strategy was developed using the 1-(4,4-dimethyl-2,6-dioxocyclohex-1-ylidene)ethyl (Dde) group as **B** (cleaved by hydrazine/DMF) and the linker was HMP [90]. Due to the mild conditions the Fmoc/Dde combination (with ivDde replacing Dde) is now the most preferred. For example, Farndale, Barnes, and colleagues reported the synthesis of THPs, based on the bovine α_1(III)CB4 fragment, via this *C*-terminal covalent attachment strategy [113–115].

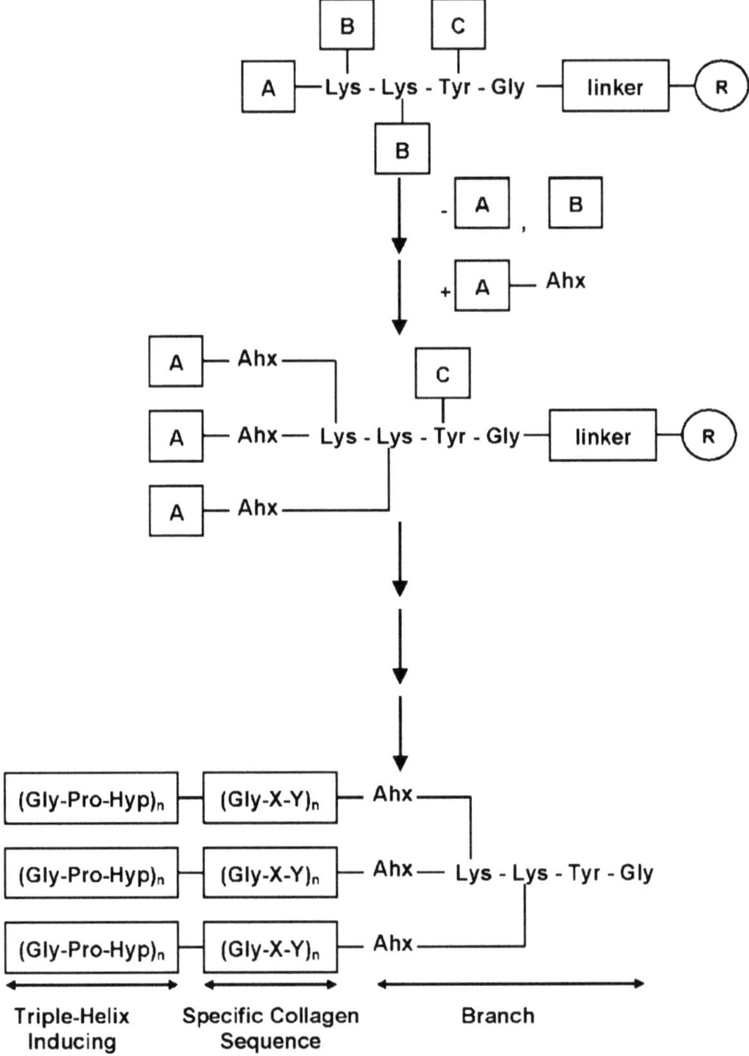

Fig. 8 General scheme for the synthesis of branched, triple-helical peptides. Ahx is 6-aminohexanoic acid. Reprinted with permission from *Biopolymers*, copyright 1993, Wiley and Sons

A double-disulfide "knot" (cystine knots) either at the *C*-terminal region or at the *N*-terminal region of three peptide strands is another frequently used strategy to stabilize THPs [3, 92, 94–96]. One representative example of a *C*-terminal cystine knot is construction of branched heterotrimeric THP **15** (Fig. 9). For this peptide all three individual chains are constructed by solid-phase methodology using Fmoc chemistry. The α1 chain contained either 3 or 5 Gly-Pro-Hyp repeats on the *N*-terminus of Gly-Pro-Gln-Gly-Ile-Ala-Gly-Gln-Arg-Gly-Val-Val-Gly-Cys(Acm)-Gly-Gly-OH, where Acm was acetamidomethyl, the α2 chain contained

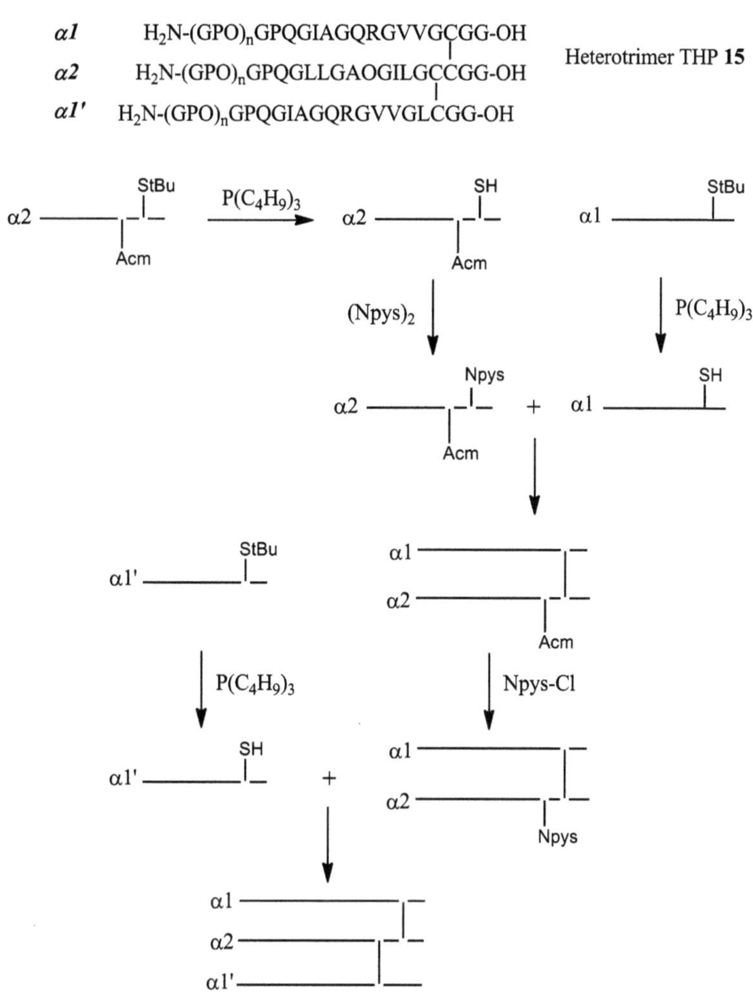

Fig. 9 Heterotrimer THP **15** and general scheme for regioselective assembly of the α1, α2, and α1′ cysteine peptides into heterotrimers with the α1α2α1′ register

3 or 5 Gly-Pro-Hyp repeats on the *N*-terminus of Gly-Pro-Gln-Gly-Leu-Leu-Gly-Ala-Hyp-Gly-Ile-Leu-Gly-Cys(Acm)-Cys(S*t*Bu)-Gly-Gly-OH, and the α1' chain contained either 3 or 5 Gly-Pro-Hyp repeats on the *N*-terminus of Gly-Pro-Gln-Gly-Ile-Ala-Gly-Gln-Arg-Gly-Val-Val-Gly-Leu-Cys(S*t*Bu)-Gly-Gly-OH. The three chains are assembled into the heterotrimer by stepwise regioselective cross-linking. Peptide α1 is treated with 3-nitro-2-pyridine-sulfenyl chloride (Npys-Cl) to convert Cys(Acm) to Cys(Npys). Peptide α2 is treated with $P(C_4H_9)_3$ to remove the S*t*Bu group. Peptides α1 and α2 are reacted at pH 4.5 to form a dimer. The dimer is treated with Npys-Cl to convert the α2 chain Cys(Acm) to Cys(Npys). The α1' chain is treated with $P(C_4H_9)_3$ to remove the S*t*Bu group. Peptide α1' and the dimer are reacted at pH 4.5 to form a trimer.

Another example of an *N*-terminal cystine knot is construction of branched THPs **16–19** (Fig. 10) using the same protocol described for the *C*-terminally branched cystine knot heterotrimeric THP [53]. Heterologous trimerization is achieved by stepwise disulfide bond-forming reactions (Fig. 9). First, the free thiol groups of the biscysteinyl peptides (**20a**, **20b**) are converted to 2-pyridylthio (PyS) groups by treatment with PySSPy. The purified PyS-activated peptides (**21a**, **21b**) are then mixed with monocysteinyl peptides (**22a**, **22b**) to yield corresponding heterodimeric peptides (**23a**, **23b**). Formation of small amounts of homodimers is also observed in this reaction. The S-Acm groups in the heterodimers are converted to S-3-nitro-2-pyridinesulfenyl (Npys) groups by treating them with NpysCl. Even under optimized conditions, this reaction generated some by-products, including homodimers. Finally, the Npys-activated dimers (**24a**, **24b**) were

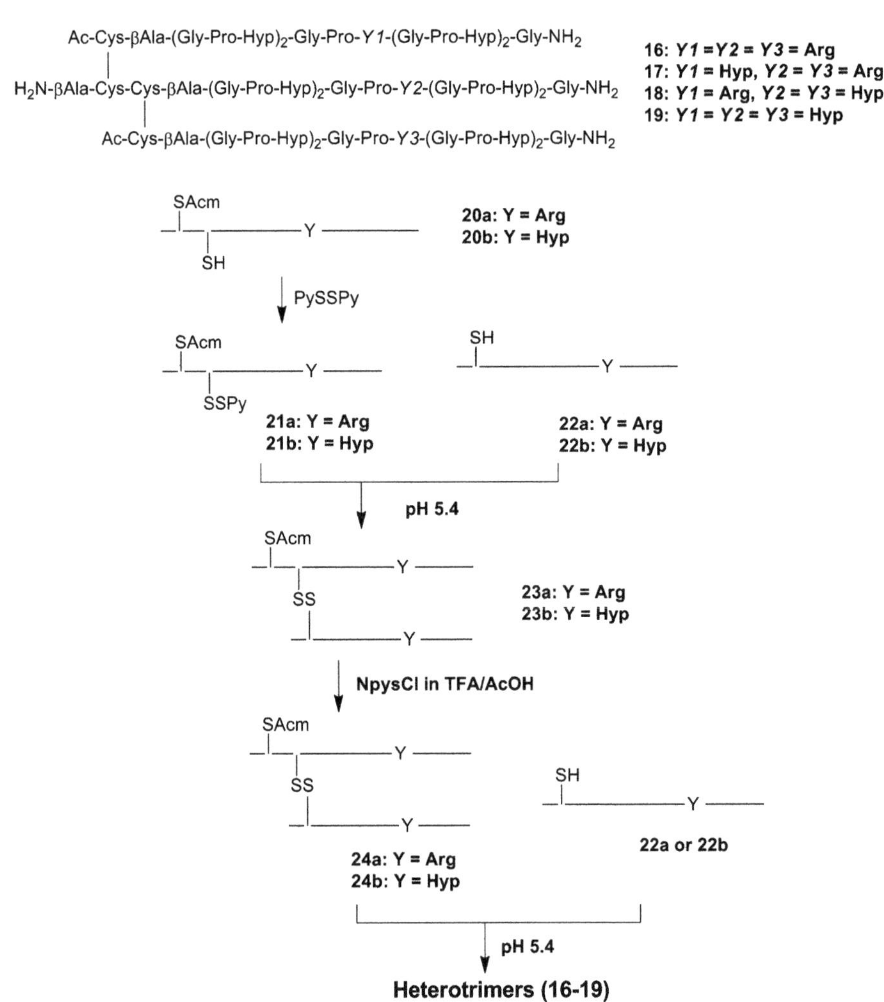

Fig. 10 Synthesis of heterotimeric THPs utilizing an *N*-terminal cystine knot

mixed with monocysteinyl peptides (**22a**, **22b**) to yield desired heterotrimers.

Stabilization of THPs can also be achieved by cross-linking both the *C*- and *N*-terminal ends of three peptide chains. This strategy required synthesis of three different fragments (Fig. 11) [88]. The first synthesis was of linear peptides of the sequence Cys-Gly-(Gly-Pro-Hyp)$_n$-Lys(Ser)-NH_2 ($n = 3$–10) **25** using Fmoc chemistry, starting from Fmoc-Lys[Boc-Ser(*t*Bu)]-OH, which in turn was prepared from reaction of Boc-Ser(*t*Bu)-OSu with Fmoc-Lys-OH. The second synthesis was of two different trivalent anchor molecules: tris-bromoacetylated Lys-Lys dimer **26** and tris-aminooxyacetylated branch peptide **27** (synthesized from Boc-aminooxyacetic acid). Chemoselective ligation of **25** and **26** followed by oxidation of the Lys(Ser) residue created an aldehyde group for the next ligation site. Final ligation of the *N*-terminus-coupled peptide with the tris-aminooxyacetyl group of anchor molecule **27** in aq. buffer produced di-cross-linked peptide **30**.

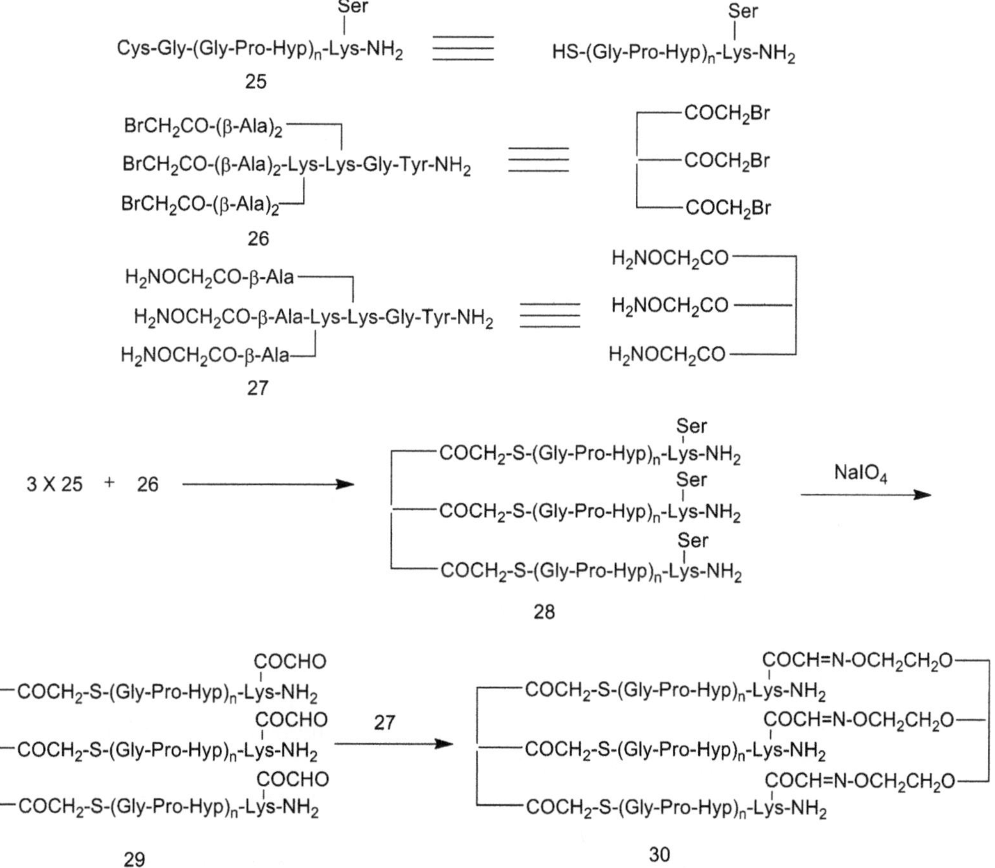

Fig. 11 Structures of the collagen-model peptide **25** and the linker peptides **26** and **27**, and the synthetic scheme for the assembly of cross-linked THP **30**

Use of the conformationally constrained organic template *cis,cis*-1,3,5-trimethylcyclohexane 1,3,5-tricarboxylic acid (Kemp triacid, KTA, Fig. 7) is another strategy to induce or stabilize triple-helical structures [97–101]. This template possesses three carboxyl groups which can be coupled to the *N*-termini of three peptide chains. Attachment of a Gly residue as a spacer between each peptide chain and each carboxyl group on KTA is vital to compensate for the difference in diameters between the KTA and the collagen triple-helix and to facilitate the synthesis [97]. Two synthetic routes have been used to prepare KTA template-assembled collagen-based structures. The first method exclusively utilized solid-phase methodology [101], whereas in the second method formation of the peptide bond between KTA and peptide–peptoid chains was in solution [99]. In the solid-phase method Boc-(Gly-Pro-Hyp)$_n$ MBHA resin is prepared and then the Boc group is removed. KTA-(Gly-OH)$_3$ is then coupled to the *N*-termini of peptide chains using DIPCDI and HOBt. In the second method peptide–peptoid chains are synthesized by solid-phase methods and cleaved from the resin. The free amine peptide–peptoid chains are then coupled with KTA-(Gly-OH)$_3$ in solution using EDC and HOBt.

TREN (Fig. 7) is another beneficial scaffold for the synthesis of template THPs [102]. The TREN molecule is a flexible tripodal structure with three aminoethyl groups. TREN-(suc-OH)$_3$ template is prepared by coupling TREN to monobenzylated succinate, and removing the benzyl ester groups by hydrogenation. The succinic acid groups extend the flexibility of the scaffold and provide terminal carboxylates for attachment of peptide chains. The succinic acid spacers release any steric hindrance by extending the reactive sites. TREN-(suc-OH)$_3$ is acylated to three peptide strands simultaneously on the solid phase using DIC and HOBt.

Another synthetic scaffold molecule, which can be attached to both model and native collagen sequences, is cone-shaped CVT (Fig. 7) [103]. Attachment to the *N*-terminus of the collagen peptides with CVT is done in solution using BOP as a coupling reagent.

Stable collagen triple-helices are also synthesized by using a macrocyclic scaffold [104]. This scaffold belongs to a class of 18-membered cyclic hydropyran oligolides with alternating ester and ether linkages. In this scaffold, three ligand attachment sites form an equilateral triangle on one face. Macrocyclic template is coupled to the *N*-terminus of collagen-model peptides in solution using PyBOP and DIEA.

The use of carbohydrates as potential templates is another strategy for the *de novo* design of triple-helical peptides (Fig. 7) [105, 106]. Carbohydrates are multifunctional molecules having comparative rigidity of ring forms, simplicity in regioselective manipulation of functional groups, and access to mono- and di-saccharide stereoisomers. The primary and secondary hydroxyl of mono- and di-saccharides provide a more flexible control of the directionality and distances among attachment points of the peptide chains. This carbopeptide strategy starts with aminooxy-acetyl (Aoa) functionalization of the methyl α-D-galactopyranose (d-Gal*p*) derivative to obtain template molecules followed by *C*-terminal peptide aldehyde coupling to the template via oxime formation.

The simplified *N*-terminal di-Glu template Gly-Phe-Gly-Glu-Glu-Gly was assembled by solid-phase Fmoc methodology, isolated as Fmoc-Gly-Phe-Gly-Glu-Glu-Gly, and acylated to three peptide strands simultaneously on the solid phase using 2-(1H-benzotriazole-1-yl)-1,1,3,3-tetramethyluronium hexafluorophosphate (HBTU)/1-hydroxybenzotriazole (HOBt) [89].

As discussed earlier, heterotrimeric triple-helical peptides have usually been prepared using (1) disulfide-mediated cross-linking or (2) electrostatic-assisted self-assembly. More recently, "2 + 1 strand click synthesis" has been utilized to synthesize discrete homo- and heterotrimeric triple-helical peptides whereby the *C*-termini of the peptides are chemically stapled using the Huisgen Cu(I)-catalyzed azide-alkyne cycloaddition reaction [116]. The "2 + 1 strand click synthesis" requires assembly of a two-stranded branched peptide possessing a *C*-terminal 6-azidelysine residue, which has the capacity to "click" to a single-stranded peptide bearing a (2S)-propargylglycinamide at the *C*-terminus.

Metal chelation also plays an important role in the construction of templated triple-helices [107–109]. As an example the addition of a dopamine residue on the *N*-terminus of (Gly-Nleu-Pro)$_6$ via a flexible succinic acid linker followed by incubation with Fe^{3+} converted a non-triple-helical peptide to a triple-helical peptide with a T_m value of 28 °C [107]. In similar fashion, addition of hydroxamic acid on the *C*-terminus of Boc-β-Ala-TRIS-[(Gly-Pro-Nleu)$_6$]$_3$, followed by incubation with Fe^{3+}, resulted in a stabilized triple-helical structure (7 °C increase in T_m) [108]. Based on metal ion chelation, Chiemelewski et al. reported the self-assembly of THPs to form microflorettes [109]. To obtain collagen-based macromolecular assemblies with temporal control, the peptide design consisted of a central collagen-based core composed of nine repeating units of the tripeptide Pro-Hyp-Gly and distinct metal-binding ligands at each termini (**NCoH**, Fig. 12). The metal-binding ligands were a His$_2$ moiety on the *C*-terminus and a nitrilotriacetic acid unit on the *N*-terminus. The core triple-helical unit has been used in the formation of collagen-like peptide fibers in solution upon the introduction of Zn^{2+}, Cu^{2+}, Ni^{2+}, or Co^{2+}. This assembly process was found to be fully reversible using EDTA as a metal ion chelator. The assembly process proceeds under mild conditions using neutrally buffered aqueous solution at room temperature.

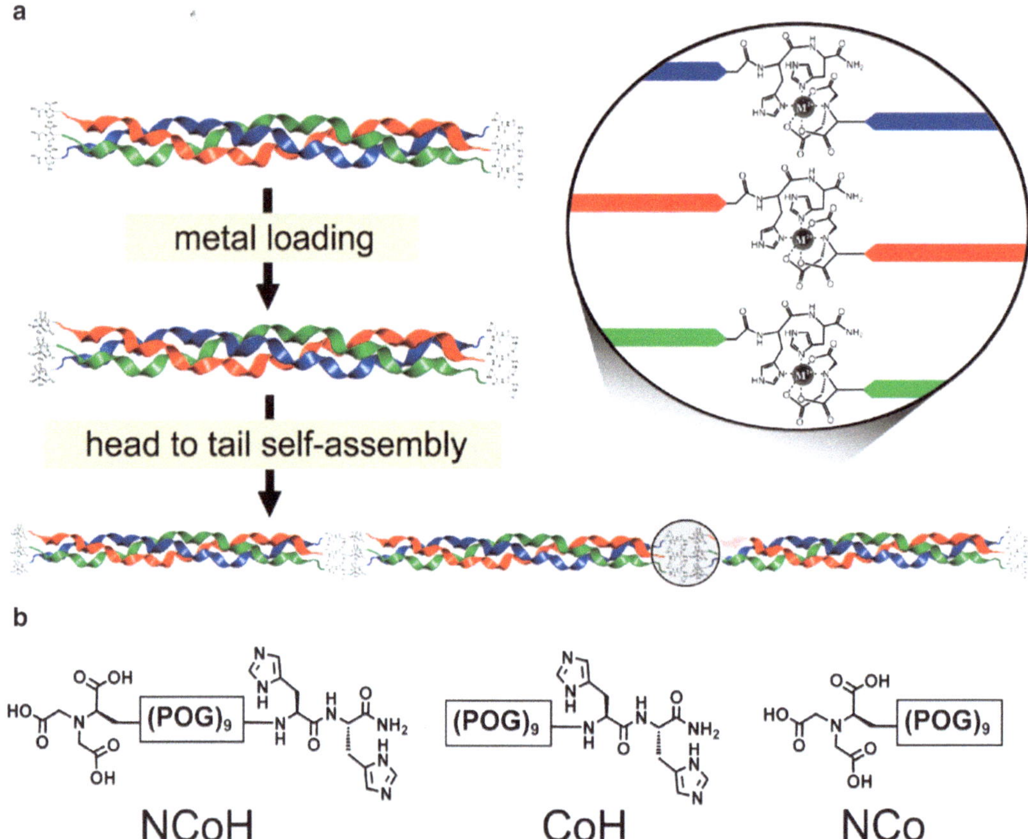

Fig. 12 (**a**) Schematic representation of the design of the **NCoH** THP and assembly into higher order structures. Following triple-helix formation, the addition of metal ions triggers an initial assembly directed by the NTA and His_2 moieties. (**b**) Structures of peptides **NCoH**, **CoH**, and **NCo**. Reprinted with permission from J Am Chem Soc, copyright 2009, American Chemical Society

The effect of templates is often to enhance the thermal stability of collagen-like sequences. For example, the THP acetyl-Gly-Gly-(Pro-Hyp-Gly)$_5$-NH_2 has a T_m value of 9.2 °C, while template versions of the same sequence, using either (+)CTV or KTA, have T_m values of 58 and 62 °C, respectively [103]. Similarly, (Gly-Pro-Hyp)$_6$ has a $T_m = 25.4$ °C, (Pro-Hyp-Gly)$_6$ with a *C*-terminal branch has a $T_m = 39.4$ °C, (Gly-Pro-Hyp)$_6$ with an *N*-terminal branch has a $T_m = 56.2$ °C, and (Gly-Pro-Hyp)$_6$ with both an *N*- and *C*-terminal branch has a $T_m = 69.7$ °C [88, 117]. An *N*-terminal macrocyclic template induced triple-helical structure of the sequence Gly-(Pro-Pro-Gly)$_7$-NH_2 ($T_m = 39.9$ °C), whereas the non-templated sequence was not triple-helical [104]. Triple-helices containing peptoid residues, such as *N*-isobutylglycine (Nleu), have been stabilized by templates. Acetyl-(Gly-Nleu-Pro)$_6$-NH_2 had a $T_m = 26$ °C, while KTA-[Gly-(Gly-Nleu-Pro)$_6$-NH_2]$_3$, TREN-[suc-(Gly-Nleu-Pro)$_6$-NH_2]$_3$, and Boc-β-Ala-TRIS-[(Gly-Nleu-Pro)$_6$-OCH_3]$_3$ had T_m values of 36, 46, and 33 °C, respectively [100, 102, 118].

The thermal stability of the α2β1 integrin-binding sequence (Gly-Pro-Hyp)$_3$-Gly-Phe-Hyp-Gly-Glu-Arg-(Gly-Pro-Hyp)$_3$ was enhanced by an *N*-terminal Gly-Phe-Gly-Glu-Glu-Gly template, resulting in an increase in T_m from 25 to 44 °C [89]. Acetyl-(Pro-Hyp-Gly)$_5$-Pro-Cys(S*t*Bu)-Cys(S*t*Bu)-Gly-Gly-Gly-NH_2 had a T_m = 20.3 °C, while the double-disulfide-linked [acetyl-(Pro-Hyp-Gly)$_5$-Pro-Cys-Cys-Gly-Gly-Gly-NH_2]$_3$ had a T_m = 68.1 °C [119]. Formation of the cystine knot in (Gly-Pro-Pro)$_3$-Gly-Pro-Arg-Gly-Glu-Lys-Gly-Glu-Arg-Gly-Pro-Arg-(Gly-Pro-Pro)$_3$-Gly-Pro-Cys-Cys-Gly increased T_m from 35 to 43 °C [120].

2 Materials

1. Dimethylformamide (DMF), *N*-methylpyrrolidone (NMP), *N*,*N*-diisopropylethylamine (DIEA), trifluoroacetic acid (TFA, peptide synthesis grade), methyl-*tert*-butyl ether (MTBE), 1,8-diaza-bicyclo[5.40.]undec-7-ene (DBU), dimethyl sulfoxide (DMSO), acetonitrile (peptide synthesis or HPLC grade), sodium acetate, ammonium acetate, NH_4HCO_3, isopropanol (Optima/HPLC grade), β-mercaptoethanol (molecular biology grade), thioanisole, ethanedithiol, phenol, Tris–HCl, NaCl, $CaCl_2$, and brij-35 are purchased from Fisher Scientific (Atlanta, GA).
2. Piperidine is purchased from AnaSpec (San Jose, CA).
3. Fmoc-amino acids, *N*-[(1H-benzotriazol-1-yl)(dimethylamino)methylene]-*N*-methylmethanaminium hexafluorophosphate *N*-oxide (HBTU), 1-H-benzotriazolium-1-[*bis*(dimethylamino)methylene]-5-chloro-hexafluorophosphate-(1-), 3-oxide (HCTU), 1-hydroxybenzotriazole (HOBt), Fmoc-Gly-Sasrin resin, Fmoc-Rink Amide 4-methylbenzhydrylamine (MBHA) resin, and NovaSyn TGR PEG resin (*see* **Note 1**) are purchased from EMD Biosciences (San Diego, CA).
4. 5,5'-Dithiobis(2-nitrobenzoic acid), α-cyano-4-hydroxycinnamic acid (MS grade), and 2-hydroxyethyl disulfide are purchased from Sigma-Aldrich (St. Louis, MO).
5. Alamar Blue is obtained from BioSource International (Camarillo, CA).
6. The monoalkyl chains hexanoic acid [CH_3–$(CH_2)_4$–CO_2H, designated C_6], octanoic acid [CH_3–$(CH_2)_6$–CO_2H, designated C_8], decanoic acid [CH_3–$(CH_2)_8$–CO_2H, designated C_{10}], dodecanoic acid [CH_3–$(CH_2)_{10}$–CO_2H, designated C_{12}], tetradecanoic acid [CH_3–$(CH_2)_{12}$–CO_2H, designated C_{14}], palmitic acid [CH_3–$(CH_2)_{14}$–CO_2H, designated C_{16}], and stearic acid [CH_3–$(CH_2)_{16}$–CO_2H, designated C_{18}] are purchased from Sigma-Aldrich.

3 Methods

3.1 Synthesis and Characterization of Peptide-Amphiphiles

1. Incorporation of individual amino acids is by Fmoc solid-phase methodology using cycles described previously [90, 91].
2. For Fmoc removal, a solution of DBU–piperidine–NMP (1:5:44) is used for 5–10 and 10–15 min.
3. All couplings are performed with HCTU/HOBt.
4. Acylation of the peptide–resin involves condensing a fourfold molar excess of alkyl acid to the N^{α}-deprotected resin, with a 3.8-fold molar excess each of HBTU and HOBt in DMF for 2 h. The reaction is initiated by the addition of an eightfold molar excess of DIEA and proceeds for 1 h.
5. Cleavage and side-chain deprotection of the lipidated peptide–resins is accomplished by treating the resin for 1–2 h with either H_2O–TFA (1:19) or ethanedithiol–thioanisole–phenol–H_2O–TFA (2.5:5:5:5:82.5) [121, 122].
6. Resins are filtered and rinsed with TFA, and the combined filtrate and wash reduced under vacuum at room temperature to ~0.5 mL and precipitated with MTBE.
7. The precipitate is centrifuged and washed three times with MTBE, dissolved in acetonitrile–H_2O (1:9–1:4), and purified by preparative reversed-phase HPLC.
8. Preparative reversed-phase HPLC is performed on a Rainin AutoPrep System with a Vydac 218TP152022 C_{18} or 214TP152022 C_4 column (15–20 μm particle size, 300 Å pore size, 250 mm × 22 mm) at a flow rate of 10 mL/min with 0.1 % TFA in H_2O (A) and 0.1 % TFA in acetonitrile (B), and detection at $\lambda = 220$ nm. For very hydrophobic samples, it is advantageous to use 0.1 % TFA in isopropanol instead of acetonitrile [73, 74]. Analytical HPLC is performed on a Hewlett-Packard 1100 Liquid Chromatograph equipped with an ODS Hypersil C-18 column (5 μm particle size, 100 mm × 2.1 mm).
9. PA composition can be determined by MALDI-TOF mass spectrometry (MS) [64, 73, 84, 85, 123–125]. PA homogeneity can also be evaluated by diphenyl and nonporous C18 reversed-phase HPLC [124] and/or hydrophobic interaction HPLC.
10. Triple-helicity is monitored by circular dichroism (CD) spectroscopy in the far UV wavelengths [64, 73, 74, 83–85, 90, 91, 125]. Peptides are dissolved in the appropriate buffer at concentrations of 2–500 μM. Conditions should reflect final relevant assay conditions as much as possible based on limitations imposed by sample availability and/or instrumentation. For example, if the peptides are to be used in a cell-binding assay at 10 μM in PBS, then the CD spectra should be acquired

at that peptide concentration in a similar buffer. Since chloride ions interfere with the acquisition of data, a phosphate buffer can be substituted for PBS. A typical spectrum for a triple-helix shows a positive molar ellipticity at $\lambda \sim 225$ nm and a negative molar ellipticity at $\lambda \sim 205$ nm. In addition to the far UV wavelength scan of $\lambda \sim 195$–250 nm, one can determine the melting temperature by monitoring the change in molar ellipticity at $\lambda = 225$ nm with a constant change in temperature from 5 to 85 °C. For samples exhibiting sigmoidal melting curves, the inflection point in the transition region (first derivative) is defined as T_m. Alternatively, T_m is evaluated from the midpoint of the transition.

11. NMR spectroscopy can also be used to study the relative thermal stability, alignment, and flexibility (backbone mobilities) of triple-helical PAs [73, 74, 126].

3.2 Template-Assembled Synthetic THP Methods

3.2.1 Preparation of [N-tris(Fmoc-Ahx)-Lys-Lys]-Tyr-Gly Branched Template

1. Deprotect 1 g of Fmoc-Gly-Sasrin resin with 20 % piperidine in DMF for 30 min followed by a second 5-min treatment.
2. Wash the resin three times with DMF. If the substitution level is higher than 0.5 mmol/g, add 0.2 mmol benzoic anhydride dissolved in 25 ml DMF and 0.4 mmol DIEA, mix for 1 h, and wash the resin three times with DMF.
3. Dissolve 3 eq of Fmoc-Tyr(*t*Bu) in approximately 10 mL DMF. Add 2.7 eq of HBTU and 3 eq of HOBt, and then add the mixture to resin. Add 6 eq of DIEA, mix for 30–60 min, and wash the resin with DMF three times.
4. Remove Fmoc as indicated above and wash the resin three times with DMF.
5. Dissolve 3 eq of Fmoc-Lys(Dde) (*see* **Note 1**) in approximately 10 mL DMF. Add 2.7 eq of HBTU and 3 eq of HOBt and then add the mixture to the resin. Add 6 eq of DIEA, shake for 30–60 min, and wash with DMF three times.
6. Repeat the Fmoc removal and Fmoc-Lys(Dde) coupling steps such that the resin contains two adjacent Lys(Dde) residues (*see* **Note 2**).
7. If the desired construct is homotrimeric (*see* **Note 3**), the Dde and Fmoc groups can be simultaneously removed by treatment with 2 % hydrazine in DMF for 30 min followed by three rinses with DMF. An optional spacer of Fmoc-Ahx acid can be added to reduce the steric hindrance of the branch.
8. Keeping in mind that the chain is now trimeric, dissolve 3 eq of Fmoc-Ahx in approximately 10 mL DMF, add 2.7 eq of HBTU and 3 eq of HOBt, and add the mixture to the resin. Add 6 eq of DIEA, mix for 60 min, and wash three times with DMF.

9. A small volume of template is cleaved from the resin using H_2O–TFA (5:95) for 1 h and analyzed by MALDI-TOF mass spectrometry to confirm proper branch assembly. This scale of template is sufficient for three 0.25 mmol syntheses.

3.2.2 Peptide Synthesis, Side-Chain Deprotection, and Peptide–Resin Cleavage

1. Incorporation of individual amino acids is by Fmoc solid-phase methodology [90, 91].
2. For Fmoc removal, a solution of DBU–piperidine–NMP (1:5:44) is used for 5–10 and 10–15 min.
3. All couplings are performed with HCTU/HOBt.
4. Peptide–resins are cleaved and side-chain deprotected by treatment with H_2O–TFA (5:95), H_2O–thioanisole–TFA (5:5:90), or ethanedithiol–H_2O–thioanisole–phenol–TFA (2.5:5:5:5:82.5) for ≥2 h depending upon the peptide sequence [121, 122].
5. Resins are filtered and rinsed with TFA, and the combined filtrate and wash reduced under vacuum at room temperature to ~0.5 mL and precipitated with MTBE.
6. The precipitate is centrifuged, washed three times with MTBE, dissolved in acetonitrile–H_2O (1:9–1:4), and purified by preparative reversed-phase HPLC.

3.2.3 Purification and Characterization of Peptide-Template

1. Preparative reversed-phase HPLC is performed with a Vydac 218TP152022 C18 column (15–20 μm particle size, 300 Å pore size, 250 mm × 22 mm) at a flow rate of 10 mL/min with 0.1 % TFA in H_2O (A) and 0.1 % TFA in acetonitrile (B), and detection at $\lambda = 220$ nm.
2. The eluent is lyophilized to a powder and repurified using a semipreparative Vydac 219TP54 diphenyl column (5 μm particle size, 300 Å pore size, 250 mm × 4.6 mm) at a flow rate of 2 mL/min with detection at $\lambda = 220$ nm.
3. Analytical HPLC is performed with an ODS Hypersil C-18 column (5 μm particle size, 100 mm × 2.1 mm).
4. Chain assembly is confirmed by Edman degradation and MALDI-TOF mass spectrometry using a specialized matrix mixture containing 2,5-dihydroxybenzoic acid/2-hydroxy-5-methoxybenzoic acid (9:1, v/v) [84, 117].

4 Notes

1. By utilizing the array of readily available orthogonally protected Lys derivatives, including, but not limited to, Fmoc-Lys(Dde), Dde-Lys(Fmoc), Fmoc-Lys(Fmoc), Fmoc-Lys(Mmt), Fmoc-Lys(Mtt), or Fmoc-Lys(ivDde), an array of topological designs can be created. The dilute acid utilized for side-chain deprotection

of the Mmt and Mtt groups will cleave the template from the Gly-SASRIN resin. Thus, a more acid-stable resin such as Rink amide MBHA should be utilized in combination with Fmoc-Lys(Mmt) and Fmoc-Lys(Mtt).

2. One can create a variety of templates based on the simple Lys-branching method. The template can accommodate 2–8 branches with ease and can also create a number of homo- or hetero-multimeric sequences. However, as the number of branching points increases care must be taken during *N*-α-amino or *N*-ε-amino deprotection and subsequent couplings. The resin becomes increasingly "sticky" as the number of free amino groups increases, which can complicate subsequent deprotection and coupling reactions. Ensure that a proper solution volume or resin mixing action is taken to avoid deletions during peptide chain assembly.

3. One can also utilize the branching strategy to construct triple-helical peptides [112]. The synthesis of triple-helical peptides with one chain of different sequence from the other two requires a four-dimensional orthogonal scheme, where HMP is the linker and Dde or 1-(4,4-dimethyl-2,6-dioxocyclohex-1-ylidene)-3-methylbutyl (ivDde) [127, 128] side-chain protection of Fmoc-Lys is used. Synthesis by Fmoc strategy of one chain (i.e., the α2 collagen chain sequence) proceeds through the α-amino of Lys; thus, the Dde/ivDde and Fmoc groups are *not* removed simultaneously. The α2 chain sequence is assembled by Fmoc chemistry as described above. After incorporation of the α2 chain sequence, the peptide chain is "capped" by using allyloxycarbonyl (Aloc) chloride or Aloc-Gly, creating an *N*-terminal Aloc group. The Lys α-amino Dde or ivDde groups are then removed with hydrazine (see above), which does not remove Aloc, and synthesis by Fmoc strategy proceeds for the other two chains (i.e., α1 collagen chain sequence). Once all three chains are equivalent in terms of number of residues incorporated, the Aloc group is removed by treatment with $(Ph_3P)_4Pd$ in $CHCl_3$–acetic acid–*N*-methyl morpholine (20:1:0.5) [129] and the Fmoc groups removed as described above. Finally, Fmoc-Gly, -Pro, and -Hyp(*t*Bu) are incorporated into all three chains. Cleavage and side-chain deprotection are by TFA as described above.

References

1. Myllyharju J, Kivirikko KI (2001) Collagens and collagen-related diseases. Ann Med 33:7–21

2. Hashimoto T, Wakabayashi T, Watanabe A, Kowa H, Hosoda R, Nakamura A, Kanazawa I, Arai T, Takio K, Mann DM, Iwatsubo T (2002) CLAC: a novel Alzheimer amyloid plaque component derived from a transmembrane precursor, CLAC-P/collagen type XXV. EMBO J 21:1524–1534

3. Boulegue C, Musiol HJ, Gotz MG, Renner C, Moroder L (2008) Natural and artificial cystine knots for assembly of homo- and heterotrimeric collagen models. Antioxid Redox Signal 10:113–125
4. Gordon MK, Hahn RA (2010) Collagens. Cell Tissue Res 339:247–257
5. Shoulders MD, Raines RT (2009) Collagen structure and stability. Annu Rev Biochem 78:929–958
6. Cole WG (1994) Collagen genes: mutations affecting collagen structure and expression. Prog Nucleic Acid Res Mol Biol 47:29–80
7. Brodsky B, Shah NK (1995) Protein motifs. 8. The triple-helix motif in proteins. FASEB J 9:1537–1546
8. Fields GB, Prockop DJ (1996) Perspectives on the synthesis and application of triple-helical, collagen-model peptides. Biopolymers 40:345–357
9. Jenkins CL, Raines RT (2002) Insights on the conformational stability of collagen. Nat Prod Rep 19:49–59
10. Koide T (2005) Triple helical collagen-like peptides: engineering and applications in matrix biology. Connect Tissue Res 46: 131–141
11. Koide T (2007) Designed triple-helical peptides as tools for collagen biochemistry and matrix engineering. Philos Trans R Soc Lond B Biol Sci 362:1281–1291
12. Brodsky B, Thiagarajan G, Madhan B, Kar K (2008) Triple-helical peptides: an approach to collagen conformation, stability, and self-association. Biopolymers 89:345–353
13. Fields GB (2010) Synthesis and biological applications of collagen-model triple-helical peptides. Org Biomol Chem 8:1237–1258
14. Ramachandran GN, Kartha G (1954) Structure of collagen. Nature 174:269–270
15. Ramachandran GN, Kartha G (1955) Structure of collagen. Nature 176:593–595
16. Rich A, Crick FH (1955) The structure of collagen. Nature 176:915–916
17. Rich A, Crick FH (1961) The molecular structure of collagen. J Mol Biol 3:483–506
18. Bella J, Eaton M, Brodsky B, Berman HM (1994) Crystal and molecular structure of a collagen-like peptide at 1.9 A resolution. Science 266:75–81
19. Riddihough G (1998) Structure of collagen. Nat Struct Biol 5:858
20. Okuyama K, Miyama K, Mizuno K, Bachinger HP (2012) Crystal structure of (Gly-Pro-Hyp)(9): implications for the collagen molecular model. Biopolymers 97:607–616
21. Woodhead-Galloway J (1980) Collagen: the anatomy of a protein. Edward Arnold Limited, London
22. Jenkins CL, Vasbinder MM, Miller SJ, Raines RT (2005) Peptide bond isosteres: ester or (E)-alkene in the backbone of the collagen triple helix. Org Lett 7:2619–2622
23. Dai N, Wang XJ, Etzkorn FA (2008) The effect of a trans-locked Gly-Pro alkene isostere on collagen triple helix stability. J Am Chem Soc 130:5396–5397
24. Dai N, Etzkorn FA (2009) Cis-trans proline isomerization effects on collagen triple-helix stability are limited. J Am Chem Soc 131:13728–13732
25. Holmgren SK, Taylor KM, Bretscher LE, Raines RT (1998) Code for collagen's stability deciphered. Nature 392:666–667
26. Holmgren SK, Bretscher LE, Taylor KM, Raines RT (1999) A hyperstable collagen mimic. Chem Biol 6:63–70
27. Bretscher LE, Jenkins CL, Taylor KM, Derider ML, Raines RT (2001) Conformational stability of collagen relies on a stereoelectronic effect. J Am Chem Soc 123: 777–778
28. Eberhardt ES, Panisik N Jr, Raines RT (1996) Inductive effects on the energetics of prolyl peptide bond isomerization: implications for collagen folding and stability. J Am Chem Soc 118:12261–12266
29. Sakakibara S, Inouye K, Shudo K, Kishida Y, Kobayashi Y, Prockop DJ (1973) Synthesis of (Pro-Hyp-Gly) n of defined molecular weights. Evidence for the stabilization of collagen triple helix by hydroxypyroline. Biochim Biophys Acta 303:198–202
30. Kotch FW, Guzei IA, Raines RT (2008) Stabilization of the collagen triple helix by O-methylation of hydroxyproline residues. J Am Chem Soc 130:2952–2953
31. Nagarajan V, Kamitori S, Okuyama K (1999) Structure analysis of a collagen-model peptide with a (Pro-Hyp-Gly) sequence repeat. J Biochem 125:310–318
32. Inouye K, Sakakibara S, Prockop DJ (1976) Effects of the stereo-configuration of the hydroxyl group in 4-hydroxyproline on the triple-helical structures formed by homogenous peptides resembling collagen. Biochim Biophys Acta 420:133–141
33. Bann JG, Bachinger HP (2000) Glycosylation/hydroxylation-induced stabilization of the collagen triple helix. 4-trans-hydroxyproline in the Xaa position can stabilize the triple helix. J Biol Chem 275:24466–24469
34. Mizuno K, Hayashi T, Bachinger HP (2003) Hydroxylation-induced stabilization of the collagen triple helix. Further characterization of peptides with 4(R)-hydroxyproline in the Xaa position. J Biol Chem 278: 32373–32379

35. Derider ML, Wilkens SJ, Waddell MJ, Bretscher LE, Weinhold F, Raines RT, Markley JL (2002) Collagen stability: insights from NMR spectroscopic and hybrid density functional computational investigations of the effect of electronegative substituents on prolyl ring conformations. J Am Chem Soc 124:2497–2505
36. Vitagliano L, Berisio R, Mastrangelo A, Mazzarella L, Zagari A (2001) Preferred proline puckerings in cis and trans peptide groups: implications for collagen stability. Protein Sci 10:2627–2632
37. Shoulders MD, Satyshur KA, Forest KT, Raines RT (2010) Stereoelectronic and steric effects in side chains preorganize a protein main chain. Proc Natl Acad Sci USA 107:559–564
38. Shoulders MD, Hodges JA, Raines RT (2006) Reciprocity of steric and stereoelectronic effects in the collagen triple helix. J Am Chem Soc 128:8112–8113
39. Lee SG, Lee JY, Chmielewski J (2008) Investigation of pH-dependent collagen triple-helix formation. Angew Chem Int Ed Engl 47:8429–8432
40. Babu IR, Ganesh KN (2001) Enhanced triple helix stability of collagen peptides with 4R-aminoprolyl (Amp) residues: relative roles of electrostatic and hydrogen bonding effects. J Am Chem Soc 123:2079–2080
41. Umashankara M, Babu IR, Ganesh KN (2003) Two prolines with a difference: contrasting stereoelectronic effects of 4R/S--aminoproline on triplex stability in collagen peptides [pro(X)-pro(Y)-Gly]n. Chem Commun (Camb) 20:2606–2607
42. Shah NK, Ramshaw JA, Kirkpatrick A, Shah C, Brodsky B (1996) A host-guest set of triple-helical peptides: stability of Gly-X-Y triplets containing common nonpolar residues. Biochemistry 35:10262–10268
43. Persikov AV, Ramshaw JA, Brodsky B (2000) Collagen model peptides: sequence dependence of triple-helix stability. Biopolymers 55:436–450
44. Persikov AV, Ramshaw JA, Kirkpatrick A, Brodsky B (2000) Amino acid propensities for the collagen triple-helix. Biochemistry 39:14960–14967
45. Berisio R, DS A, Ruggiero A, Improta R, Vitagliano L (2008) Role of side chains in collagen triple helix stabilization and partner recognition. J Pept Sci 15:131–140
46. Persikov AV, Ramshaw JA, Kirkpatrick A, Brodsky B (2003) Triple-helix propensity of hydroxyproline and fluoroproline: comparison of host-guest and repeating tripeptide collagen models. J Am Chem Soc 125: 11500–11501
47. Persikov AV, Ramshaw JA, Brodsky B (2005) Prediction of collagen stability from amino acid sequence. J Biol Chem 280: 19343–19349
48. Venugopal MG, Ramshaw JA, Braswell E, Zhu D, Brodsky B (1994) Electrostatic interactions in collagen-like triple-helical peptides. Biochemistry 33:7948–7956
49. Persikov AV, Ramshaw JA, Kirkpatrick A, Brodsky B (2002) Peptide investigations of pairwise interactions in the collagen triple-helix. J Mol Biol 316:385–394
50. Fallas JA, Dong J, Tao YJ, Hartgerink JD (2012) Structural insights into charge pair interactions in triple helical collagen-like proteins. J Biol Chem 287:8039–8047
51. Persikov AV, Ramshaw JA, Kirkpatrick A, Brodsky B (2005) Electrostatic interactions involving lysine make major contributions to collagen triple-helix stability. Biochemistry 44:1414–1422
52. Yang W, Chan VC, Kirkpatrick A, Ramshaw JA, Brodsky B (1997) Gly-Pro-Arg confers stability similar to Gly-Pro-Hyp in the collagen triple-helix of host-guest peptides. J Biol Chem 272:28837–28840
53. Koide T, Nishikawa Y, Takahara Y (2004) Synthesis of heterotrimeric collagen models containing Arg residues in Y-positions and analysis of their conformational stability. Bioorg Med Chem Lett 14:125–128
54. Hudlicky M, M. aJM (1990) New stereospecific syntheses and x-ray diffraction structures of (−)-D-erythro- and (+)-L-threo-4-fluoroglutamic. Tetrahedron Lett 31: 7403–7406
55. Hudlicky M (1993) Stereospecific syntheses of all four stereoisomers of 4-fluoroglutamic acid. J Fluorine Chem 60:193–210
56. Panasik N Jr, Eberhardt ES, Edison AS, Powell DR, Raines RT (1994) Inductive effects on the structure of proline residues. Int J Pept Protein Res 44:262–269
57. Kronenthal DR, Mueller RH, Kuester TP, Kissick TP, Johnson EJ (1990) Stereospecific friedel-crafts alkylation of benzene with 4-mesyloxy-L-prolines. A new synthesis of 4-phenylprolines. Tetrahedron Lett 31:1241–1244
58. Gottlieb AA, Yoshimasa F, Undenfriend S, Witkop B (1965) Incorporation of cis- and trans-4-fluoro-L-prolines into proteins and hydroxylation of the trans isomer during collagen biosynthesis. Biochemistry 4: 2507–2513
59. Shirota FN, Nagasawa HT, Elberling JA (1977) Potential inhibitors of collagen biosynthesis. 5,5-Difluoro-DL-lysine and 5, 5-dimethyl-DL-lysine and their activation by

lysyl-tRNA ligase. J Med Chem 20: 1623–1627
60. Hart BP, Coward JK (1993) The synthesis of DL-3,3-Difluoroglutamic acid from a 3-oxoprolinol derivative. Tetrahedron Lett 34:4917–4920
61. Avent AG, Bowler AN, Doyle PM, Marchand CM, Young DW (1992) Stereospecific synthesis of 4-fluoroglutamic acids. Tetrahedron Lett 33:1509–1512
62. Demange L, Menez A, Dugave C (1998) Practical synthesis of Boc and Fmoc protected 4-fluoro and 4-difluoroprolines from trans-4-hydroxyproline. Tetrahedron Lett 39:1169–1172
63. Tran TT, Patino N, Condom R, Frogier T, Guedj R (1997) Fluorinated peptides incorporating a 4-fluoroproline residue as potential inhibitors of HIV protease. J Fluorine Chem 82:125–130
64. Malkar NB, Lauer-Fields JL, Borgia JA, Fields GB (2002) Modulation of triple-helical stability and subsequent melanoma cellular responses by single-site substitution of fluoroproline derivatives. Biochemistry 41: 6054–6064
65. Martinez J, Tolle JC, Bodanszky M (1979) Side reactions in peptide synthesis. 12. Hydrogenolysis of the 9-fluorenylmethyloxycarbonyl group. J Org Chem 44:3596–3598
66. Atherton E, Bury C, Sheppard RC, Williams BJ (1979) Stability of fluorenylemthoxycarbonylamino groups in peptide synthesis. Cleavage by hydrogenolysis and by dipolar aprotic solvents. Tetrahedron Lett 20:3041–3042
67. Vnek V, Budesinsky M, Rinnová M, Rosenberg I (2009) Prolinol-based nucleoside phosphonic acids: new isosteric conformationally flexible nucleotide analogues. Tetrahedron 65:862–876
68. Doi M, Nishi Y, Kiritoshi N, Iwata T, Nago M, Nakano H, Uchiyama S, Nakazawa T, Wakamiya T, Kobayashi Y (2002) Simple and efficient syntheses of Boc- and Fmoc-protected 4(R)- and 4(S)-fluoroproline solely from 4(R)-hydroxyproline. Tetrahedron 58:8453–8459
69. Hodges JA, Raines RT (2003) Stereoelectronic effects on collagen stability: the dichotomy of 4-fluoroproline diastereomers. J Am Chem Soc 125:9262–9263
70. Bhowmick M, Sappidi RR, Fields GB, Lepore SD (2011) Efficient synthesis of Fmoc-protected phosphinic pseudodipeptides: building blocks for the synthesis of matrix metalloproteinase inhibitors. Biopolymers 96:1–3
71. Goodman M, Del Valle JR (2003) Asymmetric hydrogenations for the synthesis of Boc-protected 4-alkylprolinols and prolines. J Org Chem 68:3923–3931
72. Murakami Y, Nakano A, Yoshimatsu A, Uchitomi K, Mastsuda Y (1984) Characterization of molecular aggregates of peptide amphiphiles and kinetics of dynamic processes performed by single-walled vesicles. J Am Chem Soc 106:3613–3623
73. Yu Y-C, Brendt P, Tirrell M, Fields GB (1996) Self-assembling amphiphiles for construction of protein molecular architecture. J Am Chem Soc 118:12515–12520
74. Yu YC, Berndt P, Tirrell M, Fields GB (1998) Minimal lipidation stabilizes protein-like molecular architecture. J Am Chem Soc 120: 9979–9987
75. Pakalns T, Haverstick KL, Fields GB, Mccarthy JB, Mooradian DL, Tirrell M (1999) Cellular recognition of synthetic peptide amphiphiles in self-assembled monolayer films. Biomaterials 20:2265–2279
76. Forns P, Lauer-Fields JL, Gao S, Fields GB (2000) Induction of protein-like molecular architecture by monoalkyl hydrocarbon chains. Biopolymers 54:531–546
77. Borgia JA, Fields GB (2000) Chemical synthesis of proteins. Trends Biotechnol 18: 243–251
78. Mammen M, Choi SK, Whitesides GM (1998) Polyvalent interactions in biological systems: implications for design and use of multivalent ligands and inhibitors. Angew Chem Int Ed 37:2754–2794
79. García M, Alsina M, Reig F, Haro I (2000) Liposomes as vehicles for the presentation of a synthetic peptide containing an epitope of hepatitis A virus. Vaccine 18:276–283
80. Gore T, Dori Y, Talmon Y, Tirrell M, Bianco-Peled H (2001) Self-assembly of model collagen peptide amphiphiles. Langmuir 17:5352–5360
81. Malkar NB, Lauer-Fields JL, Juska D, Fields GB (2003) Characterization of peptide-amphiphiles possessing cellular activation sequences. Biomacromolecules 4:518–528
82. Fields GB, Lauer JL, Dori Y, Forns P, Yu YC, Tirrell M (1998) Protein-like molecular architecture: biomaterial applications for inducing cellular receptor binding and signal transduction. Biopolymers 47:143–151
83. Lauer-Fields JL, Tuzinski KA, Shimokawa K, Nagase H, Fields GB (2000) Hydrolysis of triple-helical collagen peptide models by matrix metalloproteinases. J Biol Chem 275:13282–13290
84. Lauer-Fields JL, Nagase H, Fields GB (2000) Use of Edman degradation sequence analysis

and matrix-assisted laser desorption/ ionization mass spectrometry in designing substrates for matrix metalloproteinases. J Chromatogr A 890:117–125
85. Lauer-Fields JL, Broder T, Sritharan T, Chung L, Nagase H, Fields GB (2001) Kinetic analysis of matrix metalloproteinase activity using fluorogenic triple-helical substrates. Biochemistry 40:5795–5803
86. Kusebauch U, Cadamuro SA, Musiol HJ, Moroder L, Renner C (2007) Photocontrol of the collagen triple helix: synthesis and conformational characterization of bis-cysteinyl collagenous peptides with an azobenzene clamp. Chemistry 13:2966–2973
87. Hojo H, Akamatsu Y, Yamauchi K, Kinoshita M, Miki S, Nakamura Y (1997) Synthesis and structural characterization of triple-helical peptides which mimic the ligand binding site of the human macrophage scavenger receptor. Tetrahedron 53:14263–14274
88. Tanaka Y, Suzuki K, Tanaka T (1998) Synthesis and stabilization of amino and carboxy terminal constrained collagenous peptides. J Pept Res 51:413–419
89. Khew ST, Tong YW (2008) Template-assembled triple-helical peptide molecules: mimicry of collagen by molecular architecture and integrin-specific cell adhesion. Biochemistry 47:585–596
90. Fields CG, Lovdahl CM, Miles AJ, Hagen VL, Fields GB (1993) Solid-phase synthesis and stability of triple-helical peptides incorporating native collagen sequences. Biopolymers 33:1695–1707
91. Fields CG, Mickelson DJ, Drake SL, Mccarthy JB, Fields GB (1993) Melanoma cell adhesion and spreading activities of a synthetic 124-residue triple-helical "mini-collagen". J Biol Chem 268:14153–14160
92. Ottl J, Moroder L (1999) Disulfide-bridged heterotrimeric collagen peptides containing the collagenase cleavage site of collagen type I. Synthesis and conformational properties. J Am Chem Soc 121:653–661
93. Ottl J, Gabriel D, Murphy G, Knauper V, Tominaga Y, Nagase H, Kroger M, Tschesche H, Bode W, Moroder L (2000) Recognition and catabolism of synthetic heterotrimeric collagen peptides by matrix metalloproteinases. Chem Biol 7:119–132
94. Ottl J, Battistuta R, Pieper M, Tschesche H, Bode W, Kuhn K, Moroder L (1996) Design and synthesis of heterotrimeric collagen peptides with a built-in cystine-knot. Models for collagen catabolism by matrix-metalloproteases. FEBS Lett 398:31–36
95. Muller JC, Ottl J, Moroder L (2000) Heterotrimeric collagen peptides as fluorogenic collagenase substrates: synthesis, conformational properties, and enzymatic digestion. Biochemistry 39:5111–5116
96. Sacca B, Fiori S, Moroder L (2003) Studies of the local conformational properties of the cell-adhesion domain of collagen type IV in synthetic heterotrimeric peptides. Biochemistry 42:3429–3436
97. Goodman M, Feng Y, Melacini G, Taulane JP (1996) A template-induced incipient collagen-like triple-helical structure. J Am Chem Soc 118:5156–5157
98. Goodman M, Melacini G, And Feng Y (1996) Collagen-like triple helices incorporating peptoid residues. J Am Chem Soc 118: 10928–10929
99. Feng Y, Melacini G, Taulane JP, Goodman M (1996) Collagen-based structures containing the peptoid residue N-isobutylglycine (Nleu): synthesis and biophysical studies of Gly-Pro-Nleu sequences by circular dichroism, ultraviolet absorbance, and optical rotation. Biopolymers 39:859–872
100. Feng Y, Melacini G, Goodman M (1997) Collagen-based structures containing the peptoid residue N-isobutylglycine (Nleu): synthesis and biophysical studies of Gly-Nleu-Pro sequences by circular dichroism and optical rotation. Biochemistry 36:8716–8724
101. Feng Y, Melacini G, Taulane JP, Goodman M (1996) Acetyl-terminated and template-assembled collagen-based polypeptides composed of Gly-Pro-Hyp sequences. 2. Synthesis and conformational analysis by circular dichroism, ultraviolet absorbance, and optical rotation. J Am Chem Soc 118:10351–10358
102. Kwak J, De Capua A, Locardi E, Goodman M (2002) TREN (Tris(2-aminoethyl)amine): an effective scaffold for the assembly of triple helical collagen mimetic structures. J Am Chem Soc 124:14085–14091
103. Rump ET, Rijkers DT, Hilbers HW, De Groot PG, Liskamp RM (2002) Cyclotriveratrylene (CTV) as a new chiral triacid scaffold capable of inducing triple helix formation of collagen peptides containing either a native sequence or Pro-Hyp-Gly repeats. Chemistry 8: 4613–4621
104. Horng JC, Hawk AJ, Zhao Q, Benedict ES, Burke SD, Raines RT (2006) Macrocyclic scaffold for the collagen triple helix. Org Lett 8:4735–4738
105. Brask J, Jensen KJ (2001) Carboproteins: a 4-alpha-helix bundle protein model assembled on a D-galactopyranoside template. Bioorg Med Chem Lett 11:697–700
106. Thulstrup PW, Brask J, Jensen KJ, Larsen E (2005) Synchrotron radiation circular dichroism spectroscopy applied to metmyoglobin

and a 4-alpha-helix bundle carboprotein. Biopolymers 78:46–52
107. Cai W, Kwok SW, Taulane JP, Goodman M (2004) Metal-assisted assembly and stabilization of collagen-like triple helices. J Am Chem Soc 126:15030–15031
108. Kinberger GA, Taulane JP, Goodman M (2006) Fe(III)-binding collagen mimetics. Inorg Chem 45:961–963
109. Pires MM, Chmielewski J (2009) Self-assembly of collagen peptides into microflorettes via metal coordination. J Am Chem Soc 131:2706–2712
110. Thakur S, Vadolas D, Germann HP, Heidemann E (1986) Influence of different tripeptides on the stability of the collagen triple helix II. An experimental approach with appropriate variations of a trimer model oligotripeptide. Biopolymers 25:1081–1086
111. Germann HP, Heidemann E (1988) A synthetic model of collagen: an experimental investigation of the triple-helix stability. Biopolymers 27:157–163
112. Fields CG, G B, Lauer JL, Miles AJ, Yu Y-C, Fields GB (1996) Solid-phase synthesis of triple-helical collagen-model peptides. Lett Peptide Sci 3:3–16
113. Barnes MJ, Knight CG, Farndale RW (1996) The use of collagen-based model peptides to investigate platelet-reactive sequences in collagen. Biopolymers 40:383–397
114. Morton LF, Peachey AR, Knight CG, Farndale RW, Barnes MJ (1997) The platelet reactivity of synthetic peptides based on the collagen III fragment alpha1(III)CB4. Evidence for an integrin alpha2beta1 recognition site involving residues 522–528 of the alpha1(III) collagen chain. J Biol Chem 272:11044–11048
115. Knight CG, Morton LF, Onley DJ, Peachey AR, Ichinohe T, Okuma M, Farndale RW, Barnes MJ (1999) Collagen-platelet interaction: Gly-Pro-Hyp is uniquely specific for platelet Gp VI and mediates platelet activation by collagen. Cardiovasc Res 41:450–457
116. Byrne C, Mcewan PA, Emsley J, Fischer PM, Chan WC (2011) End-stapled homo and hetero collagen triple helices: a click chemistry approach. Chem Commun (Camb) 47:2589–2591
117. Henkel W, Vogl T, Echner H, Voelter W, Urbanke C, Schleuder D, Rauterberg J (1999) Synthesis and folding of native collagen III model peptides. Biochemistry 38:13610–13622
118. Cai W, Wong D, Kinberger GA, Kwok SW, Taulane JP, Goodman M (2007) Facile and efficient assembly of collagen-like triple helices on a TRIS scaffold. Bioorg Chem 35: 327–337
119. Barth D, Kyrieleis O, Frank S, Renner C, Moroder L (2003) The role of cystine knots in collagen folding and stability, part II. Conformational properties of (Pro-Hyp-Gly) n model trimers with N- and C-terminal collagen type III cystine knots. Chemistry 9: 3703–3714
120. Krishna OD, Kiick KL (2009) Supramolecular assembly of electrostatically stabilized, hydroxyproline-lacking collagen-mimetic peptides. Biomacromolecules 10:2626–2631
121. King DS, Fields CG, Fields GB (1990) A cleavage method which minimizes side reactions following Fmoc solid phase peptide synthesis. Int J Pept Protein Res 36: 255–266
122. Fields CG, Fields GB (1993) Minimization of tryptophan alkylation following 9-fluorenylmethoxycarbonyl solid-phase peptide synthesis. Tetrahedron Lett 34: 6661–6664
123. Berndt P, Fields GB, Tirrell M (1995) Synthetic lipidation of peptides and amino acids: monolayer structure and properties. J Am Chem Soc 117:9515–9522
124. Fields CG, Grab B, Lauer JL, Fields GB (1995) Purification and analysis of synthetic, triple-helical "minicollagens" by reversed-phase high-performance liquid chromatography. Anal Biochem 231:57–64
125. Grab B, Miles AJ, Furcht LT, Fields GB (1996) Promotion of fibroblast adhesion by triple-helical peptide models of type I collagen-derived sequences. J Biol Chem 271:12234–12240
126. Yu YC, Roontga V, Daragan VA, Mayo KH, Tirrell M, Fields GB (1999) Structure and dynamics of peptide-amphiphiles incorporating triple-helical proteinlike molecular architecture. Biochemistry 38:1659–1668
127. Chhabra SR, Hothi B, Evans DJ, White PD, Bycroft BW, Chan WC (1998) An appraisal of new variants of Dde amine protecting group for solid phase peptide synthesis. Tetrahedron Lett 39:1603–1606
128. Rohwedder B, Mutti Y, Dumy P, Mutter M (1998) Hydrazinolysis of Dde: complete orthogonality with Aloc protecting groups. Tetrahedron Lett 39:1175–1178
129. Kates SA, Daniels SB, Albericio F (1993) Automated allyl cleavage for continuous-flow synthesis of cyclic and branched peptides. Anal Biochem 212:303–310

Chapter 12

Identification of Adipokine Receptor Agonists and Turning Them to Antagonists

Laszlo Otvos Jr.

Abstract

Receptor–ligand interactions represent one of the most basic processes in biological systems. Receptor activation and deactivation induce or prevent a series of downstream signaling events that ultimately result in normal or abnormal cellular functions. Contemporary biology is in continuous search for the identification of novel receptors and their ligands. The adipose tissue participates in the regulation of energy homeostasis as an important endocrine organ that secretes a number of biologically active adipokines, including leptin and adiponectin. A recent discovery and design process for leptin and adiponectin receptor response modifier peptides can be generalized to a series of transmembrane receptor ligands. A family of 11–13 amino acid residue-long leptin receptor (ObR) agonists has been identified by analyzing the effect of peptides corresponding to the three presumed active sites of leptin on the growth of leptin-responsive cancer cells. In the case of adiponectin, overlapping peptides were walked across the entire globular domain of the protein to identify the active site and derive adiponectin receptor (AdipoR) agonist peptides. In both sets, native residues were replaced by nonnatural analogs to improve the pharmacological properties including stability, efficacy and targeting. Later the ObR analogs were converted into true ObR antagonists that show antagonist–agonist selectivity of 1,000 in cellular assays. The design process of ObR antagonists included shortening of the peptide length and incorporating additional nonnatural residues. Here I take a look into this receptor agonist and antagonist discovery process from a practical point of view.

Key words Adiponectin, Leptin, Cell proliferation, Nonnatural residues, Peptide length, Receptor agonist, Receptor antagonist, Signaling

1 Introduction

Ligand recognition and the ensuing signal transduction cascade are currently the most frequently studied phenomena to understand and modify cellular processes [1]. Activated receptors directly or indirectly regulate cellular biochemical processes including protein phosphorylation, nucleic acid transcription, ion transport and a series of enzyme activities. Ligands (e.g., drugs, hormones, neurotransmitters) that bind to receptors may activate or inactivate their cognate receptor and accelerate or inhibit a particular cellular function.

Predrag Cudic (ed.), *Peptide Modifications to Increase Metabolic Stability and Activity*, Methods in Molecular Biology, vol. 1081,
DOI 10.1007/978-1-62703-652-8_12, © Springer Science+Business Media New York 2013

Adipokines are cytokines (cell-to-cell signaling proteins) secreted by the adipose tissue. Since the discovery of leptin (obesity hormone, [2]), the search for other adipokines with unknown or altered function represents a major topic in obesity research [3]. The clinical significance of adipokines comes from their role in fat distribution, adipose tissue function, liver fat content, insulin sensitivity and chronic inflammation. The two most extensively studied adipokines are leptin and adiponectin, presumably due to their role in promoting certain cancer types and inhibiting tumor growth, respectively [4]. In addition to leptin's historic role as a potential therapy agent in obesity, adiponectin is one of the most abundant proteins in circulation [5]. It is just natural that modifiers of these receptor responses are the focus of current research, frequently targeting multiple clinical conditions and opposing signal transduction pathways [6].

Leptin appears to interact with its receptor at three discontinuous surfaces [7]. When screening the pharmacological properties of all three potential leptin receptor (ObR)-binding sites, we identified the C-terminal site III that can be most likely used to prepare therapeutically relevant ObR ligands [8]. Armed with knowing the active site of leptin, our groups developed 11–13 amino acid residue-long leptin receptor ObR agonists that, when administered systemically, control weight and restore deranged metabolic panel results as well as correct infertility findings in diet-induced obese mouse and normal rat models [9]. In contrast to leptin, the active site of globular adiponectin had not been known. In a different line of research, we identified the active site of adiponectin and developed a highly potent, first-in-class adiponectin receptor (AdipoR) agonist peptide [10]. Using known medicinal chemistry techniques, the ObR agonist peptides were converted into nine-residue true ObR antagonists that inhibit leptin-dependent cellular functions when exogenously added leptin is present in the assay milieu but have no cellular effects without leptin [11, 12]. These highly active ObR antagonists promote weight gain in rodents, inhibit leptin-induced STAT3 and ERK1/2 signaling in cancer cells at pM or low nM concentrations, and significantly reduce ObR-dependent cancer xenograft growth in mouse models.

The jury is still out regarding the required peptide length for pharmacologically acceptable level of receptor activation. The many variables include the size and accessibility of ligand binding surfaces, ligand stability and receptor residency time, and potential for induced fit, just to name a few. Clearly the old dogma asking for submicromolar binding constant has to be revised for low nano- or picomolar cellular responses. In our hands, ObR agonist peptides have to be at least 11 amino acid residues in length [8], and this observation is in sync with parathyroid hormone [13] and dopamine [14] receptor activators (for adiponectin agonists ten residue-long peptides were large enough). It is safe to say that for an ordinary transmembrane receptor, 12-amino acid long agonists will induce cellular responses characteristic for receptor activation. Cell and tissue penetration is usually enhanced by incorporating positively charged amino acids, especially at terminal positions [15, 16]. In general, nonnatural residues at terminal positions inhibit exopeptidase cleavage and thus increase stability en route to the site of action [17].

For turning peptide agonists into antagonists, the following techniques are proposed by either surveying the literature or by personal experience [18–22]:

- Peptide library based on the agonist.
- Truncation of the sequence (from either the N- or C-termini or from mid-sequence).
- Nonnatural residue replacement.
- Conformational restriction.

Our nine-residue first generation ObR antagonist Aca1 follows these design rules as does the optimized superantagonist Allo-aca [11, 12]. For antagonist design, the pico- or low nanomolar cellular activities I suggested for agonists are equally required. Moreover, the agonists and antagonists have to show at least a 1,000-fold selectivity in their in vitro activities, that is, agonists cannot inhibit cellular functions at a concentration at least 1,000-fold over the observed EC_{50} value and antagonists (acting in the presence of an agonist) must not stimulate cellular functions in the absence of an agonist at a concentration at least 1,000-fold over the observed IC_{50} figure.

In the following sections I show examples of the development of adipokine receptor agonists and antagonists using leptin and adiponectin as examples. In all cases, incorporation of nonnatural residues improves the pharmacological properties of the peptides. The modified synthetic peptides (or peptidomimetics if you will) can be made in-house or ordered from commercial vendors. For simplicity, I use breast cancer cell proliferation assays for measuring the primary endpoint, i.e., cellular activity of the peptides. Users are encouraged to apply the design strategies to their own receptors and assay readouts. For secondary screening measures (in vitro and in vivo stability, conformation, biodistribution, etc.) readers are referred to an earlier volume of this Methods of Molecular Biology series [23].

2 Materials

2.1 Equipment

Automated peptide synthesizer (*see* **Note 1**).

Gradient high performance liquid chromatograph (HPLC) consisting of two pumps operated by a gradient controller, interchangeable analytical/preparative columns, ultraviolet detector and integrator.

Tabletop lyophilizer.

MALDI-TOF matrix-assisted laser desorption/ionization time of flight mass spectrometer.

Shaker.

Hemocytometer.

Micropipets.

Laboratory microscope with 400× capability.

Hand held counter.

2.2 Reagents

2.2.1 Solvents and Media

N,N'-Dimethyl formamide (DMF).

Piperidine (*see* **Note 2**).

N-Methyl morpholine (NMM).

Chloroform.

Acetic acid.

Acetonitrile.

Mammalian epithelial growth serum-free medium (SFM).

Dulbecco's Modified Eagle Medium (DMEM).

2.2.2 Resin and Amino Acids

TentaGel S RAM resin.

Standard Fmoc-protected amino acids (*see* **Note 3**).

Unusual amino acids used frequently in our laboratory: D-amino acids; Nva—norvaline; Pip—pipecolic acid; Nle—norleucine; Hle—homoleucine; AlloThr—allo-threonine; Orn—ornithine; β-Ala—β-alanine; Aca—1-amino caproic acid; MeArg—*N*-methyl arginine; Hyp—hydoxy-proline; Ser(Glc)—serine carrying a β-linked glucose moiety; Thr(GalNAc)—threonine carrying an α-linked *N*-acetyl glucosamine moiety; PhSer—*O-phospho* serine; Dap(Ac)—2-acetamido-3-amino propionic acid; Tyr(I_2)—2′,4′-diiodo threonine; Nal—1-naphthyl alanine; Gla—γ-carboxy glutamic acid; pGlu—pyroglutamic acid (*see* **Note 4**).

2.2.3 Other Reagents, Cells, and Supplies

O-(7-Azabenzotriazole-1-yl)-*N,N,N' N'*-tetramethyluronium hexafluorophosphate (HATU) (*see* **Note 5**).

Trifluoroacetic acid (TFA).

Alpha-cyano-4-hydroxycinnamic acid.

100 ng/mL cholera toxin solution.

MCF-7 cells (*see* **Notes 6** and 7).

Recombinant human leptin (*see* **Note 8**).

Recombinant human adiponectin globular domain (*see* **Note 8**).

$FeSO_4$ solution (10 μM).

Bovine serum albumin (BSA) solution (0.5 %).

Trypan blue solution (0.4 %).

Micropipet tips.

3 Methods

The main goal of this chapter is to describe the identification of pharmacologically acceptable and pure receptor agonists and antagonists using leptin and adiponectin as examples. Thus, the major emphasis is placed on peptide design and general activity screening process. The technology for peptide synthesis, purification, and cancer cell proliferation assays are straightforward and these methods will be just briefly described. In most likelihood various laboratories will use different receptors and screening protocols.

When studying the effects of peptides on cellular responses, one should look for the activity of the peptides without a known (frequently the native ligand protein) agonist present in the assay milieu as well as in the presence of an agonist. When the peptides activate the receptor alone and have no effect on the cells with the native ligand present, we have a pure agonist; when in addition to stimulation alone the peptide inhibits receptor responses in the presence of a native agonist we are talking about partial agonists (partial antagonists). Pure receptor antagonists inhibit the receptor activation of a known agonist but have no effect on the cells alone. Table 1 shows how we initially identified a glycopeptide called E1 as a leptin receptor agonist [8]. Further lead optimization lead to a picomolar agonist peptide termed E1/Aca (Table 2). Table 2 also shows the pharmacological profiles of our leptin receptor antagonists Aca1 (nanomolar potency) and Allo-aca (picomolar potency) [9]. Please note how an increasing number of nonnatural residues improved the activity of the peptides.

The active site of adiponectin was identified by using a 66-member unmodified peptide library and another 330-peptide array in which 1–4 natural amino acids were replaced by nonnatural analogs. Each peptide was ten amino acid-long. The length of the peptides was reduced from the optimal 11–13 to 10 to maintain high quality peptides in spite of the large number of

Table 1
Initial screening of leptin fragments leading to a first generation leptin receptor agonist peptide. The effect of different peptides on proliferation of MCF-7 and control cells was studied in the presence or absence of leptin. The pure agonist peptide corresponds to the third receptor binding domain of leptin, and contains three unnatural residues. For agonistic activity and targeting Ser120 has to be glycosylated (we used a β-Glc moiety attached to the side-chain hydroxyl group). The natural N-terminal Tyr119 is replaced with Tyr(I_2), and the C-terminal Gln130 is replaced with acetylated diamino-propionic acid to improve stability

Leptin fragments	Effect on MCF-7 growth without leptin	Effect on MCF-7 growth with 6 nM leptin added	Effect on control MCF-10 cells	Pharmacology conclusions
36–49, Site I, no unnatural residues	Unchanged up to 1 μM	Variable, mostly unchanged	Not studied	Inactive
3–18, Site II/a, no unnatural residues	Increase at 100 nM	Reduction at 1 μM	Not studied	Partial agonist
117–132, Site III, no unnatural residues	Unchanged up to 1 μM	Reduction at 1 μM	No effect	Partial agonist
119–130, Site III, 3 unnatural residues	Increase at 100 nM	Unchanged at 1 μM	No effect	Agonist

Table 2
Sequences and pharmacological profiles of leptin site III peptides. Full antagonist is defined as having antagonistic selectivity of at least 1,000 and super agonist/antagonist is defined as acting on the leptin receptor in lower concentration than native leptin protein

Designation	Sequence	Pharmacology
E1/Aca	H-Tyr(I_2)-Ser-Thrα(GalNAc)-Glu-Val-Val-Ala-Leu-Ser-Arg-Aca-NH_2	IC_{50}: 100 nM; EC_{50}: 100 pM; super agonist
Aca1	H-Thr-Glu-Nva-Val-Ala-Leu-Ser-Arg-Aca-NH_2	IC_{50}: 5 nM; EC_{50}: >1 μM; full antagonist; no effect on ObR-negative cells
Allo-Aca	H-alloThr-Glu-Nva-Val-Ala-Leu-Ser-Arg-Aca-NH_2	IC_{50}: 200 pM; EC_{50}: >1 μM; super antagonist No effect on ObR-negative cells

nonnatural residues that are known to cause synthetic difficulties compared to their natural counterparts during peptide synthesis [24]. Currently hundreds of appropriately protected nonnatural amino acid derivatives are available commercially. These can be incorporated into synthetic peptides but their reduced coupling and deprotection rates have to be always considered when designing the peptide analogs and the synthetic protocols. Indeed, as Fig. 1a shows, five overlapping peptides (each at 100 nM concentration) equally inhibited cancer cell proliferation, suggesting that the ideal binding site comprises of more than ten residues, as expected. Nevertheless, the incorporation of nonnatural moieties increased the anti-tumor properties of the peptides (Fig. 1a). From all 396 peptides in the array, peptide ADP355 with the sequence of H-DAsn-Ile-Pro-Nva-Leu-Tyr-DSer-Phe-Ala-DSer-NH_2 emerged as the most active analog. This peptide is based on fragment 25 of the globular domain of adiponectin: Asn-Ile-Pro-Gly-Leu-Tyr-Tyr-Phe-Ala-Tyr corresponding to adiponectin residues 153–162. When a lead peptide is found, a concentration-activity study has to indicate a dose-dependent single binding site curve, desirably with pM or low nM activity figures (Fig. 1b).

While it is our desire to identify pure receptor agonists or antagonists, in reality a pharmacological switch takes place at higher concentrations. If this switch occurs at a concentration at least 1,000-fold over the desired activity, we can expect predictable pharmacological activity in in vivo models. E1/Aca and Allo-aca (Table 2) fulfill this in vitro requirement and indeed exhibit clear anorexigenic and orexigenic (as well anti-tumor) properties in rodent models [9, 11, 12].

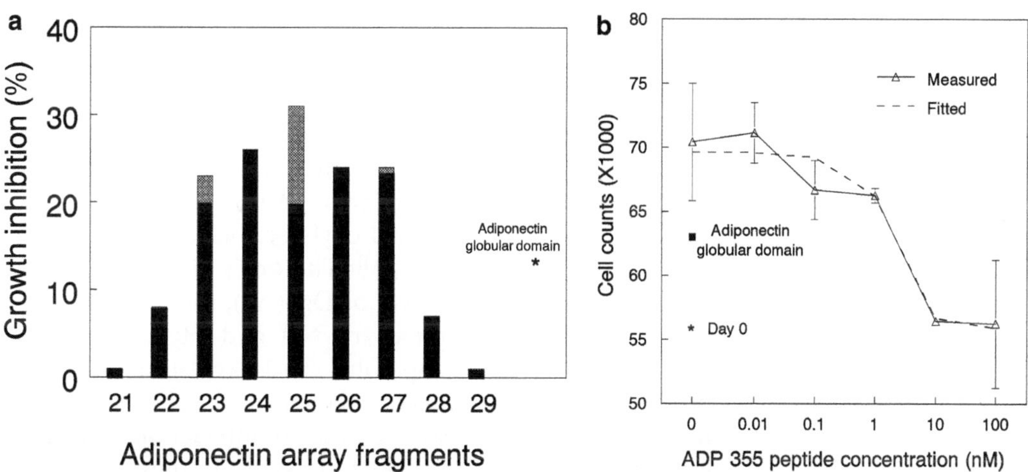

Fig. 1 Identification of a first-of-class adiponectin receptor agonist. The entire globular domain of adiponectin was synthesized from 10-mer peptide fragments, overlapping by two residues. One-to-four residues in each fragment were also replaced by nonnatural moieties. (**a**) At 100 nM concentration, peptide fragments 23–27, corresponding to adiponectin 149–166 (*solid bars*) inhibited the proliferation of MCF-7 human breast cancer cells better than the entire adiponectin globular domain (*asterisk*). For peptides 23, 27, and especially 25 amino acid replacements further improved the biological activity (*cross-hatched bars*). (**b**) Concentration dependent activity of peptide, ADP355 which corresponds to peptide 25 (residues 153–162), and in which Asn1, Gly4, Tyr7, and Tyr10 were replaced by D-Asn, Nva, D-Ser, and D-Ser, respectively. The EC_{50} of peptide ADP355 is approximately 2 nM

3.1 Receptor Agonist Development When the Receptor Binding Sites Are Known

While some ligands bind their receptors with only one site, some others have multiple receptor binding sites. While leptin binds to ObR via three bivalent fragments, AP-2β, another protein expressed in adipocytes has a single target binding site [25]. Mutational analyses of the ligand proteins or truncation studies frequently reduce the potential active site(s) to 15–20 residue-long fragments.

3.1.1 Peptide Design and Nonnatural Residue Enhancement

Reducing the Active Site

Step 1: Locate the middle 12-residue fragment of the putative active site of the protein.

Step 2: Synthesize the middle dodecapeptide, as well as peptides shifted by a single residue to the N- and C-termini. If the provisional active site consists of 20 residues, you will need nine peptides: 5–16, 4–15, 6–17, 3–14, 7–18, 2–13, 8–19, 1–12, 9–20.

Step 3: Screen the biological activity of the native fragments against the target cell line (*see* Subheading 3.3.1 as an example of cancer cell growth inhibition). Depending on the biological assay conditions, include no more than five peptides in one assay. Five test peptides, a blank, a positive control (usually the native protein), and an unrelated negative control peptide in triplicates adds up to 24 wells (1 or 2 plates) which is roughly the upper limit of non-high throughput screening by single investigators. When the most active dodecapeptide is identified, move to nonnatural amino acid replacements described in the next section.

Nonnatural Amino Acid Residue Replacements

Step 1: Start with the N- and C-terminal residues. Preferably, replace each terminal residue with homologous moieties. Our general choices for aliphatic amino acids (Gly, Ala, Val, Leu, Ile) are Nva or Nle, for hydroxyl- or thio-amino acids (Ser, Thr, Tyr, Cys) are D-serine, AlloThr, or Tyr(I_2), for amide-side chains (Asn, Gln) are D-Asn or Dap(Ac), for aromatics (Phe, Tyr, Trp) and for proline are Nal and Hyp, respectively. Carboxy amino-acids (Asp, Glu) can be replaced with D-Asp or Gla, but in many occasions the terminal residues should be neutral or downright basic to improve interactions with cellular membranes and enter cells and tissues. In these cases, pGlu is a regular choice, especially at the N terminus, where Glu or Gln → pGlu transitions occur in mildly acidic conditions [26]. To keep the negative charge at the C terminus, β-Ala or Aca are frequently used as C-terminal moieties in their free acid (not amidated) forms. Positively charged residues at pH 6–7 (Lys, Arg, His) are usually replaced with Orn, Pip or MeArg. Users are encouraged to try nonnatural residues containing the native amino-terminal charge or with an extra positively charged side-chain at terminal positions to improve solubility and interaction with the cellular environment.

Step 2: Internal modifications. To maintain pure agonistic properties, nonnatural residue replacements in mid-chain positions are not recommended; these modifications can lead to an agonist → antagonist switch (*vide infra*, Subheading 3.4). For receptor agonists, amino acid replacements are usually limited to close to terminal positions, and using naturally occurring post-translationally modifications, such as phosphorylation or glycosylation [PhSer, Ser(Glc), Thr(GalNAc)]. These modifications can target cell surface, blood–brain barrier or other receptors independent from the primary target, and can significantly improve both in vitro and in vivo properties of peptide agonists [8, 9].

Step 3: Rescreen the biological activity of the peptides as in Subheading "Reducing the Active Site," **step 3**.

Step 4: Fully purify the most active peptide(s) and identify the EC_{50} value by analyzing the activity figures after serial dilutions (*see* Fig. 1b) (*see* **Note 9**).

3.2 Identification of the Active Site and Further Modifications (the Active Site Is Unknown)

3.2.1 Overlapping Ligand Fragments

Step 1: If the binding site of a protein ligand is not known, it has to be identified before any unnatural amino acid replacement can take place. For this purpose design a series of overlapping peptides that covers the entire protein sequence. The following formula tells you how many peptides you will need:

$$\frac{a-b}{c}+1,$$

where a is the number of total residues in the given protein (domain), b is the amino acid length of the overlapping peptides and c is the step size in moving the b-mer sequences from the N terminus to the C terminus. For example, if we want to identify the active size of a protein domain that has 120 amino acid residues by using 12-amino acid long fragments and every peptide in the array is shifted by two residues, you will need (120–12)/2 + 1 = 55 peptides. In a practical point of view, while the individual synthesis of 15–20 peptides is not a difficult task, when the library grows over 30 residues it is worth preparing it in a multiple synthesis array format (*see* **Note 1**). In general, is not recommended to design less than ten residue-long peptides because most agonists are longer than this size (*vide supra*), but the quality of the peptides made in a multiple format significantly decreases over 15 residues especially if a larger array is designed which contains not only the base sequences but already has analogs containing nonnatural amino acid replacements. Twelve amino acid-long peptides with two or three residue shifts (e.g., peptide 1: ADEFGHIKLMNP, peptide 2: EFGHIKLMNPRS or FGHIKLMNPRST, peptide 3: GHIKLMNPRSTV or IKLMNPRSTVWY) appear to be a good compromise between accuracy and affordability.

Step 2: Synthesize or order the peptide library in a semi-purified form. 1 mg of each peptide or less is sufficient for most initial cellular screening assays. The final products have to be cleaned through some sort of reversed-phase treatment (chromatography or at least cartridges) to eliminate all organic contaminants originating from the solid-phase synthesis and cleavage. Cross-contamination among the array members has to be completely avoided as the presence of residual protecting groups. These can significantly bias the activity data. When we deal with 10–13-mer peptides we do not expect too many deleted sequences (present in the preparations due to inefficient coupling or deprotection steps during the synthesis procedure) in individual quantities larger than 1–2 %.

Step 3: Screen the biological activity of the peptide library as described in Subheading "Reducing the Active Site," **step 3**.

Step 4: Make amino acid replacements (if the array has not been synthesized already containing these modified peptides) as described in Subheading "Nonnatural Amino Acid Residue Replacements."

Step 5: Rescreen the biological activity of the peptides as in Subheading "Reducing the Active Site," **step 3**.

Step 6: Fully purify the most active peptide(s) and identify the EC_{50} value by analyzing the activity figures after serial dilutions (*see* Fig. 1b).

3.3 Screening the Biological Activity of the Peptides

Adipokine receptors are expressed in many cancer cells, so for simplicity we screen our designer leptin and adiponectin receptor agonists and antagonists in an MCF-7 breast cancer cell proliferation assay. As this is a quite general protocol, here I outline the basic steps. Each laboratory has to use their specialized assay, provided it is suitable for semi-high-throughput analysis.

3.3.1 Initial Screening

Step 1: Grow MCF-7 cells in a standard medium DMEM:F12 plus 5 % fetal bovine serum in 12-well plates (approximately 50,000 cells/well).

Step 2: Synchronize the cells in SFM (DMEM plus 10 μM $FeSO_4$, plus 0.5 % BSA) for 24 h.

Step 3: Add peptides to the wells in a final concentration of 100 nM (for 10-mer peptides this is approximately 100 ng/mL) in triplicate. To separate wells add a known agonist of the given receptor (we used 6 nM =100 ng/mL leptin and 50 ng/mL globular domain of adiponectin), an unrelated peptide synthesized in an identical manner and leave at least 3 wells untreated.

Step 4: Incubate the cells with the peptides and controls for 3 days.

Step 5: Count the viable cell numbers before and after treatment with trypan blue exclusion.

3.3.2 Determination of the EC_{50}

Steps 1 and 2 are identical to those in Subheading 3.3.1.

Step 3: Add the selected peptide(s) to the wells in a final concentrations 10 pM–1 μM (for 10-mer peptides this is approximately 10 pg/mL–1 μg/mL) in triplicate.

Steps 4 and 5 are identical to those in Subheading 3.3.1.

Step 6: Draw a concentration-activity curve and identify the peptide concentration that corresponds to 50 % of full activity (in our case 50 % of the cell number of untreated cells minus cells treated with 100 nM peptides, *see* Fig. 1b).

3.4 Receptor Antagonist Design and Development

The general rules of switching receptor agonist peptides to antagonists were described in Subheading 1. Here is an example by using our leptin receptor antagonists Aca1 and Allo-aca (*see* Table 2). Please observe that full pharmacological evaluation and concentration-dependent characterization of ligand functions are more labor intensive for antagonists than for agonists (*vide infra*, **step 4**, **Note 10**.)

Step 1: Design a library of shorter fragments of the agonist by gradually deleting single residues from the N- and C terminus keeping the center 6–8 residues. For most receptors, the agonistic activity of peptide ligands goes above 1 μM below 6–8 residues in length. For simplicity, let's assume that an octapeptide (especially if it contains nonnatural residues in mid-chain position) is short enough for pure antagonist activities. Then from a 12-mer ADEFGHIKLMNP agonist we will have a library of 14 pieces of 8-mer peptides, A-K, A-L, A-M, A-N, D-L, D-M, D-N, D-P, E-M, E-N, E-P, F-N, F-P and G-P.

Step 2: Design analogs of the truncated agonists by replacing natural amino acid residues with unnatural moieties. First make N- and C-terminal substitutions for improving the stability and cell surface targeting, just like for the agonists described in Subheading "Nonnatural Amino Acid Residue Replacements," **step 1**. Second, walk single homologous unnatural amino acid replacements through the entire peptide (unlike for the agonists where we restricted the modifications to close-to-terminal positions). Third, for antagonist design we can make single non-conservative replacements (such as hydroxyl-amino acid→cationic amino acid). Fourth, make multiple internal natural–unnatural amino acid mutations or deletions of selected internal residues. In most cases you can stay with the nonnatural moieties listed in Subheading "Nonnatural Amino Acid Residue Replacements," and can limit to the number of nonnatural residues to 3–5 depending upon the peptide length. Based on a 14-mer non-modified library (**step 1**), you will not need to exceed 96 peptide (14 base peptides, 14 N- and C-terminal modifications, 14 N- and C-terminal + 1 internal homologous, 14 N- and C-terminal + 1 internal non-conservative; 14 N- and C-terminal + 2 selected internally and 14 N- and C-terminal + 3 selected internally modified and 12 N- and C-terminally modified + 1 internally deleted) analogs.

Step 3: Synthesize (or order) and semi-purify the antagonist peptide library as described in Subheading 3.2.1, **step 2**. Because of the large number of peptides, most likely you will need to order or prepare a multiple peptide array.

Step 4: Screen the receptor antagonist properties of the library. Once again, all laboratories have to use their favorite cell lines and assay procedures. The screening process is identical to those I described in Subheadings "Reducing the Active Site" and 3.3 for the agonist with one major exception. The peptides have to be probed against the given cell lines both alone and in the presence of an agonist. For this purpose we add our antagonists to cells at 100 nM during the rough screening and at 100 nM–1 μM final concentrations alone at the more detailed characterization stage, as well as incubate the cells with 6 nM leptin plus 100 nM antagonist candidate during the initial screening and 6 nM leptin plus the test peptides in 10 pM–100 nM concentrations at the final characterization stage. Pure antagonists should inhibit the activity of the agonist but should not influence the cellular responses when added alone.

4 Notes

1. Currently a large number of automated peptide synthesizers available commercially. These differ in operating temperature, mixing techniques and the availability of on-line resin cleavage process. Many commercial facilities offer peptide synthesis services for very reasonable price. Peptides can be ordered individually or synthesized on an array format. The arrays are either cleaved at the contract facilities (e.g., New England Peptide provides a cleaved array in 96-well plates) or provided in a cleavable format, either attached to resins or cellulose carriers. Ordering peptide arrays can be indeed beneficial if a large number of peptides are needed. We synthesized the leptin-related peptides in our laboratory and ordered the adiponectin peptide array (attached to cellulose sheets) from the Peptide Array Facility of the University of British Columbia (396 peptides attached to a cellulose sheet and cleavable with mild base treatment). In either way, the peptides have to be at least 95 % pure and should not contain organic contaminants (i.e., should be purified by reversed-phase HPLC, or at least run through a reversed-phase cartridge). Organic contaminants kill cells and will make the evaluation of biological activity very difficult, if

not impossible. Regarding active peptide content, a 1 % active peptide contamination with EC_{50} of 1 nM will suggest that an otherwise inactive (EC_{50} = 1 μM) peptide agonist has an EC_{50} value of 100 nM, and will be considered a good lead compound for further development!

2. Piperidine, used for repetitive removal of the N-terminal Fmoc protecting group during peptide synthesis is listed as a controlled substance, and suppliers are required to obtain written information about the end-user and indented application for regulatory agencies.

3. For economical reasons, we purchase bulk amino acids and mix them with the coupling reagents prior starting the synthesis. For our specific peptide synthesizer, Protein Technologies sells pre-packaged Fmoc-protected amino acids mixed with coupling reagents but these are limited to those with standard side–side protection.

4. Unusual amino acids, particularly with positively charged side-chains, are frequently placed to the amino and carboxy termini of peptides to avoid exopeptidase cleavage. Hydrophobic unusual amino acids are regularly incorporated into mid-chain position to turn peptide agonists into antagonists or as spacers between solid carriers used peptide synthesis during array preparation. For most efficacy screening applications Aca- or β-Ala-conjugated peptides work just fine. D-amino acids not only improve proteolytic stability but also increase peptide solubility [27]. In all cases we design peptides for which appropriately protected unnatural amino acid residues, ready for solid-phase peptide synthesis, can be purchased from commercial sources. Our choice of supplier is Bachem.

5. HATU is the most powerful yet most expensive peptide coupling reagent on the market. It is preferentially used for the coupling of sterically hindered amino acids (such as many nonnatural moieties) or during the synthesis of complex peptides [28].

6. MCF-7 cells are arguably the most frequently studied cancer cells in oncology research. These cells (just like many other cell types) change their surface cancer markers upon multiple passages [29]. Since many MCF-7 preparations are currently in use, the proliferation efficacy can vary significantly upon stimulation with ligands. We found 10- or even 100-fold differences in EC_{50} and IC_{50} values when various MCF-7 batches were treated with identical leptin fragments. It is advisable that after initial purchase (I recommend the American Type Culture Collection, ATCC) or acquisition the cells are split, and multiple vials of the same passage stage are stored.

7. While it is not absolutely required for the identification of receptor agonists and antagonists, in most cases in the initial

screening process we include a control cell line that does not express the receptor. For breast cancer studies, the most frequently used control is MCF-10, a normal mammalian epithelial cell line that is void of cancer markers.

8. Leptin, and especially adiponectin, as well as their receptors are hard to handle proteins. They are not easily soluble, stick to glass surfaces, etc. Now recombinant adipokines are available from multiple sources. For reproducibility of the results, it is a good idea to buy these from one manufacturer and stick to those preparations.
9. At very high concentrations some agonists switch pharmacology and become inhibitory for the cellular functions. This phenomenon is explained by the negative feedback on receptor signaling at extreme stimulatory conditions [30].
10. Full characterization of receptor antagonists is more complex than that of agonists. If high throughput screening is available then this is not a major issue. However, if the products have to be screened individually, then it is a good idea to select a few peptides from the various modification groups outlined in Subheading 3.4, **step 1** and screen them first to see the general trend of agonist to antagonist change after a certain family of peptide modifications, and extend the study to further select peptide analogs based on the initial input.

References

1. Yan Z, Wang J (2012) Specificity quantification of biomolecular recognition and its implication for drug discovery. Sci Rep 2:309
2. Zhang Y, Proenca R, Maffei M, Barone M, Leopold L, Friedman JM (2004) Positional cloning of the mouse obese gene and its human homologue. Nature 372:425–432
3. Bluher M (2012) Clinical relevance of adipokines. Diabetes Metab J 36:317–327
4. Lang K, Ratke J (2009) Leptin and adiponectin: new players in the field of tumor cell and leukocyte migration. Cell Commun Signal 7:27
5. Kelesidis I, Kelesidis T, Mantzoros CS (2006) Adiponectin and cancer: a systematic review. Br J Cancer 94:1221–1225
6. Inui A, Meguid MM (2003) Cachexia and obesity: two sides of one coin? Curr Opin Clin Nutr Metab Care 6:395–399
7. Iserentant H, Peelman F, Defeau D, Vandekerckhove J, Zabeau L, Tavernier J (2005) Mapping the interface between leptin and the leptin receptor CHR2 domain. J Cell Sci 118:2519–2527
8. Otvos L Jr, Terrasi M, Cascio S, Cassone M, Abbadessa G, di Pascali F, Scolaro L, Knappe D, Stawikowski M, Cudic P, Wade JD, Hoffmann R, Surmacz E (2008) Development of a pharmacologically improved peptide agonist of the leptin receptor. Biochim Biophys Acta 1783:1745–1754
9. Kovalszky I, Surmacz E, Scolaro L, Cassone M, Sztodola A, Olah J, Hatfield MPD, Lovas S, Otvos L Jr (2010) Leptin-based glycopeptide induces weight loss and simultaneously restores fertility in animal models. Diabetes Obes Metab 12:393–402
10. Otvos L Jr, Haspinger E, La Russa F, Maspero F, Graziano P, Kovalszky I, Lovas S, Nama K, Hoffmann R, Knappe D, Cassone M, Wade JD, Surmacz E (2011) Design and development of a peptide-based adiponectin receptor agonist for cancer treatment. BMC Biotechnol 11:90
11. Otvos L Jr, Kovalszky I, Scolaro L, Sztodola A, Olah J, Cassone M, Knappe D, Hoffmann R, Lovas S, Hatfield MPD, Olah G, Wade JD, Surmacz E (2011) Peptide-based receptor antagonists for cancer treatment and appetite regulation. Biopolymers 96:117–125
12. Otvos L Jr, Kovalszky I, Riolfi M, Ferla R, Olah J, Sztodola A, Nama K, Molino A, Piubello Q, Wade JD, Surmacz E (2011) Efficacy of a leptin

receptor antagonist peptide in a mouse model of triple-negative breast cancer. Eur J Cancer 47:1578–1584

13. Shimizu M, Carter PH, Khatri A, Potts JT Jr, Gardella TJ (2001) Enhanced activity in parathyroid hormone-(1–14) and -(1–11): novel peptides for probing ligand-receptor interactions. Endocrinology 142:3068–3074
14. Demchyshyn LL, McConkey F, Niznik HB (2000) Dopamine D5 receptor agonist high affinity and constitutive activity profile conferred by carboxy-terminal tail sequence. J Biol Chem 275:23446–23455
15. Teesalu T, Sugahara KN, Kotamraju VR, Ruoslahti E (2009) C-end rule peptides mediate neuropilin-1-dependent cell, vascular, and tissue penetration. Proc Natl Acad Sci USA 106:16157–16162
16. Li ZJ, Cho CH (2012) Peptides as targeting probes against tumor vasculature for diagnosis and drug delivery. J Transl Med 10(Suppl 1):S1
17. Otvos L Jr, Wade JD, Feng J-Q, Condie BA, Hanrieder J, Hoffmann R (2005) Designer antibacterial peptides kill fluoroquinolone-resistant clinical isolates. J Med Chem 48:5349–5359
18. Hruby VJ (2002) Designing peptide receptor agonists and antagonists. Nat Rev 1:847–858
19. Sillerud LO, Larson RS (2005) Design and structure of peptide and peptidomimetic antagonists of protein-protein interaction. Curr Protein Pept Sci 6:151–169
20. Vessey KA, Lencses KA, Rushforth DA, Hruby VJ, Stell WK (2005) Glucagon receptor agonists and antagonists affect the growth of the chick eye: a role for glucagonergic regulation of emmetropization? Invest Opthalmol Vis Sci 46:3922–3931
21. Lempainen H, Molnar F, Gonzalez MM, Perakyla M, Carlberg C (2005) Antagonist- and inverse agonist-driven interactions of the vitamin D receptor and the constitutive androsterone receptor with corepressor protein. Mol Endocrinol 19:2258–2272
22. Higginbottom A, Cain SA, Woodruff TM, Proctor LM, Madala PK, Tyndall JD, Taylor SM, Fairlie DP, Monk PN (2005) Comparative agonist/antagonist responses in mutant human C5a receptors define the ligand binding site. J Biol Chem 280: 17831–17840
23. Otvos L Jr (ed) (2008) Peptide-based drug design, vol 494, Methods in molecular biology. Humana Press, Totowa, NJ
24. Otvos F, Murphy RF, Lovas S (1999) Coupling difficulty following replacement of Tyr with HOTic during synthesis of an analog of an EGF β-loop fragment. J Pept Res 53: 302–307
25. Meng X, Kondo M, Morino K, Fuke T, Obata T, Yoshizaki T, Ugi S, Nishio Y, Maeda S, Araki E, Kashiwagi A, Maegawa H (2010) Transcription factor AP-2β: a negative regulator of IRS-1 gene expression. Biochem Biophys Res Commun 392:526–532
26. Liu YD, Goetze AM, Bass RB, Flynn GC (2011) N-terminal glutamate to pyroglutamate conversion in vivo for human IgG2 antibodies. J Biol Chem 286:11211–11217
27. Quintana FJ, Gerber D, Bloch I, Cohen IR, Shai Y (2007) A structurally altered D, L-amino acid TCRα transmembrane peptide interacts with the TCRα and inhibits T-cell activation in vitro and in an animal model. Biochemistry 46:2317–2325
28. Angell YM, Garcia-Echeverria C, Rich DH (1994) Comparative studies of the coupling of N-methylated sterically hindered amino acids during solid-phase peptide synthesis. Tetrahedron Lett 35:5891–5894
29. Wu G, Fan RS, Li W, Srinivas V, Brattain MG (1998) Regulation of transforming growth factor-β type II receptor expression in human breast cancer MCF-7 cells by vitamin D3 and its analogues. J Biol Chem 273: 7749–7756
30. Kawano H, Katayama Y, Minagawa K, Shimoyama M, Henkemeyer M, Matsiu T (2012) A novel feedback mechanism by Ephrin-B1/B2 in T-cell activation involves a concentration-dependent switch from costimulation to inhibition. Eur J Immunol 42: 1562–1572

Chapter 13

Peptide Detection and Structure Determination in Live Cells Using Confocal Raman Microscopy

Andrew C. Terentis and Jing Ye

Abstract

Peptides are an important class of bioactive compounds that continue to be developed for a variety of therapeutic uses. The bioactivity of peptides stems in most cases from their ability to enter or bind to the surface of cells to elicit a cellular response, and the primary sequence and secondary structure of the peptide determine this. Therefore, experimental methods that can provide structural information on peptides in live cells are useful for exploring peptide structure–activity relationships and metabolism directly within the targeted cellular environment. In this chapter we describe an experimental methodology for the detection and structure determination of exogenous peptides within living cells using confocal Raman microscopy (CRM). CRM is Raman spectroscopy performed under a confocal microscope. Raman spectroscopy itself has been applied to the study of peptides for several decades and provides a wealth of information, including secondary structure via the amide backbone vibrational modes, cysteine redox status via the S–S and S–H stretches, and disulfide conformation via the S–S stretch. The Raman spectra of peptides are dominated by intense bands associated with the aromatic ring vibrations of Phe, Tyr, and Trp. The positions and intensities of some of these bands are sensitive to the hydrophobicity and pH of the peptide environment and thus can potentially be used as intracellular probes. Heavy-isotope labeling of aromatic ring side chains shifts the spectral positions of the aromatic ring vibrations and enables unambiguous detection of the peptide within cells. We employ this method primarily for the study of cell penetrating peptides in live cells. However, the method could in principle be applied to the study of any type of peptide within any type of cell if the intracellular concentration of the peptide reaches high enough levels to enable detection.

Key words Raman spectroscopy, Confocal Raman microscopy, Microspectroscopy, Cell penetrating peptide, Secondary structure

1 Introduction

Peptides are an important class of bioactive compounds that continue to be developed for a variety of therapeutic uses, including as antimicrobials and analgesics [1]. In order for a peptide to be bioactive it typically must enter cells, or bind to the surface of cells and elicit a cellular response. The primary sequence and the secondary structure(s) that the peptide can adopt are the principal

Predrag Cudic (ed.), *Peptide Modifications to Increase Metabolic Stability and Activity*, Methods in Molecular Biology, vol. 1081,
DOI 10.1007/978-1-62703-652-8_13,

determinants of the function of the peptide. Therefore, any experimental method that can detect exogenous peptides and provide information on their structure within living cells will be useful for the development of new therapeutic peptides, allowing peptide structure–activity relationships to be explored directly within the cellular environment.

Fluorescence microscopy has been the standard method for studying peptide–cell interactions. While it remains an essential tool, it does not provide information on the structure of the peptide. Recently we developed a new experimental approach for the detection of exogenous peptides within living cells using confocal Raman microscopy (CRM) [2]. The main advantage of the approach is that it can provide information on the structure of the peptide in addition to its distribution within cells. While we have applied it to the study of cell penetrating peptides in melanoma cells, the technique can in principle be used to study many other different types of peptides and cell types as well. In this chapter we describe experimental methods for the study of peptides in live cells and how information on the structure of the peptide can be obtained. In a previous volume of this series Schweitzer-Stenner et al. provided a detailed outline of methods for measuring Raman spectra of peptides in aqueous solution using a confocal Raman microscope [3]. Therefore, the main focus of this chapter will be on the methods and considerations relating specifically to the detection and structure determination of peptides in live cells using CRM. We begin with a brief introduction to Raman spectroscopy.

1.1 Raman Spectroscopy and Microscopy

Raman spectroscopy is a type of vibrational spectroscopy that involves the inelastic scattering of light by matter. When a sample is irradiated with a monochromatic light source, such as a laser, the vast majority of the scattered photons possess the same energy as the incident photons. This is called elastic or Rayleigh scattering. A very small fraction of photons (about 1 in every 10^7–10^9) can be energy-shifted relative to the incident photons, which is inelastic or Raman scattering. When the energy of the scattered photons is less than those of the incident light the scattered light is described as being Stokes-shifted. Conversely, when the energy of the scattered photons is greater than those of the incident light the scattered light is anti-Stokes-shifted (*see* Fig. 1). In both cases the magnitudes of the energy shifts, which are conventionally reported in wavenumber (cm^{-1}) units, reflect the energies of the fundamental vibrations of the molecules being probed. Only those vibrational modes that involve a change in the polarizability of the molecule during the course of the vibration will be Raman-active [4]:

$$I = \mathrm{K}l\alpha^2 v^4 \tag{1}$$

where I is the non-resonant Raman scattering intensity, K is a constant, l is the power density and ν the frequency of the incident laser,

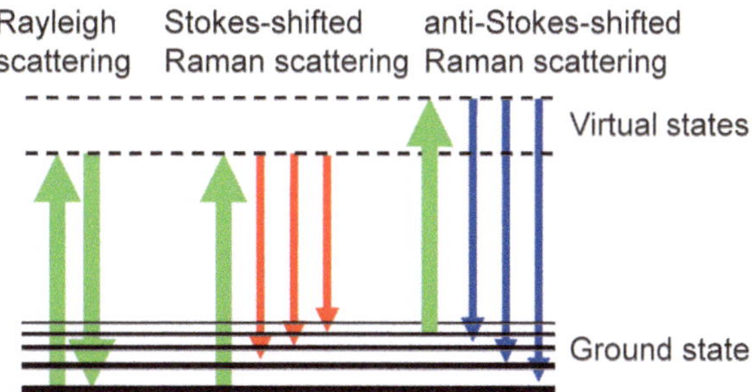

Fig. 1 Molecular energy level schematic of Rayleigh scattering, Stokes Raman scattering, and anti-Stokes Raman scattering. The *solid black lines* represent vibrational levels of the molecule in the ground electronic state. The *dashed black lines* represent virtual energy states that lie between the ground and the first excited electronic state. The *green upward arrows* represent the absorption of photons from the excitation laser. Rayleigh scattered photons (*green downward arrow*) have the same energy as the laser photons. Stokes-shifted photons (*red arrows*) are lower in energy than the laser photons while the anti-Stokes-shifted photons (*blue arrows*) are higher in energy

and α is the polarizability term for the vibration. Raman scattering intensity is therefore proportional to the square of the polarizability—vibrations involving the displacement of bulky atoms (e.g., S) or chemical groups (e.g., aromatic rings) that are readily polarizable will exhibit the strongest Raman intensities. Many detailed reviews of the theory of Raman scattering can be found in the literature (e.g., ref. 5).

Raman microscopy is simply Raman spectroscopy performed with a microscope. The microscope stage is used as a convenient means to mount and visualize the sample (e.g., a cell culture dish containing a monolayer of adhered cells) and the excitation laser is directed through the microscope objective, which focuses the beam onto the sample. The scattered light is detected through the same microscope objective. High quality, high numerical aperture objectives can achieve close to a diffraction-limited laser beam diameter at the focal point, which can roughly be estimated as half the wavelength of the light being employed. In the *z*-direction along the focal axis the spatial resolution is controlled by the confocal hole in the optical detection path (*see* Fig. 2). Thus, a Raman microscope enables Raman spectra to be measured from very small probe volumes ($\geq 1\ \mu m^3$), such as single points within living cells, as well as enabling the acquisition of 2D or 3D Raman images or "maps" of samples when a motorized scanning stage is employed.

1.2 Raman Spectroscopy of Peptides

1.2.1 Amino Acid Side-Chain Vibrations

Many of the intense vibrational bands that appear in the Raman spectra of peptides are associated with the aromatic ring vibrations of Phe, Trp, and Tyr, which appear throughout the fingerprint spectral region (~500–1,600 cm^{-1}). Another particularly strong band can arise from the S–S stretch of cystine in the 430–550 cm^{-1} region [6]. The S–H stretch of cysteine, while not as intense, is also

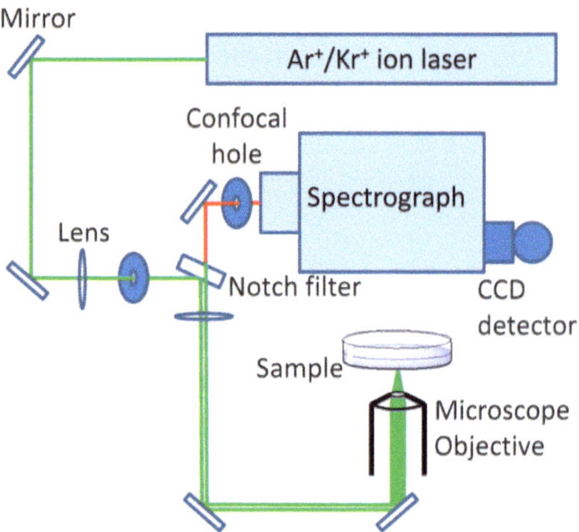

Fig. 2 Simplified schematic of the confocal Raman microscope system. The laser beam is directed into the microscope and focused onto the sample by the microscope objective. Raman- and Rayleigh-scattered light is gathered by the microscope objective, which travels back down the same optical path. At the notch filter the Rayleigh/laser light is rejected while the Raman scattered light is transmitted and enters the spectrograph, where it is dispersed and imaged on the CCD detector

a useful marker band because it appears in an uncongested region of the spectrum at 2,550–2,600 cm^{-1}. The C–S stretches of cysteine and methionine side chains appear in the 630–790 cm^{-1} region but are weaker than the S–S stretch. Vibrational bands associated with C–H bends (~1,400–1,470 cm^{-1}) and stretches (~2,800–3,000 cm^{-1}) are generally amongst the most intense bands, whereas bands associated with C–C stretches of aliphatic side chains are generally weaker but may be detectable in the spectrum in the range 600–1,300 cm^{-1}, especially if the peptide contains a large number of a particular amino acid in its sequence to create a cumulative intensity effect for a particular mode.

Raman spectra of penetratin (RQIKIWFQNRRMKWKK-NH_2) and oxytocin (CYIQNCPLG-NH_2) are shown in Fig. 3. The band assignments above the spectra in the figure highlight the prominence of the aromatic side-chain vibrations. Thus, when studying peptide–cell interactions using CRM, it is advantageous if not essential for the peptide to contain at least one Phe, Trp, or Tyr in the sequence to enable the detection of the peptide within cells (*vide infra*). Some of the aromatic side-chain vibrations can also provide information on the local environment of the peptide. For example, the intensity ratio of the tyrosine 830/850 cm^{-1} Fermi doublet is sensitive to the nature of the hydrogen bonding or ionization of the phenolic hydroxyl group [7, 8]. The intensity of the 880 cm^{-1} tryptophan band reflects the strength of hydrogen bonding at the N_1H site of the indole ring while the intensity of the

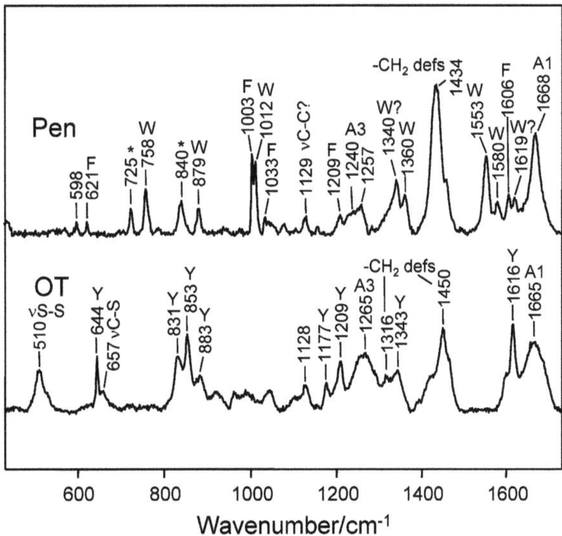

Fig. 3 Drop deposition Raman spectra of penetratin (Pen) and oxytocin (OT). Vibrational bands are assigned to aromatic ring vibrations of phenylalanine (F), tryptophan (W), and tyrosine (Y). A1 and A3 refer to amide I and amide III vibrations, respectively. Bands at 725 and 840 cm^{-1} marked by an *asterisk* in the penetratin spectrum are due to residual TFA from the sample preparation [34]. Other vibrational assignments include C–H deformations (–CH_2 defs), C–C stretches (νC–C), and the disulfide stretch (νS–S). Band assignments are derived from literature refs. 21, 35–39

1,360 cm^{-1} band is a marker of the hydrophobicity of the environment of the indole ring [9].

The intense S–S stretching mode that appears in the uncongested 430–550 cm^{-1} region of the Raman spectrum offers a potentially viable experimental method for detecting disulfide-containing peptides in cells using CRM. The S–S stretch appears at 510 cm^{-1} in the Raman spectrum of oxytocin (*see* Fig. 3). The peak position of the disulfide stretch is sensitive to the –S–S– dihedral angle and therefore can be used to monitor conformational changes in the peptide around the disulfide bond [10–13]. The appearance/disappearance of the S–S stretching mode in conjunction with the disappearance/appearance of the S–H stretching mode in the 2,550–2,600 cm^{-1} region provides a means for monitoring the redox status of the peptide cysteines.

1.2.2 Peptide Amide Backbone Vibrations

The amide bond constitutes the principal repeat unit in peptides and therefore the amide backbone vibrations (i.e., the "amide modes") generate intense bands in the Raman spectra of peptides. The amide modes are conventionally labeled amide I through VII and have varying degrees of IR and Raman activity [14]. In Raman spectra the amide I and III vibrational modes are the most important due to their high intensities and their high sensitivity to the secondary structure of peptides. Thus, amides I and III have been extensively used to determine the secondary structure composition of proteins and peptides by both IR and Raman spectroscopy [14–16]. The amide I mode possesses mainly C=O stretching character, whereas the amide III mode involves mainly C–N stretching and N–H in-plane bending [17–19]. In Raman spectra the amide I mode is easier to interpret because it is more intense, not significantly overlapped with other vibrational bands in the spectrum, and displays a more straightforward correlation between band peak position and peptide secondary structure. For these reasons the amide I band has received the most attention for secondary structure studies of peptides and proteins using Raman spectroscopy.

At the simplest level of interpretation, the peak position of the amide I band can be interpreted is indicating a predominant secondary structure of the peptide. This approach will obviously be most appropriate for peptides that possess uniform secondary structure, and it has generally proven reliable for distinguishing between predominantly α-helix, β-sheet/strand, and unstructured conformations. Thus, Raman and IR spectroscopy and theoretical calculations of model polypeptides with uniform secondary structures established that α-helices can be characterized by the appearance of amide I bands in the range 1,650–1,655 cm^{-1}, which can be distinguished from β-sheets/strands that produce amide I bands in the range 1,665–1,680 cm^{-1}, where β-sheets correspond to the lower end of this range and with narrower bandwidths than β-strands. PP_{II} structures manifest similar amide I band features as β-strands. Unstructured ("random coil") conformations are characterized by a high degree of structural heterogeneity with respect to the peptide backbone dihedral angles and give rise to broad amide I bands, typically peaking around ~1,665 cm^{-1} [14, 17–22].

To illustrate this interpretational approach we show Raman spectra of penetratin, a 16 residue cell penetrating peptide, in water/TFE solvent of varying composition (*see* Fig. 4). In 100 % water solvent the Raman spectrum of penetratin displays a rather broad, asymmetric amide I band with a peak at 1,675 cm^{-1}. This high peak wavenumber value is indicative of a significant β-strand contribution. With increasing TFE solvent, a shift in the amide I peak position from 1,675 to 1,655 cm^{-1} occurs (*see* Fig. 4). This change is consistent with a transition of the peptide toward a more α-helical structure. The drop in intensity in the amide III region at 1,239 cm^{-1} is similarly consistent with this structural change [19, 22]. Nevertheless, the amide I band remains relatively broad, and retains a shoulder on its high wavenumber edge that together is indicative of the retention of some β-strand and unstructured components. CD measurements under the same set of solvent conditions were in good agreement with these qualitative structural interpretations [2].

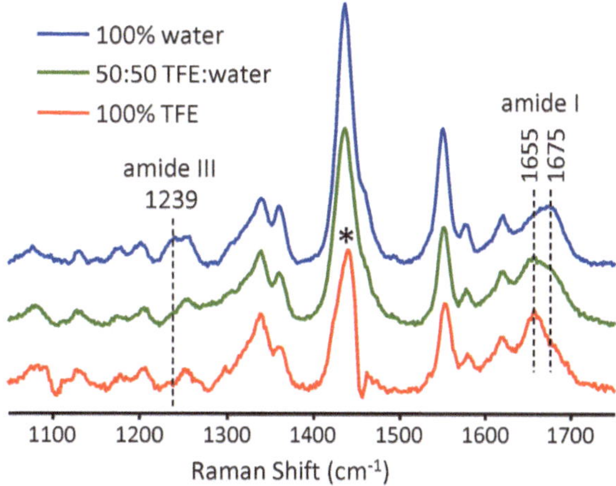

Fig. 4 Raman spectra of penetratin-Fh in three different water/TFE solvent conditions. Solvent peaks were removed by subtracting solvent spectra measured under matching experimental conditions. The profile of the band marked by an *asterisk* was not fully recoverable due to partial overlap with a strong solvent band (reproduced from ref. 2)

A more quantitative approach to obtaining secondary structure information from the Raman spectrum is to deconvolve the amide I band profile into a sum of either Lorentzian or Gaussian–Lorentzian (i.e., Voigt) component functions, each of which represents a distinct secondary structure component in the peptide [22]. Before illustrating this approach, let us first ponder the problem briefly. The amide I mode can be described as predominantly (83 %) C=O stretching [14]. Furthermore, the wavenumber of this vibration is sensitive to the local amide bond conformation, which can be characterized by the dihedral angles (ψ, ϕ). Thus, for an *n*-residue peptide containing n backbone CO groups associated with n amide bonds, one might anticipate a maximum of n distinct amide I band components if the peptide is highly unstructured. (The opposite extreme would be for all the dihedral angles in the peptide to be equal, leading to only one band component.) However, even an unstructured peptide can conceivably have some coincidental overlap of dihedral angles and therefore band components, making the number of distinct components less than n. The problem is further complicated by the fact that there is significant coupling between CO groups of the peptide, both through-bond via covalent and H-bonding interactions, and through-space via transition dipole coupling [15, 23–25]. As a result, there is not a simple one-to-one correlation between amide I band components and individual amide bond conformations. Instead, an *n*-residue peptide contains n distinct amide I modes, each of which is characterized by the concerted motions of several CO bonds along the peptide (*see* Fig. 5). Therefore, in reality, the number of resolvable

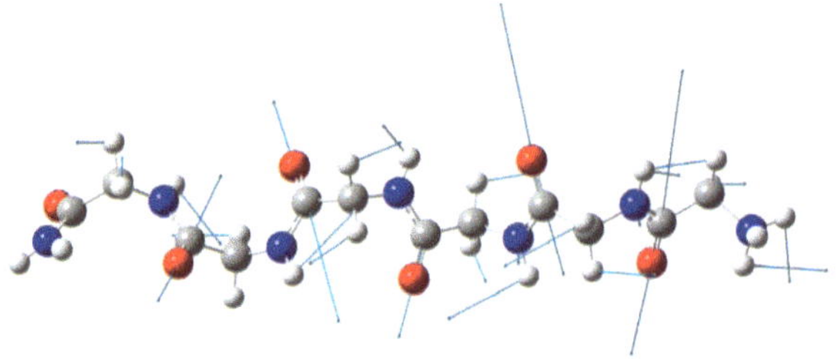

Fig. 5 Example of an amide I normal mode for the model peptide GGGGGG-NH_2 in an extended β-strand conformation. The relative amplitudes of displacement of the atoms for the vibration are represented by the lengths of the vectors. The structure and normal mode displacement vectors were calculated with Gaussian 09 using the DFT B3LYP method and 6-31G(d,p) basis set. Atom coloring: *grey* = carbon, *white* = hydrogen, *blue* = nitrogen, *red* = oxygen

amide I band components is considerably less than the number n, and in many cases the amide I band profile can be well fitted with just three components.

We illustrate the method of analysis for penetratin, using three independent Voigt functions to model the amide I band profile in the Raman spectrum of penetratin in water and TFE solutions (*see* Fig. 6). According to the procedure prescribed by Maiti et al. [22], the first Voigt function is constrained to be in the range 1,650–1,656 cm^{-1} and represents an α-helix secondary structure component for the peptide. The second Voigt function is constrained within the range 1,664–1,670 cm^{-1} and represents a β-sheet component if the function is narrow (~25 cm^{-1} FWHM) or an unstructured component if the function is broad. The third Voigt function is constrained within the range 1,674–1,685 cm^{-1} and represents a β-strand and/or a PP_{II} component (these two cannot be distinguished). According to this model then, the secondary structure composition of penetratin in 100 % water solvent is 24 % α-helical, 43 % unstructured, and 33 % β-strand/PP_{II} (*see* Fig. 6). The broadness of the Voigt components suggests poorly formed helical and strand components. In TFE solvent, the secondary structure composition of the peptide changes to 54 % α-helix, 32 % unstructured, and 14 % β-strand/PP_{II} (*see* Fig. 6).

The above amide I band fitting procedure illustrates that Raman spectroscopy can match CD spectroscopy in terms of the level of secondary structure information that each provide. The key advantage of Raman spectroscopy over CD is that it can be implemented in the living cell environment. Additionally, it can provide primary sequence and environmental insights by way of the side-chain vibrations, as mentioned earlier. Schweitzer-Stenner and coworkers have developed a method for analyzing the amide I band profiles of polarized Raman spectra in conjunction with the respective profiles in the IR and VCD spectra to obtain dihedral angle distributions for short peptides [3, 15]. This level of information exceeds that

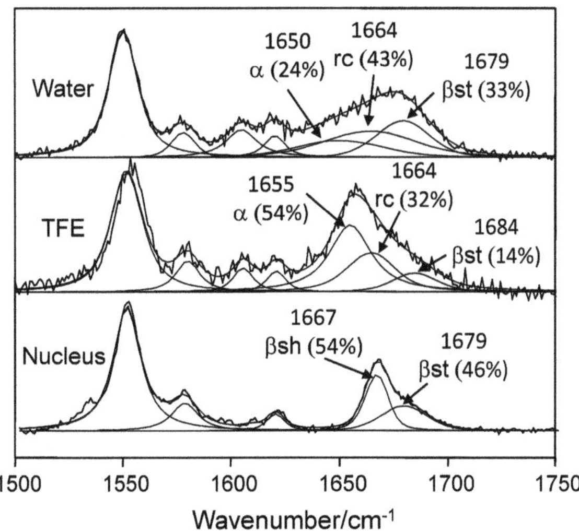

Fig. 6 Raman spectra of penetratin-Fh in the amide I region. The spectra were obtained from solutions of the peptide in 100 % water solvent, 100 % TFE solvent, and from within the nucleolus of an SK-Mel-2 cell ("Nucleus") using CRM. The amide I band profiles are modeled with a constrained three component Voigt functions model, as described previously [22]. Adjacent bands associated with Trp and Phe ring modes are also fitted with Voigt functions in the model. The component Voigt functions are shown in the figure as well as the summative fit to the experimental data. Also shown are the peak wavenumber values for each Voigt function contributing to the amide I band profile, the secondary structure assignments for each component (α = α-helix, rc = unstructured, βst = β-strand, βsh = β-sheet), and the percent contribution of each component to the total area of the amide I band

provided by CD but still does not match the level of structural detail that can be obtained from X-ray crystallography or NMR. Nevertheless, Raman spectroscopy offers some advantages over these two techniques, including comparative ease of measurement and analysis, applicability to peptides of any size or structure, applicability to aqueous solution studies, and adaptability to ultrafast dynamics studies as well as live cell studies.

1.3 Peptide Detection in Living Cells

Thus far we have discussed the main features of peptide Raman spectra and the types of structural information that can be obtained from them. In relation to the aim of measuring Raman spectra of exogenous peptides in live cells, the main practical challenge will be to distinguish the peptide Raman signals from those of the endogenous cellular constituents. Vibrational bands associated with proteins dominate the Raman spectra exhibited by cells [26–30], which can mask the signals from the peptide. Indeed, if the concentration of the exogenous peptide does not reach a high enough level within the cells then this problem may not be surmountable. However, for cell-penetrating peptides, the local concentrations of peptide can reach levels high enough inside cells that the spectrum of the peptide can be successfully distinguished from the cellular constituents. To aid the detection of the peptide in cells, one or more residues of the peptide are labeled with heavy isotopes in order to shift the position of one or more characteristic marker bands, which can then be unambiguously identified within cells. The intense aromatic ring modes of Phe, Trp, or Tyr are the best candidates for this. For our study of penetratin, we labeled the Phe residue with ^{13}C and ^{15}N atoms (*see* Fig. 7), which had the effect of shifting the 1,003 cm^{-1} ring mode down to a peak position of 967 cm^{-1} (*see* Fig. 8). This band was easily identifiable in the raw Raman spectra from peptide-treated cells. Subtraction of the Raman spectrum obtained prior to peptide treatment yielded the Raman spectrum of the peptide (*see* Fig. 9). The Raman spectrum of the peptide may also be filtered from the background Raman signals using Principal Components Analysis [2].

The amide I band profile in the Raman spectrum of the peptide obtained from cells may be modeled to estimate the secondary structure composition of the peptide (*see* Fig. 6, bottom spectrum). The 3-Voigt-function model described earlier indicates that the peptide is exclusively β-form within the nucleoli of SK-Mel-2 cells, specifically being 54 % β-sheet and 46 % β-strand.

We have shown that under favorable conditions, specifically when the peptide concentration reaches high enough levels within cells, the Raman spectrum of the peptide may be obtained following removal of the background (cell) Raman signals. Furthermore, the amide I band in the resulting Raman spectrum furnishes information on the secondary structure of the peptide. However, some caution must be exercised when interpreting this amide I band profile because in the raw Raman spectrum it contains contributions not only from the exogenous peptide but also the endogenous cell proteins and the water bending mode that peaks around 1,640 cm^{-1}. The reliability of the peptide amide I band obtained by spectral

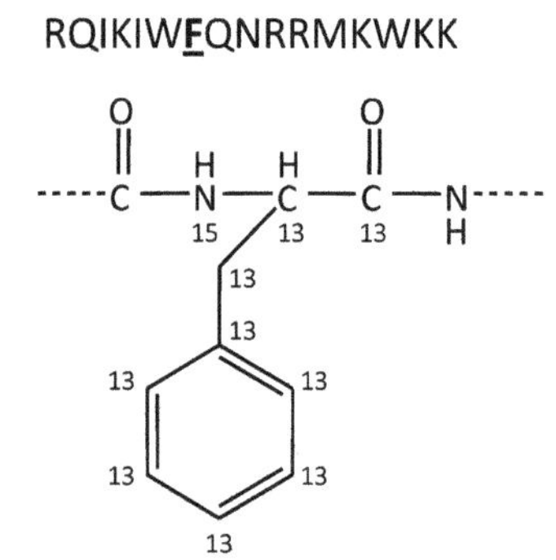

Fig. 7 Primary sequence of penetratin-Fh and heavy isotope (^{13}C, ^{15}N) labeling scheme of the Phe residue (reproduced from ref. 2)

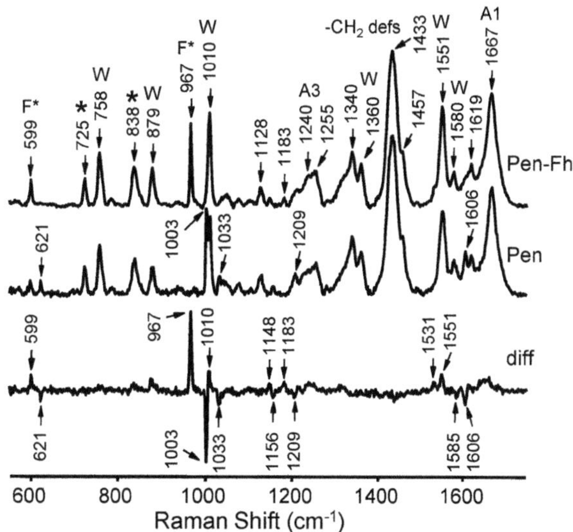

Fig. 8 Raman drop deposition spectra of heavy-labeled penetratin (Pen-Fh), unlabeled penetratin (Pen), and Pen-Fh—Pen difference spectrum (diff) showing the 967 cm^{-1} $^{13}C/^{15}N$-Phe (F*) mode and other Phe-associated modes. Other peak assignments: Trp (W) [17, 37, 38, 40], $-CH_2$ deformations ($-CH_2$ defs) [17, 21], amide I (A1) and amide III (A3) bands [14, 19]. Peaks at 725 and 838 cm^{-1} marked by an *asterisk* are due to residual TFA from the peptide preparation [34] (reproduced from ref. 2)

subtraction (or by PCA) critically relies on the quality of the control spectrum, which is measured under matching experimental conditions and must reliably represent the background signals present when the spectrum of the peptide is measured. This tends to be easier when probing the nucleus of cells compared with the cytoplasm, because the cytoplasm is more heterogeneous and dynamic [2]. Another important consideration is the degree of overlap between the peptide amide I band and the amide I band of the cell proteins. Raman spectra of cell proteins generally exhibit an amide I band peaking near 1,655 cm^{-1}, which is reflective of a high α-helical content. If the secondary structure of the peptide is predominantly helical then its amide I band will be strongly overlapped with that of the cell proteins, making data subtraction more difficult and perhaps less reliable. On the other hand, if the peptide secondary structure is predominantly β-form, its amide I peak should appear in the range 1,665–1,680 cm^{-1} and data subtraction should be more reliable (*see* Fig. 9). The overall reliability of the CRM method for determining the secondary structure of peptides in cells by way of amide I band processing will become clearer as more experiments on a wider range of peptides are performed. This work is ongoing in our laboratory.

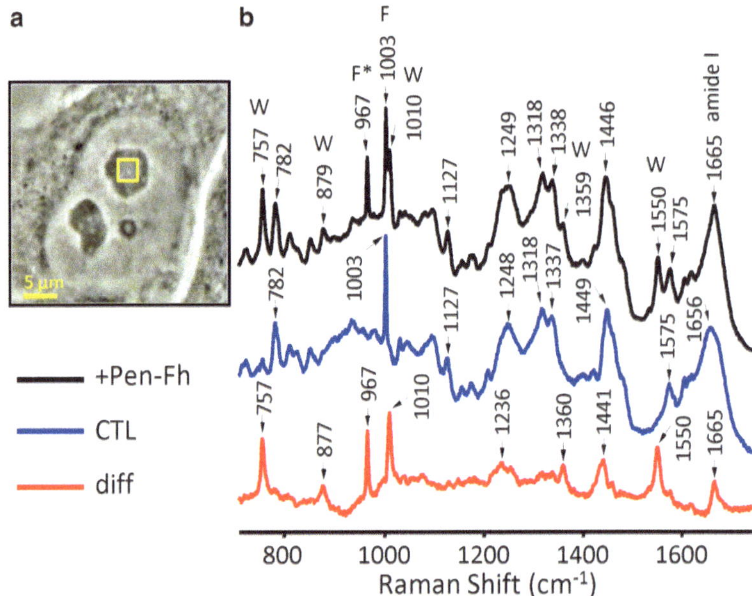

Fig. 9 (**a**) Phase contrast image of a melanoma (SK-Mel-2) cell captured with 100× magnification. (**b**) Averaged Raman spectra derived from the area indicated by the *yellow box* within the nucleolus of the cell shown in (**a**). Spectra were measured either ~120 min after treatment with 50 μM penetratin-Fh (+Pen-Fh) or before treatment (control, CTL). Pen-Fh—CTL difference spectrum (diff) is also shown. Bands due to Trp (W), Phe (F) and $^{13}C/^{15}N$-labeled Phe (F*) are indicated above the spectra. Scan parameters: 3.5 μm × 3.5 μm, 0.5 μm step-size, 2 × 10 s dwell-time (reproduced from ref. 2)

2 Materials

1. Heavy-isotope-labeled Phe, Fmoc-[U-^{13}C9, ^{15}N]-Phe-OH from Cambridge Isotope Laboratories (Andover, MA). The labeled penetratin peptide with sequence RQIKIWF[^{13}C9,^{15}N]QNRRMKWKK-NH_2 ("penetratin-F-heavy" = penetratin-Fh) is synthesized using standard Fmoc-chemistry [2]. Peptide stocks are made up as 20 mM solutions in water (pH 7) and frozen, or kept at 4 °C for no longer than 10 days. Stock concentrations are determined by the absorbance at 280 nm using the extinction coefficient ε_{280} = 11,000 $M^{-1}cm^{-1}$ (5,500 $M^{-1}cm^{-1}$ per Trp residue).
2. Human metastatic melanoma cells (Sk-Mel-2) from American Type Culture Collection (ATCC) (Manassas, VA).
3. DMEM from CellGro (Herndon, VA).
4. Opti-MEM, trypsin, antibiotics from Invitrogen (Carlsbad, CA).

5. Trypsin–EDTA solution, 1 L, pH 7.4 (all chemicals except trypsin from Fisher Scientific): 80.0 g NaCl, 4.0 g KCl, 5.8 g $NaHCO_3$, 2.0 g Na_2EDTA, 10.0 g d-glucose, 5.0 g Difco™ Trypsin 250, dissolved in MilliQ water to a total volume of 1 L and sterile (0.2 μm) filtered. This is a 10× stock for storage at −20 °C. Use a diluted 1× stock on cells.
6. Adhesion medium, 250 mL, pH 7.4 (all chemicals except DMEM from Fisher Scientific): 0.375 g $NaHCO_3$, 3.195 g DMEM powder (without phenol red, l-glutamine, sodium pyruvate, sodium bicarbonate, Cellgro 90-013-PB), 12.5 mL of 400 mM HEPES, 237.5 mL MilliQ water. Solution is sterile (0.2 μm) filtered.
7. Trypan Blue (0.4 % in PBS) from Fisher Scientific.
8. Sterile 100 mm polystyrene cell culture dishes from VWR Scientific.
9. 35 mm collagen-coated glass coverslip (#1.5) bottom dishes from MatTek Corporation.
10. All Raman spectroscopy measurements are performed with a 0.8 m dispersive spectrograph (LabRam HR800, Horiba Scientific) equipped with 600 groove/mm and 1,800 groove/mm gratings, liquid-N_2-cooled 1,024×256 pixel CCD detector, and motorized slit and confocal hole adjustments. CRM measurements are performed with an Olympus IX-71 inverted microscope coupled to the spectrograph (*see* **Note 1**). The microscope is equipped with a Marzhauser motorized X–Y scanning stage (112×74 mm range, 0.1 μm minimum step-size), 100× phase contrast oil immersion objective (Olympus UPlanFL N, NA 1.3), and digital video camera for cell visualization and image capture. Raman spectroscopy of peptide solutions is performed through a separate port of the spectrograph, bypassing the microscope (*see* **Note 2**). Laser beams (514 and 647 nm) are generated by a Spectra Physics Beamlok 2060 mixed argon/krypton ion laser.
11. Premium #1 glass coverslips (Fisher Scientific, 12-544-10).
12. 50 mm×0.6 mm square I.D. glass capillary tubes (VitroCom, 8250).
13. Critoseal® capillary tube sealant (215003).
14. C_{18} Ziptips (Millipore, ZTC18S960).
15. α-cyano-4-hydroxycinnamic acid (Proteochem, P9100). Make a saturated solution in 50/50 acetonitrile/water, centrifuge and use the supernatant for MALDI.
16. Applied Biosystems Voyager MALDI-TOF mass spectrometer.
17. Origin® 8 software (OriginLab Corporation, Northampton, MA).

3 Methods

3.1 Drop Deposition Raman Spectroscopy

A necessary precursor to the cell experiments is measurement of the Raman spectrum of the pure peptide so that the features of the spectrum are known. It also serves to confirm any band-shifts resulting from heavy isotope labeling of the peptide. A convenient sampling technique that requires very little peptide sample is drop deposition [31]. With this technique a drop of the peptide solution is placed on a glass coverslip and allowed to dry at room temperature (RT). The residual solid is then examined under the Raman microscope. The peptide solution concentration can be as low as 1–10 μM in some favorable cases, but high quality spectra are much easier to obtain with peptide solutions in the low millimolar range.

1. Place a 2 μL droplet of 1 mM peptide solution at the center of a glass coverslip using a 10 μL capacity micropipette. Allow the droplet to dry at RT (~30 min for aqueous solutions).
2. A small spot or ring of peptide residue should be visible on the coverslip (*see* **Note 3**). Locate the spot under the microscope using regular bright field illumination and 40× objective. Scan the microscope stage to find the edge of the ring.
3. Target the laser beam a small distance (100–200 μm) inside the edge of the ring of peptide residue, which usually gives better results than when the beam is focused near the center of the sample spot.
4. Optimize the focus of the laser beam on the sample by adjusting the microscope objective focal control knob and continuously monitoring the Raman signal at 1–2 s per CCD exposure. If the focus is set too low, too much glass interference will occur from the coverslip, which manifests as a broad and distorted baseline in the low wavenumber region of the spectrum. If the focus is set too high, the Raman signals from the sample will be too low. Optimize the focus for maximum Raman signal from the peptide while minimizing glass interference. For very thin residual films of peptide, a higher N.A. objective, such as a 100× oil-immersion objective, may be required to achieve a good focus within the peptide film without interference from the underlying glass coverslip. The confocal hole diameter may also be decreased to improve the Z-resolution.
5. Record Raman spectra using 5 mW of 514 nm laser light through the 40× objective with 2 × 60 s collection time (*see* **Note 4**).
6. Record Raman spectra from several *different* positions within the sample to check for consistency. Dried samples can sometimes be heterogeneous with some points of the sample giving

a spectrum that is not representative of the peptide. Multiple spectra from the sample are averaged later for better S/N.

7. If a difference spectrum is the goal (e.g., *see* Fig. 8) then it is desirable to have carefully matching experimental conditions for the two samples. In this case, prepare the different samples at the same concentrations and drop each sample as close as possible to each other (without touching) on the same coverslip. Raman spectra of the samples are then recorded under closely matching measurement conditions, one after the other. Additionally, a background spectrum is measured by moving the laser a short distance (<1 mm) away from the samples (*see* **Note 5**). The background spectrum is later subtracted from the sample spectra to improve the baselines.

3.2 Raman Spectroscopy of Peptide Solutions

Measuring the Raman spectrum of the pure peptide in solution provides an important reference for the spectra obtained from cells. Spectra can be measured in different solvents to change the structure of the peptide and therefore show how the spectral features change in response to the structure. Non-resonant Raman scattering is a very weak phenomenon and thus the measurement of the peptide spectrum in solution is difficult due to the low number of scattering centers present within the laser focal volume; concentrations in the 5–20 mM range are typically required as well as laser powers at the sample in the 200–400 mW range. Raman scattering is also more efficient for shorter laser wavelengths (Eq. 1). High laser power is not typically obtainable under a Raman microscope and so a separate external sample setup may be required for these measurements (*see* **Note 2**).

1. Load 5 μL of 20 mM peptide solution into a 50 mm, 0.6 mm square I.D. glass capillary tube using a 10 μL micropipette. Keep the tube horizontal when doing this.
2. With the tube horizontal, seal both ends of the tube with Critoseal. Try to avoid having the solution inside the tube come in contact with the sealant (*see* **Note 6**).
3. Place the capillary tube on an appropriate holder or mount and focus the aligned laser beam onto the sample (*see* **Note 7**).
4. Record the Raman spectrum of the peptide solution using 250 mW of 514 nm laser light at the sample and 5 min acquisition time (*see* **Note 8**).
5. Repeat the above procedure for the solvent alone using an identical tube and without changing the measurement parameters. The solvent spectrum is subtracted from the peptide solution spectrum to remove the solvent background (*vide infra*).

3.3 Cell Preparation for Raman Measurements

Cell culture conditions are strongly dependent on the cell type. We describe the conditions that are conducive to growing SK-Mel-2 cells. In principle, any adherent cell type can be studied using CRM.

1. Upon arrival from ATCC immediately passage the cells four times (splitting at 70–80 % confluence) and store at –80 °C in DMEM containing 10 % FBS and 5 % DMSO.
2. For CRM experiments, thaw the cells and passage 2–3 times in 100 mm polystyrene dishes. Cells are maintained in Opti-MEM medium supplemented with 4 % fetal bovine serum (FBS), 50 U/mL penicillin, and 0.05 mg/mL streptomycin at 37 °C in a humidified atmosphere of 5 % CO_2 in air.
3. The day before CRM experiments, harvest the cells from a sub-confluent (<80 %) polystyrene dish using trypsin–EDTA solution and then resuspend the cells in fresh medium. Seed the cells into the desired number of 35 mm glass bottom dishes at an initial density of ~10,000 cells/cm^2 and place in the incubator overnight.
4. The next day the cells in the glass bottom dishes should be adhered to the bottom of the dish and healthy in appearance (no round or floating cells). The cells should not be too crowded or too sparse—50 to 75 % confluence is ideal. Immediately before CRM measurements are to be performed on a dish, gently wash the adherent cells with PBS twice and then cover them with about 2 mL of serum-free and phenol red-free DMEM (*see* **Note 9**). Gently place the dish on the microscope stage.

3.4 CRM Measurements in Live Cells

As discussed in the Introduction, a crucial requirement for obtaining a reliable spectrum of the exogenous peptide within a cell is to obtain background spectra of the cell in the absence of the peptide. There are two possible approaches to this. The first is to add the peptide and then to measure Raman spectra in a particular region of the cell before the peptide has arrived. This is possible if the peptide takes a significant amount of time to diffuse into the region being probed. Penetratin, for example, has a nuclear localization sequence that causes it to build up in the nucleus of cells (*see* Fig. 9). Under the conditions of our study it typically took around 30 min for the peptide to reach the nucleus of the cells. This enabled enough time for the peptide to be added to the medium covering the cells and then for background spectra to be recorded immediately within the nucleus before the peptide arrived. At a later time, the spectrum of the peptide was measured from the same region within the cell. In the cytoplasm, however, penetratin was detectable in as little as 5 min [2]. In this case it was necessary to first measure the background spectrum in a chosen area of the cytoplasm of a cell and then to add the peptide and measure again a short time later with the peptide present (*see* **Note 10**). Thus, the experimental approach will depend on the nature of the peptide and the location within the cell that is being probed.

All CRM measurements are performed at room temperature under normal atmosphere with a 647 nm, 6 mW excitation laser beam focused with a 100× phase contrast, oil immersion microscope objective (*see* **Note 11**). A 200 μm confocal hole-size is used and spectral collection times are 10–30 s per point. Automated X–Y area scanning is performed with a step-size of 0.5–1.0 μm.

1. Visualize the cells under the microscope using phase contrast, which is necessary in the absence of cell staining.
2. Choose a healthy cell—flat and shapely, not rounded or raised. The nucleoli within the nucleus should be clearly visible.
3. Target the desired region of the cell and program a scan across a 3–5 μm square region. For example, a 3 × 3 μm scan region with 1.0 μm stepsize will generate 16 spectra in total. If a 2 × 10 s dwell-time is employed the total run time will be ~5 min (*see* **Note 12**).
4. Optimize the laser focus (*see* Subheading 3.1, **step 4**) within the scan region to maximize cell Raman signals while minimizing glass interference.
5. Scan the cell to collect the background spectra, either before or immediately after peptide addition (see comments at the beginning of this section).
6. The peptide is added as a concentrated aqueous solution (5–10 mM) directly to the medium in the dish while placed on the microscope stage. The total volume of medium in the 35 mm dish is 2 mL and the final concentration of the added peptide can be in the range 10–100 μM (*see* **Note 13**). Best results are obtained when gently adding the concentrated peptide solution to the center top of the medium in the dish and mixing with the same pipette many times around the center surface of the medium. It is crucial to be gentle so that no cells are disturbed and the dish position on the stage is not moved.
7. Measure the Raman spectra again in the presence of the peptide at the desired time after treatment, without changing any of the measurement conditions or scan parameters used for the before-peptide measurements in **step 5**.
8. In addition to measurements within individual cells, control measurements are performed where the laser is focused in unoccupied areas of the dish, between cells. This is to confirm that peptide Raman signals are not detectable outside of the cells.

3.5 Assessing Cell Viability During and After CRM Measurements

At the completion of CRM measurements the dish is kept on the microscope stage and washed of excess peptide with PBS. Trypan blue solution (250 μL) is then added to the dish to test the viability of cells by assessing Trypan blue uptake under bright-field illumination. Using this procedure it is possible to assess the viability of both laser-irradiated and non-irradiated cells.

Separate control experiments may also be performed where the cells are first incubated with the peptide at 37 °C for a certain period of time, washed with PBS, and then covered again in serum/phenol-free medium for CRM measurements. Thus, in this case the peptide is not present in the medium during the CRM measurements. Any peptide detected in the cells must have therefore entered prior to the CRM measurements. This is a good way to confirm that the peptide is not entering the cells as a result of the laser irradiation. The reason that we do not recommend this peptide "pulse-chase" approach as the standard approach for CRM experiments is that it makes it difficult to match the background (no peptide) and treatment spectra. With this approach the background spectra can be recorded on cells in the dish prior to the addition of the peptide, but then the dish must be removed and incubated with the peptide before returning it to the microscope stage. It is then virtually impossible to follow up with CRM measurements on the same cells unless a gridded coverslip is used.

3.6 Intracellular Peptide Analysis by MALDI-TOF

It is important to know whether the exogenous peptide remains intact inside the cells or whether it becomes cleaved or fragmented. This can be assessed by analyzing the intracellular contents by MALDI-TOF MS. We employed a procedure adapted from that of Fischer et al. [32] to study penetratin-Fh in SK-Mel-2 cells [2]. The following procedure is essentially the same, except that we now recommend the use of C_{18} Ziptip columns instead of PD Minitrap G10 columns for sample desalting because the Ziptips provide better peptide recovery.

1. A confluent cell layer of SK-Mel-2 cells in a 9.6 cm^2 dish (the same type of dish used for Raman measurements) is washed once with DMEM and incubated with 1.5 mL of DMEM containing 50 μM penetratin-Fh for 2 h at RT covered from light. A total of three treatment dishes are prepared. Additionally, three control dishes are prepared, in which cells are incubated with 1.5 mL DMEM *not* containing peptide.
2. The adhered cells in each dish are gently washed twice with 1 mL PBS and then detached by the addition of 1 mL of 1× trypsin–EDTA solution and incubated for 10 min at 37 °C. An additional 2 mL PBS is then added to each dish. The contents of the three treatment dishes are combined in a fresh 50 mL polystyrene BD Falcon tube and a further 4 mL PBS is added. Similarly, the cells from the three control dishes are combined

in another Falcon tube and diluted to 10 mL with PBS. Hereafter, the treatment and the control cells are processed in parallel according to the following procedure.

3. The combined cells are washed three times with PBS by centrifugation for 10 min at 150 × *g* and resuspension in 10 mL PBS.
4. After final PBS wash, cell pellet is resuspended in 200 μL of 0.1 % HCl for 10 min at RT to lyse the cells, and lysate is transferred to a 1.5 mL microcentrifuge tube.
5. The cell lysate is then centrifuged for 15 min at RT and 12,000 × *g* using a microcentrifuge (Eppendorf mini-spin). Only 10–15 μL of the supernatant is required for the remainder of the procedure. The rest of the supernatant can be frozen and stored at −20 °C for later analysis or discarded.
6. The peptide-containing supernatant is immediately desalted (*see* **Note 14**) using a C_{18} Ziptip column according to the following:
   - 6.1 Condition Ziptip by pipetting in and then discarding 10 μL 50/50 acetonitrile/0.1 % aqueous TFA. Do this three times.
   - 6.2 Hydrate the tip three times with 10 μL 0.1 % aqueous TFA, in a similar fashion to 6.1.
   - 6.3 Load sample by pipetting 10 μL of cell lysate supernatant in and out of the tip slowly several times.
   - 6.4 Wash the tip three times with 10 μL 0.1 % aqueous TFA to remove salts.
   - 6.5 Elute the peptide with 5 μL of 50/50 acetonitrile/0.1 % aqueous TFA (*see* **Note 15**).
7. The sample is immediately prepared for MALDI-TOF MS by mixing 1 μL desalted peptide with 1 μL α-cyano-4-hydroxycinnamic acid matrix solution in a microcentrifuge tube. 1 μL peptide–matrix mixture is then spotted on the MALDI plate and allowed to dry. In addition to the treatment and control cell samples, 1 μL pure α-cyano-4-hydroxycinnamic acid matrix is spotted on the MALDI plate as a further control.
8. MALDI-TOF MS analysis is performed in positive reflector mode.

3.7 Raman Spectroscopy Data Analysis

3.7.1 Spectral Calibration

Spectral calibrations are performed daily before and after the experimental measurements are completed to ensure that the calibration has not changed throughout the day. A change may happen if the grating position is moved frequently. Calibration may be performed against the known emission lines of mercury from ordinary room light (e.g., 546.074 nm). Another very common procedure is to place a small piece of silicon onto the microscope stage and to record the position of the standard 520.7 cm^{-1} Raman line of silicon. Either of these single peak calibration methods is suitable for setting the origin of the grating position. Once the constant wavenumber discrepancy for the standard is known, the constant is added to the "x" (wavenumber) values in the spectrum text file, which is viewed using a software program such as Microsoft Excel or MicroCal Origin. If the coefficient (slope) of the calibration is incorrect, spectral "stretch" or "compression" may be observed. I.e., peaks may be in the correct position at one end of the spectrum but too high or too low at the other. In this case a single peak calibrant is inadequate and multiple standard spectral lines must be measured across the spectral range of interest. For this purpose, the standard lines of polystyrene or neon may be employed. A polystyrene dish can be conveniently placed on the microscope stage and the laser focused onto it to record the Raman spectrum. Alternatively, a neon lamp may be held up to the microscope objective or spectrograph entrance slit to record the spectrum.

3.7.2 Background and Baseline Corrections

1. Drop deposition spectra—the multiple spectra recorded from different points within the sample are averaged. The matching background spectrum is then subtracted from the peptide spectrum. At this point the baseline of the corrected peptide spectrum should be quite flat if the sample and control measurement conditions were well matched. The exception is when the sample fluoresces. In this case a manual or polynomial least-squares fitted baseline may be subtracted from the spectrum to make it more aesthetically pleasing. If the intention is to model the amide I band of the corrected spectrum, caution must be exercised when performing a baseline correction so as not to unduly affect the amide I band profile. Baseline anchor points should be kept well away from the band.
2. Peptide solution spectra—the procedure is identical to **step 1** above, where the solvent spectrum is subtracted from the peptide solution spectrum.
3. Calculating the difference between two spectra—this is the case for the cell experiments, where the background cell spectrum is subtracted from the cell + peptide spectrum, as well as for the detection of isotope-sensitive bands (heavy peptide spectrum—light peptide spectrum). If the conditions employed for the acquisition of the two spectra were well matched then immediate spectral subtraction without any prior data manipulation should yield a clean difference spectrum with a level baseline. Common peaks between the two spectra should cancel out across the entire range of the spectrum. In truth, this is usually difficult to achieve. Typically it is necessary to multiply

one of the spectra by a constant intensity scaling factor prior to subtraction. In the worst case scenario, the two ends of the spectrum may require slightly different scaling factors to achieve good peak cancellation across the entire spectral range. Any baseline corrections are best performed on the resulting difference spectrum *after* spectral subtraction.

3.7.3 Amide I Band Fitting

The following amide I band fitting procedure is based on the published method of Maiti et al. [22], using the Origin 8 software program to perform the nonlinear least squares fitting.

1. Spectrum is first processed according to the procedures outlined in Subheadings 3.7.1 and 3.7.2. For amide I fitting, the spectrum is culled down to the range ~1,500–1,750 cm^{-1}. It is important to include a significant portion of the spectrum either side of the amide I band in order to ensure the best possible fit.
2. Plot the spectrum and keep the graph window active. There should not be any other data plotted in the graph besides the spectrum to be fitted.
3. In the drop-down menu go to Analysis → Fitting → Nonlinear curve fitting → open dialog.
4. Within the NLFit dialog, click on the Settings tab and then click on Function Selection. Select Category = Spectroscopy and Function = Voigt in the drop-down menus.
5. Still within the Settings tab, click on Advanced and choose the Number of Replicas. This will determine the total number of Voigt functions to be included in the fit. There should be three functions included for the amide I band, plus one function for every other band in the spectrum to be fitted. For the data in Fig. 6, the total is 7 so the number of replicas is 6.
6. Also under Advanced, check the "Plot individual peak curve" and "Plot cumulative fitted curve" boxes.
7. Click on the Bounds tab. A table of all the fitting parameters will be visible. Enter the initial values, upper and lower bounds for the fit. For example, Table 1 below lists the values employed for the fits shown in Fig. 6. Note that the area values for the fit are arbitrary as they depend on the intensity of the spectrum being fitted.
8. Click on the <►►|> "fit until converged" button that appears on the lower right of the NLFit dialog box. Check under the Messages tab for the progress of the fit. If the maximum number of iterations is reached before convergence click the "fit until converged" button again. Repeat until converged.
9. Click OK to exit the NLFit dialog.

Table 1
Initial values and constraints employed for multiple Voigt function fitting of experimental data presented in Fig. 6

Parameter	Meaning	Initial value/cm^{-1}	Lower bound	Upper bound
y0	Offset	0	–	–
xc	Center	1,550	1,548	1,552
A	Area	10,000	0	–
wG	Gaussian width	20	5	50
wL	Lorentzian width	20	5	50
xc_2	Center	1,579	1,577	1,581
A_2	Area	500	0	–
wG_2	Gaussian width	10	5	40
wL_2	Lorentzian width	10	5	40
xc_3	Center	1,605	1,603	1,607
A_3	Area	500	0	–
wG_3	Gaussian width	10	5	40
wL_3	Lorentzian width	10	5	40
xc_4	Center	1,619	1,617	1,621
A_4	Area	500	0	–
wG_4	Gaussian width	10	5	40
wL_4	Lorentzian width	10	5	40
xc_5	Center	1,655	1,650	1,656
A_5	Area	2,500	0	–
wG_5	Gaussian width	20	5	40
wL_5	Lorentzian width	20	5	40
xc_6	Center	1,665	1,664	1,670
A_6	Area	2,500	0	–
wG_6	Gaussian width	20	5	40
wL_6	Lorentzian width	20	5	40
xc_7	Center	1,680	1,674	1,685
A_7	Area	2,500	0	–
wG_7	Gaussian width	20	5	40
wL_7	Lorentzian width	20	5	40

10. The results of the fit will be overlaid on the graph with the experimental data. Tabulated data can be found in the FitNL1 sheet in the original workbook that contains the experimental data (*see* **Note 16**).

4 Notes

1. There are several commercial Raman microscopes now on the market that could be employed for the cell experiments described here. An inverted microscope is preferable over an upright microscope for cell measurements because it enables more convenient measurements on adherent cells in culture dishes.
2. This arrangement is necessary because the Raman microscope does not allow more than ~20 mW of laser power through the microscope objective onto the sample. For non-resonant Raman spectroscopy of peptides high laser power (>100 mW) is required to obtain satisfactory spectra. Our macro-sampling arrangement, built in-house, allows very high laser powers to be focused onto the sample. The scattered light is collected at 90° by a collection lens and focused directly into the spectrograph via a 125 μm fixed entrance slit.
3. The diameter of the dried spot or ring should be around 2–3 mm if an aqueous solution of peptide is used. For organic solvents (e.g., TFE) the spot will usually spread more, creating a larger, thinner film of residue. In this case use a higher concentration and a smaller volume of peptide solution.
4. Some dried peptide samples are prone to burning and/or fluorescence. If this is the case, try a longer laser wavelength and/or lower laser power with longer acquisition time. If a motorized scanning stage is available, scan the laser through a line of the sample while employing a short dwell time at each step to avoid sample burning or damage.
5. The laser focus can have a strong influence on the spectrum baseline, particularly for measurements on thin sample films where the laser focal point is set close to the underlying glass coverslip surface. Small changes in the laser focus can then have a significant effect on the spectrum baseline, making spectral subtraction more complicated. Therefore, it is best to keep the same microscope focal setting for the two samples. Also, changing the X–Y position of the laser a large amount (i.e., several millimeters) can inadvertently change the laser focus if the coverslip is not level. Therefore, it is desirable to travel only short distances when making multiple measurements on a sample.

6. Although we recommend this as a precaution, we are yet to detect any interference from the sealant, even when the tube contains organic solvents like carbon tetrachloride.
7. Place the capillary tube horizontally by sticking one end of the tube onto the top of a flat-ended steel post using double-sided tape. Ensure that the square tube is securely stuck on one of the flat sides rather than one of the corners of the tube to avoid undue scattered light.
8. Lower sample concentration and laser power and longer acquisition time is feasible if the sample is stable and not prone to heating or damage. We employ a 5 cm focal length convex lens to focus the beam onto the sample. A more powerful lens could be used to create a higher power density at the sample.
9. This is important because serum and phenol red generate too much fluorescence using visible laser light. This modified cell medium maintains the viability of the cells during the CRM experiments much better than simple buffered solutions like HBSS or PBS.
10. Some cells do not take up peptide. If such a cell is chosen for the background measurements then a different cell will need to be chosen for the peptide measurements.
11. Red or near IR laser wavelengths and low laser powers are required to ensure the viability of the cells throughout the CRM measurements [33]. The cells should be carefully monitored for any changes that may indicate laser damage.
12. Short total run times are desirable to avoid cell damage. This is highly dependent on the cell type.
13. For penetratin, we have found the concentration limits to be ~10–100 μM. For transportan, the limits are ~5–50 μM (unpublished data). Concentrations above these ranges are toxic to the cells whereas concentrations below these ranges lead to intracellular concentrations of the peptides that are not detectable using the CRM technique.
14. The samples should be placed on ice to suppress cell protease activity if a delay in sample processing is necessary.
15. The ratio of the acetonitrile and aqueous TFA solution can be adjusted according to the hydrophobicity of the peptide.
16. The results should be inspected closely to ensure a good, physically meaningful fit. The R^2 value should be close to 1 (typically better than 0.98). The standard error for a parameter should not be significantly greater than the final value for the parameter. Preferably none (or maximally 1 or 2) of the parameters will be at the upper or lower bound limit.

Acknowledgments

The authors wish to thank the following people for their contributions to this work: Ms. Sara Fox and Dr. Evonne Rezler for cell culture assistance; Dr. Mare Cudic for synthesizing the heavy-labeled penetratin peptide; Mr. Storm Stillman for measuring the oxytocin spectrum; Mr. Richard Lantz for the GGGGGG-NH_2 calculation using Gaussian.

References

1. Vlieghe P, Lisowski V, Martinez J, Khrestchatisky M (2010) Synthetic therapeutic peptides: science and market. Drug Discov Today 15:40–56
2. Ye J, Fox SA, Cudic M, Rezler EM, Lauer JL, Fields GB, Terentis AC (2010) Determination of penetratin secondary structure in live cells with Raman microscopy. J Am Chem Soc 132:980–988
3. Schweitzer-Stenner R, Soffer JB, Toal S, Verbaro D (2012) Structural analysis of unfolded peptides by Raman spectroscopy. Methods Mol Biol 895:315–346
4. Smith E, Dent G (2005) Modern Raman spectroscopy: a practical approach. Wiley, Hoboken, NJ
5. Keresztury G (2002) Raman spectroscopy: theory. In: Chalmers JM, Griffiths PR (eds) Handbook of vibrational spectroscopy. John Wiley and Sons Ltd, Chichester, pp 71–87
6. Socrates G (2001) Infrared and Raman characteristic group frequencies. John Wiley and Sons, Chichester
7. Arp Z, Autrey D, Laane J, Overman SA, Thomas GJ Jr (2001) Tyrosine Raman signatures of the filamentous virus Ff are diagnostic of non-hydrogen-bonded phenoxyls: demonstration by Raman and infrared spectroscopy of p-cresol vapor. Biochemistry 40:2522–2529
8. Siamwiza MN, Lord RC, Chen MC, Takamatsu T, Harada I, Matsuura H, Shimanouchi T (1975) Interpretation of the doublet at 850 and 830 cm^{-1} in the Raman spectra of tyrosyl residues in proteins and certain model compounds. Biochemistry 14:4870–4876
9. Miura T, Takeuchi H, Harada I (1988) Characterization of individual tryptophan side chains in proteins using Raman spectroscopy and hydrogen-deuterium exchange kinetics. Biochemistry 27:88–94
10. Hernández B n, Carelli C, Coïc Y-M, De Coninck J, Ghomi M (2009) Vibrational analysis of amino acids and short peptides in aqueous media. V. The effect of the disulfide bridge on the structural features of the peptide hormone somatostatin-14. J Phys Chem B 113:12796–12803
11. Hruby VJ, Deb KK, Fox J, Bjarnason J, Tu AT (1978) Conformational studies of peptide hormones using laser Raman and circular dichroism spectroscopy. A comparative study of oxytocin agonists and antagonists. J Biol Chem 253:6060–6067
12. Tu AT, Bjarnason JB, Hruby VJ (1978) Conformation of oxytocin studied by laser Raman spectroscopy. Biochim Biophys Acta 533:530–533
13. Tu AT, Lee J, Deb KK, Hruby VJ (1979) Laser Raman spectroscopy and circular dichroism studies of the peptide hormones mesotocin, vasotocin, lysine vasopressin, and arginine vasopressin. Conformational analysis. J Biol Chem 254:3272–3278
14. Bandekar J (1992) Amide modes and protein conformation. Biochim Biophys Acta 1120: 123–143
15. Schweitzer-Stenner R (2006) Advances in vibrational spectroscopy as a sensitive probe of peptide and protein structure—a critical review. Vib Spectrosc 42:98–117
16. Williams RW (1986) Protein secondary structure analysis using Raman amide I and amide III spectra. Methods Enzymol 130:311–331
17. Krimm S, Bandekar J (1986) Vibrational spectroscopy and conformation of peptides, polypeptides, and proteins. Adv Protein Chem 38:181–364
18. Overman SA, Thomas GJ Jr (1998) Amide modes of the alpha-helix: Raman spectroscopy of filamentous virus fd containing peptide 13C and 2H labels in coat protein subunits. Biochemistry 37:5654–5665
19. Chen MC, Lord RC (1974) Laser-excited Raman spectroscopy of biomolecules. VI. Some polypeptides as conformational models. J Am Chem Soc 96:4750–4752
20. Hansen CL, Hansen PR, Callisen TH, Bauer R, Nielsen OF (2002) Secondary structure and association of melittin during and after solid-phase synthesis: a Raman and static light scattering study. J Raman Spectrosc 33:142–146
21. Laporte L, Stulz J, Thomas GJ (1997) Solution conformations and interactions of alpha and

beta subunits of the Oxytricha nova telomere binding protein: investigation by Raman spectroscopy. Biochemistry 36:8053–8059

22. Maiti NC, Apetri MM, Zagorski MG, Carey PR, Anderson VE (2004) Raman spectroscopic characterization of secondary structure in natively unfolded proteins: alpha-synuclein. J Am Chem Soc 126:2399–2408
23. Brauner JW, Flach CR, Mendelsohn R (2005) A quantitative reconstruction of the amide I contour in the IR spectra of globular proteins: from structure to spectrum. J Am Chem Soc 127:100–109
24. Moore WH, Krimm S (1975) Transition dipole coupling in Amide I modes of betapolypeptides. Proc Natl Acad Sci USA 72:4933–4935
25. Myshakina NS, Asher SA (2007) Peptide bond vibrational coupling. J Phys Chem B 111: 4271–4279
26. Taleb A, Diamond J, McGarvey JJ, Beattie JR, Toland C, Hamilton PW (2006) Raman microscopy for the chemometric analysis of tumor cells. J Phys Chem B 110:19625–19631
27. Matthaus C, Chernenko T, Newmark JA, Warner CM, Diem M (2007) Label-free detection of mitochondrial distribution in cells by nonresonant Raman microspectroscopy. Biophys J 93:668–673
28. van Manen H-J, Kraan YM, Roos D, Otto C (2005) Single-cell Raman and fluorescence microscopy reveal the association of lipid bodies with phagosomes in leukocytes. Proc Natl Acad Sci USA 102:10159–10164
29. Krafft C, Knetschke T, Funk RHW, Salzer R (2005) Identification of organelles and vesicles in single cells by Raman microspectroscopic mapping. Vib Spectrosc 38:85–93
30. Bonnier F, Knief P, Lim B, Meade AD, Dorney J, Bhattacharya K, Lyng FM, Byrne HJ (2010) Imaging live cells grown on a three dimensional collagen matrix using Raman microspectroscopy. Analyst (Cambridge, UK) 135:3169–3177
31. Ortiz C, Zhang D, Xie Y, Ribbe AE, Ben-Amotz D (2006) Validation of the drop coating deposition Raman method for protein analysis. Anal Biochem 353:157–166
32. Fischer R, Kohler K, Fotin-Mleczek M, Brock R (2004) A stepwise dissection of the intracellular fate of cationic cell-penetrating peptides. J Biol Chem 279:12625–12635
33. Puppels GJ, Olminkhof JH, Segers-Nolten GM, Otto C, de Mul FF, Greve J (1991) Laser irradiation and Raman spectroscopy of single living cells and chromosomes: sample degradation occurs with 514.5 nm but not with 660 nm laser light. Exp Cell Res 195:361–367
34. Wei F, Zhang D, Halas NJ, Hartgerink JD (2008) Aromatic amino acids providing characteristic motifs in the Raman and SERS spectroscopy of peptides. J Phys Chem B 112:9158–9164
35. Chen MC, Lord RC (1976) Laser-excited Raman spectroscopy of biomolecules. VIII. Conformational study of bovine serum albumin. J Am Chem Soc 98:990–992
36. Lord RC, Yu NT (1970) Laser-excited Raman spectroscopy of biomolecules. II. Native ribonuclease and alpha-chymotrypsin. J Mol Biol 51:203–213
37. Lord RC, Yu NT (1970) Laser-excited Raman spectroscopy of biomolecules. I. Native lysozyme and its constituent amino acids. J Mol Biol 50:509–524
38. Overman SA, Thomas GJ Jr (1995) Raman spectroscopy of the filamentous virus Ff (fd, fl, M13): structural interpretation for coat protein aromatics. Biochemistry 34:5440–5451
39. Tuma R (2005) Raman spectroscopy of proteins: from peptides to large assemblies. J Raman Spectrosc 36:307–319
40. De Gelder J, De Gussem K, Vandenabeele P, Moens L (2007) Reference database of Raman spectra of biological molecules. J Raman Spectrosc 38:1133–1147

Index

Predrag Cudic (ed.), *Peptide Modifications to Increase Metabolic Stability and Activity*, Methods in Molecular Biology, vol. 1081, DOI 10.1007/978-1-62703-652-8, © Springer Science+Business Media New York 2013

D

E

F

G

H

I

K

L

M

N

O

P

R

S